Zoos

Also by Keekok Lee:

PHILOSOPHY AND REVOLUTIONS IN GENETICS

Zoos

A Philosophical Tour

Keekok Lee

First published 2005 by
PALGRAVE MACMILLAN
Houndmills, Basingstoke, Hampshire RG21 6XS and
175 Fifth Avenue, New York, N.Y. 10010
Companies and representatives throughout the world

PALGRAVE MACMILLAN is the global academic imprint of the Palgrave Macmillan division of St. Martin's Press, LLC and of Palgrave Macmillan Ltd. Macmillan® is a registered trademark in the United States, United Kingdom and other countries. Palgrave is a registered trademark in the European Union and other countries.

ISBN-13: 978–1–4039–8624–5 hardback
ISBN-10: 1–4039–8624–X hardback

This book is printed on paper suitable for recycling and made from fully managed and sustained forest sources.

A catalogue record for this book is available from the British Library.

Library of Congress Cataloging-in-Publication Data

Lee, Keekok, 1938-
    Zoos : a philosophical tour / Keekok Lee.
        p.   cm.
    Includes bibliographical references.
    ISBN 1-4039-8624-X (hardback)
    1. Zoos—Philosophy. I. Title.

QL76.L44 2005
590.73—dc22   2005051356

10   9   8   7   6   5   4   3   2   1
14   13   12   11   10   09   08   07   06   05

Printed and bound in Great Britain by
Antony Rowe Ltd, Chippenham and Eastbourne

Contents

Acknowledgements

I wish to thank all those kind people who gave me technical assistance in producing the typescript. Most of all, I wish to thank Daniel Bunyard, the philosophy commissioning editor of Palgrave for his extraordinary speed and efficiency in enabling the book to see the light of day.

Introduction

This book has an unusual take on zoos. It is a philosophical exploration of the concept of zoos, not, however, from the usual ethical angle of either animal welfare or animal rights, but from the ontological standpoint.[1] It demonstrates that the animals kept in zoos are, indeed, unique to zoos – they are not wild, nor are they domesticants in the classical understanding of domestication. Ontologically speaking, they are not tokens of wild species but of what may be called artefactual species. Such a provocative thesis also has controversial and radical policy implications for zoos, as it challenges those policies advocated by the World Zoo Conservation Strategy (1993) and the European Union Zoos Directive (1999). The book is written not solely for philosophers and philosophy students but in a manner which makes it readily accessible to zoo theorists and managers world-wide, as well as to anyone who has a professional or lay interest in zoos and their futures.

The central contention of this book is that the ontological status of animals in zoos is different from that of animals in the wild. The former are not wild, amounting to what may be called biotic artefacts, and constitute the *ontological foil* to the latter.

This claim is inspired by two sources: first, what may be called the zoological conception of animals; second, the commonly accepted official definition of zoos as collections of animal exhibits open to the public. The respective implications of each of these two accounts are then teased out. Such an approach leads, as already mentioned, to some surprising conclusions and to some interesting implications for policy consideration in zoo management. The ontological stance and arguments presented in this book are, in principle, not against zoos; zoos are acceptable provided they are prepared to admit that their exhibits are not wild but tame or immurated animals, that the individual immurated animal is not a token of

1

any wild species.[2] However, to admit this logic would entail that the only sound theoretical justification of zoos lies in recreation, and not in the more high-minded mission of education-for-conservation of wild animals, or of *ex situ* conservation, tasks which zoos necessarily cannot accomplish, as an ontological dissonance exists between, on the one hand, the immurated animals and their behaviour on view as exhibits, and on the other, the mistaken belief, on the part of zoos, that by looking at such exhibits, visitors would actually be learning about wild animals, their behaviour in the wild and the need to save them and their habitats in the wild. Ironically, it transpires that the zoo-visiting public may have a better intuitive grasp of the ontological issues at stake, as they seem to accept happily and simply the fact that they derive pleasure and enjoyment from looking at, indeed, a unique kind of animal – only to be found in zoos – animals which may look wild but are not wild, and which necessarily do not behave like those in the wild. For them, the zoo experience is indeed unique, but not, however, in the way that zoo theorists and professionals have in mind.

Chapter 1 sets the scene for the ontological exploration of zoos by distinguishing between five different conceptions of animals, namely, the lay person's conception, the conceptions presupposed respectively by the philosophy of animal welfare (Singer) and animal rights (Regan), the zoo's conception and the zoological conception.

Chapter 2 elaborates on the zoological conception which rests primarily on the Darwinian theory of natural evolution and its mechanism of natural selection, in the context of other equally relevant sciences, such as genetics as well as ethology and ecology. At the same time, it introduces certain key philosophical notions such as trajectory, the distinction between existing 'by themselves' and 'for themselves'. The scientific and the philosophical components combine to lay down a delineation of the ontological status of animals in the wild, which in the chapters to follow, is argued as constituting the ontological foil to zoo animals.

Chapter 3 begins the delineation of the ontological status of animals under captivity in zoos by disentangling the conceptual confusions in terms such as 'wild animals in captivity', a phrase which occurs and recurs in zoo literature. However, it is an oxymoron, as the animals under zoo captivity are *ex hypothesi* tame; 'tame' is the antonym of 'wild', in a crucial sense of the term.

Chapters 4, 5 and 6 begin to go beyond the mere conceptual contradiction to demonstrate in earnest that zoo animals are, from the ontological perspective, very different beings from animals in the wild. Chapter 4 takes seriously the definition of zoos as 'collections of animal exhibits open to the public', teasing out its logical and ontological implications; it also examines the ideas that zoo animals are necessarily exotic (in the technical sense) animals, and that the environment provided for them is also, therefore, necessarily 'exotic' in the sense that, at best, their exhibit enclosures are simulated, naturalistic environments which bear no resemblance to habitats in the wild. Chapters 5 and 6 go beyond the simulation of naturalistic environments to examine several crucial aspects of animal life under zoo management which bear no resemblance whatsoever to the lives led by animals in the wild: (a) the miniaturisation of simulated space (in some extreme cases, enclosure space is 10 000 times smaller than the home range of the animal in the wild); (b) 'hotelification' (a word coined to refer to the fact that food is not food as in the wild, which is hunted or foraged, but are substitutes prepared in zoo kitchens, served up in much the same way as one would put out food for one's pet); (c) medication to prevent suffering from discomfort or disease, but also to ward off death and prolong the lives of the animals under zoo management. All these features add up to the suspension of the mechanism of natural selection within the context of natural evolution.[3]

Chapters 7 and 8 pursue the implications of the suspension of natural evolution and natural selection in the context of zoo management. Chapter 7 argues that zoo animals are immurated animals – 'immuration' is a term coined to embody the key idea that they are, indeed, domesticated animals, even though it is true that they do not readily fall within the traditional, classical definition of domestication, and that there is a need to distinguish between two kinds of domestication. Chapter 8 reinforces this logic, arguing that immurated animals are biotic artefacts, by laying bare the ontological features of artefacts in general, and showing that the term 'biotic artefacts' is neither conceptually confused nor incoherent. It formally introduces the type/token distinction and shows that an individual immurated animal is not a token of a wild species, but of an immurated species. In other words, zoo animals are uniquely different from animals in the wild as well as from traditional domesticants such as cats and dogs.

Chapters 9 and 10 begin the task of teasing out the policy implications of the ontological thesis established in the preceding chapters by critically examining the justifications for zoos. The three so-called serious justifications – research, conservation (*ex situ*) and education – are shown to be defective and, indeed, even deeply flawed given the ontological dissonance between zoo animals and wild animals in the wild. Curiously, the so-called frivolous justification in terms of recreation seems to have survived critical scrutiny best of all.

Chapter 11 and the Conclusion draw the book to a close by reconsidering the five conceptions of animals distinguished in Chapter 1 in the light of the ontological insights yielded by the subsequent chapters. The single most important conclusion, from the policy point of view, which this exploration has come up with is that zoos do, emphatically, have a future, indeed, a thriving future but, ironically, not perhaps the future which the World Zoo Conservation Strategy (1993) and the European Union Zoos Directive (1999) envisage and strenuously advocate – zoos are neither the modern-day scientific Noah's Ark nor are they the Open University for education-in-conservation, as they are human cultural spaces which necessarily humanise the animals under their care and control. Their animal exhibits, in reality, offer clean, wholesome family fun and entertainment. That is why, world-wide, they are big draws and big business. Ironically, it follows that although zoos (via *ex situ* conservation) may not be truly relevant to the project of saving extant threatened wild species from extinction, they, unwittingly, play a role in adding to biodiversity – though not of the natural kind – by nurturing and creating, in the long run, new immurated, artefactual species.

1
What Does the Public Find in Zoos?

The question in the title of this chapter seems silly and naïve. However, there is a point in posing it, as behind the obvious answer to it stands a whole lot of complexities which must be unravelled before one could ultimately answer in a satisfactory manner the issues raised in the Introduction.[1]

Surely, even a child would be able to say that in zoos one finds animals. The child would be right in one undoubted sense. While there are trees in zoos, one does not go to a zoo specially to see plants; plants are incidental to the *raison d'être* and aims of zoos. Zoos exist to exhibit animals.[2]

However, the intriguing question is: 'What animals do zoos exhibit? All animals?' To which we may answer: 'Not at all; only some select few.' In other words, the selection presupposes a certain conception of what counts as an animal. We need to explore that conception, and compare it with at least three other conceptions, bringing out the overlapping concerns and relationships (if any) between them, making clear their respective hidden agendas and assumptions, whenever relevant.

The lay person's conception

Let us start with the ordinary lay person's conception of what counts as an animal. The first thing to notice is that it sticks to the commonsensical understanding of the traditional two-kingdom schema (as evidenced in the obvious answer to the naïve question mentioned above); they would have no difficulty classifying squirrels as animals and conifers as plants, but on the whole they would have no opinion whether bacteria count as animals, since they would have no knowledge nor would they have given the matter any thought.

In general, society's interest in certain animals was/is dictated by the roles they play in human lives – these animals have either religious/cultural, culinary, economic or personal significance for the social group or individual in question. For instance, some groups have chosen even rats (vertebrate) and snakes (invertebrate) as objects of religious worship. Some cherish the bald eagle as a symbol of national (tribal) virility, others the lion. Which animals are good to eat and which are not clearly varies according to culture and to historical period. Dogs are good to eat for the Dayaks (in Borneo), while cows, for Hindus, are not for eating at all. Tigers, rhinoceroses and some whales, today, are in danger of being hunted to extinction for economic reasons.

Beyond identifying animals which belong to these main categories of concern, most people remain in ignorance of those not encompassed. Today, in the industrialised and industrialising world, 'animals' as a generality are not all that pertinent to their lives. Particular types of animals may, outside these categories, catch their attention because they are exotic (in which case they go to the zoo to see them or watch them in a television programme), charismatic like the lion, or cuddly like the panda. As far as lay people are concerned, birds seem to be the only class of animals which commands a sizeable minority of followers dedicated to watching and studying them; and amongst ornithologists knowledge of them can be thorough and comprehensive.

As most societies have long left the hunter/gatherer mode of existence behind, the only animals which form part of their immediate experience and consciousness are domesticated animals – cows, pigs, sheep, goats, chickens (which are good to eat); horses, bullocks, camels (which are good for traction and transportation but also to eat in some cultures); and dogs and cats (which, in the West, are good for companionship, guarding the house or catching mice but not good to eat). Except for chickens, ducks and turkeys which are birds, the rest are mammals. In other words, the word 'animal' would, in the context of utility derived from domestication, typically conjure up either of these two classes of the phylum chordate. In some cultures, some species of fish have been domesticated, and increasingly today salmon and trout are also being cultivated. However, domestication, throughout world history, has been confined to only a few species of these vertebrate classes.

In contemporary consciousness, the image or idea of what counts as an animal has become even more circumscribed, as increasingly in urban contexts, domesticated animals are not directly encountered. Some children even have difficulty associating pork with an animal called the pig or milk with the cow, pork and milk being just packaged items the family purchases from the supermarket. This means that animals as domestic pets occupy centre stage; especially in developed countries, cats and dogs are the most prevalent. Children, increasingly, are taught to identify animals via these exemplars. For them, the denotation as well as the connotation of the term 'animal' is paradigmatically given and determined by the varieties of dogs and cats they find in the household. If asked whether they share their homes with animals, they would confidently say no provided they kept no cats, dogs, budgerigars, goldfish or hamsters. If reminded that most homes, and therefore, theirs, would have a mouse or two, they would feel justifiably shocked. But to them, if compelled to acknowledge their presence, mice are not animals in the way pets are animals – they are at best animals only in some technical sense. To them they are just pests. And if told that mites live in the detritus of their scalp or upon their skin, and in their carpets, they would be horrified; unlike mice or rats, they would even have difficulty accepting or understanding these as animals at all.

To sum up, the lay consciousness increasingly is confined to grasping animals in terms of a few domesticated species of mammals which are regarded as friends to humans or of a few exotic and/or charismatic animals which they see occasionally in zoos (as we shall see).

Conceptions presupposed by animal welfare/liberation and animal rights

We next turn to the conception of what counts as an animal via the defence of animals against human cruelty; it is essentially a protest against the ways in which animals are: (a) kept/caught and then slaughtered for food or other human purposes; (b) used in scientific research and experimentation whether for the serious purpose of saving human lives or the relatively more trivial one of improving the appearance of human bodies; and (c) hunted, hounded or killed for human pleasure.

The more traditional justification, derived from philosophers like Kant, is that the duty not to be cruel to animals is in reality an indirect duty to humans, as the infliction of cruelty upon animals could dispose us to be callous towards fellow human beings. But of late, this highly anthropocentric standpoint has been powerfully challenged by two contemporary philosophers – Peter Singer (1976) and Tom Regan (1983) – who, in spite of the obviously different philosophical stances each has adopted, nevertheless are united in repudiating the dominant humanist tradition of Kant and the Enlightenment, at least regarding the treatment of animals.

A minimalist reconstruction of Singer's philosophy of animal welfare includes the following:

(a) *The hedonic postulate.*   Pleasure and pain as mental states are respectively intrinsically good and evil.

(b) *The consequentialist/utilitarian postulate.*   One ought always to maximise pleasure and minimise pain in one's actions.

(c) *The boundaries of sentience postulate.*   (a) and (b) are 'blind' to the kind of being which is capable of feeling pleasure and pain. Humans clearly are sentient but empirically it can be shown that humans are not the only sentient beings. Other mammals, too, clearly are sentient. Birds are as well. Erring on the side of caution and charity, the boundary should then be drawn somewhere around shrimps and, possibly, lobsters.

(d) *The consistency postulate.*   As we, today, believe that we have a moral duty not to keep and eat fellow humans for food, to perform vivisection on them with or without their consent, to hunt, maim or slaughter fellow humans for entertainment, then equally, we have a moral duty not to do likewise to fellow sentient beings.

A minimalist reconstruction of Regan's philosophy of animal rights includes the following:

(a) *The rights postulate.*   (i) an entity is intrinsically (or inherently, in Regan's terminology) valuable if and only if it is capable of being the subject of a life, that is to say, possessing memory, beliefs and desires as well as other mental states, and (ii) an entity is a rights holder if and only if the entity is capable of being the subject of a life.

(b) *The conceptual postulate.*   To be the subject of a life, to experience mental states like beliefs and desires, conceptually speaking, it is not necessary to possess verbal language at all or human language as we understand it to be.

(c) *The boundaries of the subject-of-a-life postulate.*   (a) and (b) are 'blind' to the kind of entity which can satisfy the criterion of being the subject of a life. Humans (or at least the majority of them) are clear candidates, but empirically it can be shown (once (b) has been conceded) that mammals, too, are candidates. Erring on the side of caution and charity, the boundary of eligibility should then be drawn at birds.

(d) *The consistency postulate.*   As we, today, hold the view that human beings have a right not to be kept and eaten by fellow humans, to have vivisection performed on them with or without their consent, to be hunted, maimed or slaughtered by fellow humans for entertainment, then equally, other mammals (and possibly birds) have a right not to be treated likewise by us humans.

There are obvious differences in the philosophical foundations provided by Singer and Regan and the debate which ensues between the two sides – one is anchored in moral duties understood in the context of hedonic consequentialism, the other in moral rights, deontologically understood, in the context of certain characteristics of mental life in humans and closely related mammalian others.[3] However, these differences notwithstanding, the two do have certain things in common, apart from their agreed common goal to end cruelty to, and the suffering of, animals. Their respective implicit conceptions of what is an animal are given by the criterion they each have chosen as the most fundamental postulate of their philosophy of animal liberation – the hedonic postulate in the case of Singer and the rights postulate in the case of Regan.

In either, the paradigmatic animal is the human animal. Although Bentham, as the commonly acknowledged founding father of modern utilitarianism, had said that certain animals also come within the purview of his fundamental postulate, nevertheless, utilitarianism, as propagated and inspired by him, has chosen to concentrate on humans as the paradigmatic sentient beings. Similarly, the concept of rights – understood either as natural or contractual rights – has long been conducted, until very recently, within an exclusively human domain.

Singer uses the image of the expanding (moral) circle, in order to draw certain other beings, so far excluded by modern Western philosophy, into its orbit. Regan endorses this implicitly. However, both proceed on the assumption that there is a limit to which this circle may be enlarged – Singer's fundamental postulate allows him to redraw it with a somewhat wider radius than Regan's. But in the centre of their circles is the human. The further a being is from that centre, the more difficult it would be to make a case for bestowing on it the status of being morally considerable.[4] The human is, of course, a mammal. Hence, extending moral duties or rights to fellow mammals is their most obvious target. This has prompted some commentators to say, especially in the case of Regan's account, that it is really about mammalian rights.

In general, it might not be too unfair to say that both philosophies are underpinned by an overarching postulate, namely the search for similarities and likenesses between humans and certain animal others.[5] As such, the more an animal resembles humans in certain specified ways, the easier it is to argue for admitting them into the moral circle.

Of the mammals, the Great Apes come closest to us – this class is held to consist of the gorillas, the orang-utans, the chimpanzees and then ourselves as the long-missing fourth Great Ape.

While those animals within the pale are accorded a dignity befitting their newly acquired status of being morally considerable, those outside, as a result, are dealt a double blow – first, they are owed no moral duties and denied moral rights, and second, the term 'animal liberation' or 'animal rights' itself goes even further and serves implicitly to deny them the status of animality itself. In other words, only those beings which qualify to be the bearers of rights or to be the object of our moral duties are 'proper' or 'true' animals. The denotation and connotation of the word 'animal' has surreptitiously and subtly been revised so that even on Singer's more hospitable expansion of the moral circle, worms, molluscs and many more are debarred. The similarities postulate has forcefully challenged human chauvinism, the view which sets humans apart from other animals, assigning to themselves a superior status of privilege and domination.[6] It attempts to force human consciousness to concede that humans, as mammals, are really fellow animals. They (together with those admitted into the expanded circle) are owed duties not to be tortured and held to enjoy rights to life, and so on. Strictly speaking, in Singer's moral/political philosophy, a single hedonic consequentialist theory is postulated to embrace all sentient beings, from mammals down the evolutionary scale to possibly some crustaceans like lobsters, just as in Regan's moral/political philosophy, a single unified theory of rights is postulated, covering all mammals and possibly birds. However, the price for this revision is the construction of a new demarcation line between the ingroup and the out-group. Members of the latter are pariahs because they are unlike us in crucial respects, and therefore cannot be animals, a category to which we, ourselves, now belong. A hierarchical or class system remains in place – the franchised and the privileged against the nonfranchised and the disadvantaged. It is just that the former now includes not simply us, but those beings which are like us in certain selected aspects. Human chauvinism may have been vanquished but its spirit has not been challenged by either Singer or Regan and remains unexorcised in their respective philosophies.

The zoo's conception

We have seen how the denotation and connotation of the term 'animal' endorsed by the lay conception as well as that of Singer or Regan permit the lay public as well as animal theorists such as Singer and Regan to be selective in what they count as animals. The same selectivity holds in the case of the zoo conception of animals. However, it differs from the lay conception in that zoos largely exclude domesticated animals, while the lay public includes them in their account; otherwise, it concurs with it in all other respects.[7] In the words of one prominent apologist for zoos, zoos deal in the main with animals like 'the rhinos, the tiger, leopards, primates, parrots, the Asian elephants, many an antelope, many a bird of prey, various cranes and so on; all the creatures of our childhood; what most people mean by the word "animal" ' (Tudge, 1992, p. 46).[8] However, it may be fair to say that while both the lay public and zoos consider zoo animals to be exotic, for the former, the term 'exotic' carries the connotation of being from distant parts, uncommon, strange, or rare, whereas for the latter, the term is understood in the more technical sense of referring to the geographical origins of the animals, to the fact that their natural range is found in parts of the world other than where the zoo and its exhibits are.

The zoo image, then, powerfully reinforces the lay image of what counts as an animal. What zoos and ordinary people mean by the word 'animal' and the species the animals belong to, refers, then, only to a minuscule fraction of the total animal species known today to science; that number stands at 1 032 000 species, of which insects account for nearly three-quarters, standing at 751 000 species. Mammalian species only total 4000, with birds slightly more than double that at 8600.

The zoological conception

To get a more comprehensive conception of what counts as an animal we need to turn to zoology, which is commonly understood as the scientific study of animals. One of the Greek words composing the term itself – *zoon* – is usually translated to mean 'animal', although it may be more accurate to point out that it has a wider denotation, referring to living things.[9] Of course, zoology in turn is part of biology, the study of life itself – the Greek word *bios* means life.

So how does zoology answer the question 'what counts as an animal?' It will soon be obvious that as far as it is concerned, there is no simple and quick reply. Any systematic answer, no matter how schematic, starts with the, by no means easy, problem of first distinguishing between life and non-life. Although both the living and the non-living are subject to the same laws of physics and chemistry as well as the law of the conservation of energy, the crucial differences between them lie in the fact that the former is very differently organised and structured from the latter – unlike the latter, it is capable of metabolism, growth, adaptability, irritability and interaction with the environment.

But how in turn is animal life to be distinguished from plant life? As the terms 'zoology' and 'botany' themselves indicate, we, lay people, take for granted that there are two recognisable kingdoms to which all organisms are said to belong: plant or animal. We instinctively know to classify mosses, ferns and trees as plants on the one hand and mammals, birds and fishes as animals on the other. Yet this time-honoured Aristotelian schema may be said to have outlived its usefulness in the light of more up-to-date understanding of the various forms of life on Earth. Complexities appear straightaway. The central point to grasp is that, unfortunately, no single criterion exists which can serve to distinguish all animals from all plants. Take the presence or absence of chlorophyll as an obvious distinguishing mark. Chlorophyll is a necessary condition for photosynthesis to take place. We unhesitatingly associate chlorophyll with plants but not with animals; an oak has it but not a hedgehog. Under photosynthesis, green plants produce organic compounds from sunlight and atmospheric carbon dioxide and, at the same time, restore free energy to the biosphere. These photo-autotrophic organisms, in converting inorganic substances into organic materials, not only sustain their own functioning integrity but also provide food for heterotrophic organisms, namely animals, which live on them as they themselves lack the capability of photosynthesis. Yet some organisms, for example *Euglena*, display photosynthesis under some conditions but not others – in the light, it functions like a plant, in the dark, like an animal. So is it an animal or a plant? They are considered to be animals by zoologists and plants by phycologists. Another borderline group is the slime moulds – zoologists call them Mycetazoa and botanists Myxomycophyta. Furthermore, not all plants possess chlorophyll – the higher parasitic plants and a large plant group, Fungi, also do not have it. So the presence of chlorophyll cannot identify and include all plants; neither does its absence identify all animals.

Another distinguishing mark may be said to be motility. We commonsensically believe that animals have the ability to move about in their environment (and some even travel between very different environments depending on the season) at some stage in their life history, whereas plants are stationary. Yet movement is not restricted solely to animals – a good many of the thallophytes such as Oscillatoria, several bacteria and colonial chlorophytes are quite motile.

In biology today, scientists, as we have seen, no longer regard the two kingdom schema to be all that illuminating; instead, they attach greater significance to the prokaryote/eukaryote distinction. Prokaryotes refer to organisms which lack organelles, that is, specialised structures such as nuclei, mitochondria, chloroplasts; their DNA is not found in chromosomes but forms a coiled substance called a nucleoid. In contrast, the genetic material of eukaryotes is contained within a well-defined cell nucleus with a protein coat. However, besides this crucial difference, there are others. All living organisms, except viruses, bacteria and blue-green algae (cyanobacteria) are eukaryotic.

The problems mentioned above, amongst others, led to a proposal in 1969 for a five-kingdom system which incorporates the prokaryote/eukaryote distinction. The prokaryotes are assigned to the kingdom Monera while the eukaryotes are divided into four kingdoms. The kingdom Protist includes the unicellular eukaryotic organisms (protozoa and unicellular eukaryotic algae). The multi-cellular eukaryotic organisms are in turn organised into three kingdoms according to their mode of nutrition and other significant organisation differences. The kingdom Plantae contains multi-cellular photosynthesising organisms, higher plants and multi-cellular algae. The kingdom Fungi includes yeast, moulds and fungi that get their food through absorption. Finally, the kingdom Animalia comprises the invertebrates (except the protozoa) and the vertebrates.

Traditionally, the animal kingdom has included the unicellular protozoa, but the new schema excludes them. Yet they share many characteristics with so-called animals, such as ingestion of food, advanced locomotory systems, sexual reproduction, and so on. For this reason, books on zoology regard the protozoa as animals and have a chapter on them.

To sum up a very complex set of issues, it may be fair to say that zoologists today clearly identify all invertebrates and vertebrates as animals while agreeing that the protozoa, too, may be considered as such.

One should also be reminded that the consensus which emerges takes place against a background of theoretical ideas and assumptions which have developed since the nineteenth century, of which the most salient are: (a) the Darwinian theory of natural evolution in terms of natural selection; (b) the Mendelian theory of particulate inheritance and the gene/chromosome theory; (c) DNA genetics and molecular biology;[10] (d) developments in cell theory; (e) animal ecology; (f) ethology; and (g) an understanding of intra-ecosystemic interdependence.

The larger framework is still basically neo-Darwinian. It is informed by the notion of natural selection as the mechanism of natural evolution. Its primary object of study is therefore animals in the wild, to understand their ancestry and evolutionary history, how they come to have the characteristics they posses through certain fundamental concepts and principles that govern the understanding of organic life in general and animal life in particular.[11] This basic orientation is informed by four deep themes in the philosophy of the (wild) organism:

1. *Biology of time.* Evolution cannot be detached from time and history. As Steven Rose has put it: 'Nothing in biology makes sense except in the light of *history*, by which I mean simultaneously the history of life on Earth – evolution, Darwin's concern – and the history of the individual organism – its development, from conception to death' (Rose, 1997, 15).

2. *Biology of space.* Added to the time dimension is the space dimension, which in organisms is expressed through their structures. Organisms have forms which undeniably change over time, yet they persist as they go through their trajectories in life.[12]

3. *Biology and philosophy of reciprocal causation of organisms-in-the-environment.* Organisms cannot be considered apart from their environment.[13] Organisms are the product of a complex set of dynamic causal interrelations between themselves and their environments. As Rose has put the matter so well, it is appropriate to quote him:

> for the organism, (the environment) is the external physical, living and social worlds. Which features of the external world constitute 'the environment' differ from species to species; every organism thus has an environment tailored to its needs. ... *organisms evolve to fit their environments, and environments evolve to fit the organisms that inhabit them.* No environment is constant over time. ... Stasis is death. ... Organisms – any organism, even the seemingly simplest – and the environment – all relevant aspects of it – interpenetrate. Abstracting an organism from its environment, ignoring this dialectic of interpenetration, is a reductionist step which methodology may demand but which will always mislead. ... organisms are not passive responders to their environments. They actively choose to change them, and work to that end.
>
> (Rose, 1997, p. 140 [my emphasis])

It is not good science to regard an animal in abstraction from its evolutionary history and context or from its natural habitat which informs its very existence and with which it interacts.[14]

4. *Environmental philosophy of ecocentrism.* This philosophy focuses on the conservation of (natural) biodiversity, on the survival of animal

species rather than of individual animals, irrespective of whether or not they are charismatic, exotic, capable of suffering pain or of mental activity.[15]

The remainder of the book will tease out the implications of the zoological conception of animal in order to throw ontological light on the respective status of wild animals in the wild on the one hand, and captive animals in zoos on the other.

Conclusion

This chapter has briefly set out five conceptions of what counts or does not count as an animal. We shall have occasion, as the arguments in the book develop, to return to them in order ultimately to assess the different justifications for zoos; we shall also have to go beyond this initial exploration of the different denotations and connotations of the word 'animal' as embodied in these conceptions to unravel in greater detail the conception of zoo animals as opposed to that of wild animals (not in captivity), before attempting to answer the questions and address the issues raised in the Introduction.

2
Animals in the Wild

This book in the main is an attempt to make explicit the presuppositions and tease out the implications of the zoological conception of what counts as an animal, in order to clarify the ontological difference between wild animals on the one hand and zoo animals on the other. This chapter begins the task by looking in greater detail at animals in the wild in order to characterise their ontological status before turning attention in the chapters that follow to determining in turn the ontological status of animals in zoos; it is a fundamental contention of this book that the former is the ontological foil to the latter, a point which will become clear as the arguments unfold.

Natural evolution and natural selection

The zoological conception so far outlined of what counts as an animal has made it clear that animals in the wild are the products of natural evolution. Natural evolution means that all forms of life as we know them to be today and historically are/were descendants of the first form of life on Earth. Once upon a time, nearly 4 billion years ago, Earth was more or less devoid of life. When life did appear, it was first in water as microbial mats. The first organism was prokaryotic and single-celled. Then the 'higher' eukaryotic organisms appeared about 1.8 billion years ago, at first as single-celled, later as multi-cellular. It was not until the Cambrian explosion, 540 to 500 million years ago, that macroscopic animals appeared in abundance to give rise to the types which still exist today. Apart from the protozoans, as already observed in Chapter 1, the kingdom Animalia comprises the vertebrates and the invertebrates, dating largely from the Cambrian period.[1]

According to natural evolution, the beginning of all life forms and, in principle, the end of all life forms on Earth have nothing to do with either gods (God) or humans. No divine agency of any kind is necessary to account for the origin of life and its evolutionary history.[2] It relies on natural selection to account for the latter. Natural selection – that mechanism working upon genetic variations in individual organisms which constitute populations of organisms interacting with other components of their ecosystems – primarily explains the course of evolution.[3] Their origins pre-dated human existence by about 4 billion years and their evolution, until very recently, also occurred unhindered or relatively unhindered by human activities. However, the human species today, in virtue of its success as a species with a population of over 6 billion and still growing, and its present unconscionably extravagant mode of production and consumption in terms of exhausting Earth's resources, especially amongst the world's mature economies, is feared to be capable of destroying most of Earth's (natural) biodiversity. Under the more pessimistic scenarios, it is even capable of almost wiping out the kingdoms of Plantae and Animalia (including the human species) should some crazy state in possession of nuclear bombs release them in the spirit of *après moi, le déluge*.[4]

Trajectory and the distinction between 'by themselves' and 'for themselves'

The brief account of natural evolution and natural selection given above must be supplemented by an exploration of a new notion, namely trajectory and the distinction, inherent in it, between beings which exist 'by themselves' and beings which 'exist for themselves'.[5]

Every naturally occurring entity or process has its own trajectory. The term 'trajectory' is introduced to do the precise job of referring to the history as well as the character of its autonomous existence, whatever the entity or process may be. For example, a lake has its own trajectory. As a geological form, it is considered to be one of the most transient – it dries out in a relatively short span of geological time, first by becoming a swamp, then probably a meadow. In the case of naturally occurring processes, these may be biotic, abiotic, or an interaction involving both. It is also the case that a species (for instance, *Canis lupus*), as much as an individual member of the species, (the particular wolf roaming at a particular time in a particular forest), may be said to have their own respective trajectories, although as we shall see in the next section, the trajectory of the species is not necessarily identical to that of any of its individual members.

Similarly, the use of the term to cover both the biotic and abiotic domains does not entail that their trajectories are identical, that what is true of the one is also true of the other. We have already seen that, as far as present evidence goes, the history of Earth shows that the abiotic long preceded the biotic on Earth. Present evidence also shows that without the continuing existence of a certain combination of abiotic conditions, the biotic would not, and could not, continue to exist.

Two further points need to be emphasised. One concerns the crucial differences between the biotic and the abiotic; the other is to argue that in spite of these differences, it is appropriate to use the notion of trajectory to talk about the entities and the processes in both domains.

First, the crucial differences: in general, individual organisms go through certain recognisable stages from their beginning to their end – infancy, growth, maturity, senescence (in some cases) and death. They also possess certain characteristics, depending on the particular stage of their existence or indeed, of their sex. For instance, in the trajectory of a frog, it starts off life by being an embryo, which soon develops into a tadpole, then an adult frog. The tadpole clearly looks very different from the adult frog; yet the former is but a stage in the growth of the latter. Similarly, the peahen looks very different from the peacock; yet both are members of the same species.

By contrast, the granite mountain remains a mountain of granite. Of course, even granite wears away over a large expanse of geological time. But granite does not 'mature' to become something else, although it is true even granite might weather away to become soil. However, it does not 'grow or develop to become soil' in the same way as the tadpole becomes the adult frog – hence the need, strictly speaking, to put quotation marks around the phrase in the former but not in the latter context. Granite may end up as soil, but soil is not granite – they are two very different things. However, in the case of the frog, the tadpole and the adult frog belong to the same species; each is simply a different stage in the trajectory of the individuals belonging to the species.

Another obviously important difference between the biotic and abiotic is that the former appears to be an exception to the laws of thermodynamics, whereas the latter is not. But, of course, the appearance is only misleading. However, what misleads one is that there are certain processes at work in organisms, which are absent in the case of the abiotic. Individual organisms are autopoietic; they sustain and maintain their own functioning integrity by engaging in metabolical and other physiological activities. They, therefore, appear to produce order out of chaos, so to speak; in reality, to produce order, they need to take in ordered material from external sources, in the form of nutrients which, ultimately, are dependent on solar energy itself. They produce disorder when they respire, defecate and when they die. In contrast, abiotic entities are not autopoietic or capable of self reproduction, as they do not possess any mechanisms analogous to those found in organisms.[6]

However, in spite of the admitted differences, one would want to argue that the term 'trajectory' may meaningfully be applied to both. Only the biotic may be said to be autopoietic, yet both biotic and abiotic nature may be said to be 'self-sustaining' and 'self-generating' in the larger sense of these terms, in spite of the fact that the latter does not possess physiological/metabolic, neurological, hormonal or reproductive mechanisms which the former has. In other words, the autopoietic is but a subcategory of the self-sustaining and the self-generating.

The abiotic as much as the biotic is 'self-sustaining' and 'self-generating' in the sense that it is autonomous, though not autopoeitic.[7] But, to be more precise about the autonomy of the different trajectories in general of biotic and abiotic nature, one needs to explore an important distinction between saying that an entity exists 'by itself' on the one hand, and that it exists 'for itself' on the other. While all naturally occurring entities, whether biotic or abiotic, exist 'by themselves', only the former exist 'for themselves'.

Naturally occurring entities and processes are precisely those which have come into existence, continue to exist, and go out of existence, entirely autonomously, and therefore independently of human intentionality and agency (and of supernatural agency for that matter). They do not owe their being in any way to humankind. They are also self-generating and self-sustaining. So, in so far as they exist, they can then be said to exist 'by themselves'. However, naturally occurring biotic entities display an additional dimension of complexity in their existence. In existing 'by themselves', they also necessarily exist 'for themselves' given that they are members of organic species, unlike abiotic entities which are members of inorganic natural kinds.[8] As organisms, they strive to maintain their own functioning integrity. They develop, grow, mature, replicate (and in the case of the higher animals, they nurture their young) – their success in these activities involves active appropriation of suitable nutrients, protecting themselves against adverse circumstances which would either harm or kill them.

Implications of natural evolution/natural selection in the light of the notion of trajectory

It follows from the historical fact of natural evolution and the explanatory mechanism of natural selection which underpins it that animals in the wild are:

1.   Naturally occurring entities: both as individual organisms and as species, they have come into existence, continue to exist, and will eventually go out of existence (in principle), entirely independent of human volition, manipulation and control.[9] In contrast are those entities which have come into existence, continue to exist and go out of existence precisely as the result of human intentions and manipulation. In the category of naturally occurring entities, the term 'nature' is understood in its fundamental sense as the ontological foil to the latter category which refers to human artefacts.[10] Human artefacts may be briefly defined as the material embodiment of human intentionality; as such, they are technological products, and may be either abiotic/exbiotic (like statues made of marble or wood), or biotic (like domesticated plants and animals).[11] In the case of biotic artefacts, the technology used is initially craft-based (as in the classical case of traditionally domesticated animals) or, latterly, science-induced (first based on the classical science of Mendelian genetics in the first half of the twentieth century and then on the sciences of molecular genetics and molecular biology over roughly the last 40 years).[12]

2.   To use slightly different language, one could say that animals in the wild (both as individual organisms and as species), as characterised above, exist 'by themselves'. So do abiotic entities, such as mountains and rivers. However, unlike the latter, biotic beings, such as (individual) animals in the wild, may be said also to exist 'for themselves'. By this is meant that they maintain their own metabolism and functioning integrity in the total absence of human intervention and manipulation; they keep themselves healthy, keep disease at bay until the moment comes when they fall foul of disabilities, ailment and/or fall prey to predators because of such weaknesses. Being self-maintaining, self-regenerating and self-reproducing, they may also be said to be 'autopoietic' beings.[13]

3.   In reproduction in general, the male fights off other contenders in order to mate with the female he fancies or encounters; the female, too, up to a point, chooses to mate with the male she fancies, although she usually ends up with the victorious male who has succeeded to fight off his rivals. Their reproduction is under their own control. Depending on the species, but generally speaking, the females are responsible for bringing up their young and initiating them into the way of life which is normal for the species to engage in. For example, the young will be taught to hunt if the species is carnivorous; to identify the right plants for foraging if the species is herbivorous; to learn about what constitutes danger; to get to know the terrain within its home range, and so on. Just to cite one example of such education: the mother orang-utan teaches her offspring, from its birth till the age of 7, a whole range of survival skills. The infant learns how to navigate the forest by learning how to make mental maps of distances, of food locations. The mother teaches it to identify and to obtain up to 400 different foods, as well as to swing from tree to tree with immense dexterity, to judge finely subtle variations in the strength as well as the suppleness of branches. It is a prolonged and intensive education, as if of an only child undertaken by the mother.[14]

4.   The species determines the social composition and life of the group, whether it is a group predominantly of mother (and a few other females) and their young, excluding any mature male who normally leads either an isolated life of his own (except during the mating season when he seeks contact with females) or as member of a small group of other similar males.

5.   In other words, every individual animal in the wild (as a member of a species) unfolds its own *telos*, its own trajectory from birth till death, as well as that of its species. It may be said to exemplify the thesis of intrinsic/immanent teleology.[15] The tiger hunts, while the giraffe forages; the male lion lives in a pride with a few other sexually mature males, while the emperor penguin shares the duties of parenthood equally.[16]

6.   Biodiversity may be a recently coined term but the notion refers to something which is of fundamental concern to scientists interested primarily in evolution leading to speciation – for instance, a single species of wasps which came to Hawaii 100 000 years ago has given rise to hundreds of species as the members of the original colonising population spread out, changed and evolved in response to the distinctive environments they found themselves in, which were peculiar to a particular island, mountain ridge or valley. As such, the scientists are interested, not so much in the individual animal (or organism) but in the species

and in the mechanisms of speciation, namely how changes in a population of individual organisms lead eventually to the emergence of two or more populations which no longer exchange genetic material with one another.[17]

Evolution of species means that a population responds not merely to genetic variations but to such variations in the context of specific environments. Over time, variations which prove to be adaptive may ultimately lead to the emergence of two or more species, as we have seen. This means that ecology in general, and habitats and ecosystems in particular, play a vital part both in the emergence and the maintenance of a species – such a view, in environmental philosophy, is also referred to as holism or ecocentrism.

According to this scientific understanding of animals (in the wild), behind the individual animal necessarily stands the species as well as the habitat of which both the individual and the species are a part. What one observes of the individual animals cannot be properly comprehended except in relationship with their species and, in turn, of both in relationship with the environment in which they have their being and function. As the example of the wasps in Hawaii demonstrates, the populations in the wild which constitute different species have evolved in the particular habitats they had found themselves in and to which they had adapted. The complex interactions between the different environments and the individual/species historically had led to their different evolutionary trajectories.

For the same reason, one cannot understand what a polar bear is, and what the species *Ursus maritimus* to which the bear belongs is without having an idea of the Arctic landscape and seascape, and the rapid series of evolutionary changes which its ancestors underwent in order to survive the cold.[18] Scientists believe that the polar bear is a descendant of a group of brown bears and is the most recent of the eight bear species; this group was stranded and isolated by glaciers in Siberia during the mid-Pleistocene period, that is, 100 000 to 250 000 years ago. While brown bears hibernate in the winter, polar bears do not, strictly speaking, do so, as food is plentiful in the Arctic winter. Their bodies are more elongated, and their faces have different shapes; their necks are longer which enable them to keep their heads above the water while they swim; their teeth are different as they are carnivores whereas brown bears are, in the main, if not exclusively, herbivores; they have warm thick fur and very large paws which make it possible for them to spread their weight on thin ice as well as to facilitate swimming. A polar bear's home range can be enormous, its size being normally dependent on the amount of food available – in one case, it is twice the size of Iceland, and in another, one satellite-tracked female travelled 3000 miles. A young polar bear may travel a distance of as much as 1000 kms (600 miles) in order to set up a home range of its own when it leaves its mother, after an apprenticeship of two to three years during which its mother teaches it how to catch seals, to respond to the seasonal variations in the availability of food, and so on.[19]

7. The individual animal is but a very transient member of its species. A species, as Holmes Rolston (1988) puts it, is a historical lineage. It comes to possess the characteristics it does as the outcome of an extended period of evolution which, in the case of the polar bear, as we have seen, began 100 000 to 250 000 years ago. Hence an individual animal properly understood against the background of its species is not an *ahistorical* being, as it is the product and an embodiment of evolutionary history itself. In other words, in observing a particular animal, one is not merely observing an individual being displaying whatever characteristic it does possess, but through it one grasps the whole historical dimension of its evolutionary past. This understanding of species refers to the evolutionary-species concept.[20]

The individual animal, such as the polar bear, may be said to be a token of its type or species (both as it exists today and in its evolutionary history) *Ursus maritimus*. But what is the precise relationship in this biological context between a token and its type?[21] First of all, membership of an animal species does not demand that all tokens are absolutely homogenous; membership can and does tolerate variations between individual tokens.[22] However, the fact that one token weighs more than another, that it is taller or longer than another, or that it may even look very different from another does not prevent them from being tokens of the same species.[23] Second, certain attributes of tokens cannot intelligibly be used of their species. For instance, male polar bears may weigh around 600 kg while the females are much smaller, around 400 kg. In other words, we can catch and weigh individual polar bears and record their respective weights; however, it makes no sense for us to say that we can catch and weigh the species. Third, although it makes sense to assign both birth and death dates to tokens and their species, it is true to say that while one can (in principle, especially today with the help of telemetry) record the precise birth and death of a token, one cannot (in principle) determine with any degree of precision the emergence or the extinction of a species. Fourth, an individual elephant is likely to live up to 65 years but the life span of its species is more likely to be a million years.[24]

In spite of some of the obvious differences between tokens and their species noted above, all the same there is a tight existential relationship between them. First, if no tokens whatever exist, then the species could be said to be extinct. However, in contrast, if only a few tokens exist, the species may be said to be as good as extinct, especially if only a few exclusively male or exclusively female tokens remain in the case of a sexually reproductive species. Even if a small population of male and female tokens of a reproductive species are extant, biologists would still, nevertheless, consider the species as doomed to extinction, as the collective genetic inheritance between them is too limited to make the species viable in the long run. Second, it follows that existentially speaking, a species is not something over and above its tokens – there is no immortal, eternal species existing in some Platonic heaven apart from its transient mortal tokens. Third, the extant tokens instantiate, at once, their own identity and trajectory as well as the identity and trajectory of their species.

Conclusion

This chapter seeks to establish two main points:

1. It argues that to grasp what an animal in the wild is, one must understand not simply what constitutes its own identity (in terms of what it looks like, its full range of behaviour), its trajectory, but also its identity and trajectory in relation to its species within the habitat and ecosystems to which it reacts/reacted in a complex manner here and now, as well as in evolutionary history. In other words, an animal in the wild must be suitably *contextualised.* The individual polar bear is a polar bear not simply because it looks a certain way, has certain anatomical/metabolic characteristics, behaves in certain ways with regard to securing its livelihood, its reproductive activities, to looking after and initiating its young in survival skills, but also because it is a token of a species which has evolved some 100 000 to 250 000 years ago from its ancestor – the brown bear – in the Arctic habitat, landscape and seascape.

2. It demonstrates that from the *ontological* standpoint, animals in the wild are *naturally occurring* beings; they are not the products of human design, manipulation or control, as their very existence and survival have nothing (in principle) to do with humankind.

3
'Wild Animals in Captivity': Is This an Oxymoron?

The last chapter attempted to shed light on the ontological status of animals in the wild as naturally occurring beings. The next five chapters will try analogously to clarify the ontological status of zoo animals. However, this chapter will grapple specifically with a basic problem of terminology, whether it is conceptually appropriate to refer to animals in zoos as 'wild animals in captivity'. It will argue that the term is an oxymoron. The phrase makes perfect grammatical sense, yet the terms 'wild' on the one hand, and 'captive'/'captivity' on the other, do not yield conceptual coherence.

'Wild animals in captivity'

If posed the question 'what are zoo animals?', zoo professionals, zoo scientists and commentators on zoos in general are very likely to reply: 'Zoo animals are wild animals in captivity.' The term 'wild animals in captivity' is, therefore, not something dreamt up by this author as a 'straw notion' for the convenient purpose of exposing it as being conceptually flawed, but one which actually occurs in the titles and in the texts of many books on the subject of zoo animals – just to cite one classical example: *Wild Animals in Captivity: An Outline of the Biology of Zoological Gardens* by Heinrich Hediger, the famous director of Basle's Zoological Gardens, published in its English version in 1950.[1]

From the standpoint of the critique of this book, it is important to point out that there are three types of 'wild' animals the visitors could encounter in zoos, even though they might not necessarily be aware of the difference between them. They could be looking at a jaguar (*Panthera onca*) caught from the wild who may not have been a zoo resident for long; they may be looking at a jaguar, who has never known existence in the wild, having been born and bred in the zoo, although the offspring of one parent (or both) who has (have) been caught from the wild; they may be looking at a jaguar, the offspring of parents both of whom have themselves been born and bred in zoos, either as first generation zoo-bred jaguars or several generations down the line of zoo-bred jaguars.[2] It will be shown that the critique mounted against the term 'wild animals in captivity' applies to all these three categories of 'wild animal'. However, as the strongest case in defence of the use of the term 'wild animals in captivity' rests with the first type identified, it would be fair to focus on it and to assess the strengths of the arguments against its use even in such a *prima facie* favourable case.

Freshly caught 'wild animal in captivity'

At first sight, it seems very natural to consider the jaguar just caught from the wild and delivered to the zoo as a token of the species *Panthera onca*, characterised in Chapter 2, as a naturally occurring individual organism belonging to a naturally evolved species, by and large, within its historical habitat/environment. Surely, the very act of removing such an individual animal from the wild and relocating it within a zoo environment amounts to a peripheral matter which could not undermine the very meaning of the term 'wild'. Take an analogous case with regard to a human – the court has described the defendant in the dock as a dangerously violent man who regards women as objects to be abused, and has convicted him of a nasty rape, sentencing him in consequence to several years of imprisonment. The fact that he is now behind bars, under captivity, so to speak, does not render it (conceptually) inappropriate for society to continue to refer to him as 'that dangerously violent man who regards women as objects to be abused'. The phrase is perfectly intelligible; there is no incoherence in using it in the context of his incarceration.[3] If the analogy holds, then there should be no conceptual inappropriateness in talking about an animal being 'wild' and being 'under captivity' in the same breath.

However, the analogy fails to hold. The attribute 'wild' is not the same as the attribute 'weighing x kg', or the attribute 'large but sleek' in characterising a wild animal. The (male) jaguar weighing 48 kg (120 pounds) when caught in the wild, give or take, will continue to weigh around 48 kg when received by the zoo, although the stress of being caught and then transported to another location might make the animal lose a few kilograms in weight. But to simplify the argument: assume that the jaguar has not lost any weight, we can meaningfully talk about that 'jaguar in the wild weighing 48 kg' as well as that 'jaguar now in captivity weighing 48 kg' – in other words, being made captive or kept in captivity does not affect the meaning or the sense of 'weighing 48 kg' in any way, although as already mentioned, being captive could, as a matter of fact, affect the weight of the animal in question.[4] However, as we shall see, the logic of the words 'wild' and 'in captivity' are more complicated than 'weighing 48 kg' or 'having stripes all over the body'; we have already briefly drawn attention in the two preceding chapters to the deep themes in biology, especially of time and of the reciprocal causal relations occurring in the context of organisms-in-the-environment, which establish that an animal-in-the-wild cannot be properly and fully grasped except in the context of its history and its habitat within which it (as a species) has naturally evolved and in which it (as a member of the species) lives.[5]

'Wild' is the antonym of 'tame'[6]

We begin to unravel that complexity here by showing that the term 'captive' undermines and subverts the meaning of being 'wild' in a fundamental way. To demonstrate the subversion, let us turn our attention, in the first instance, to a meaning of 'wild' which is implicit in the exploration of animals-in-the-wild, though not explicitly dealt with in Chapter 2. An animal-in-the-wild, as a naturally occurring being, has no intimate contact with human beings, and would by instinct run away from such contact. Hediger has characterised the situation well and it may be worthwhile to quote him at some length; he talks about an

> irresistible impulse towards the continual avoidance of enemies, this flight tendency dominant in the animal's behaviour, and its manifestation in specific flight reaction. This is released by enemies, in the predator-prey relationship. Man often plays the part of the predatory animal, in fact there is hardly a species of animal that has not been hunted by him, often for centuries or even thousands of years.[7] Thus it may be said that man, with his world-wide distribution and his long-distance weapons represents the arch-enemy standing, so to speak, at the flash-point of escape reactions of animals.
>
> Yet not every approach of the enemy touches off the flight reaction, nor is every approach necessarily a threat. The situation only becomes dangerous when the enemy approaches to within a certain distance of the animal – the escape distance. Only when this specific flight distance, which differs for each species, is overstepped by an observed enemy does flight reaction follow; i.e. the animal in

a typical manner runs away from it, far enough to put at least its specific escape distance between itself and the enemy once again.[8]

(1968, pp. 40–1)

Given such a basic powerful impulse to run away from its perceived enemies, zoo visitors ought to ask themselves the searching question: why do these 'wild animals in captivity' appear not to do so? This is because they have been tamed, that is to say, they have been re-programmed by their human captors to suppress that basic capability, or even to eliminate it totally. Hediger says: 'Removal of escape tendency, i.e. taming and tameness, can only come from man. He is the only creature capable of freeing another from the magic circle of flight, from the irresistible impulse to avoid enemies continually' (ibid., p. 49).[9]

The taming process requires skills and expertise, but if properly undertaken the wild animal would gradually lose its flight tendency, permitting its human keeper/carer to approach it and under certain circumstances even to move freely in its presence. Without successful taming, zoo management is just impossible; unless anaesthetic darts are used every time to knock them out when they are examined for husbandry purposes, zoo carers/keepers must be able to approach them without provoking either a violent reaction (injuring or even killing the human carers) or a flight reaction. In any case, minimally, they have to be taught to get used to the presence of humans, as their captive environment is nothing but an environment which is run and controlled by humans.

Taming is essential not only for zoo animals, but obviously also for circus animals.[10] What is perhaps not so obvious, but which Hediger (ibid.) points out, is that it is a procedure essential to the evolution of domesticated animals or domesticants. Such animals cannot remotely be useful to us humans, unless we first tame and then ultimately breed out their tendency to escape – what good is a horse's pulling powers or the hen's capability to lay eggs, if the former cannot be harnessed and attached to a cart/carriage and the latter cannot be made to live in and lay their eggs in the hen house? In other words, taming and training are essential to the process of domestication itself.

This insight, in turn, then calls for an examination of the issue whether domestication, or at least some understanding of the notion, can be said to be appropriately applied to animals in zoos.[11] However, this matter is raised here only *en passant*; a later chapter will deal with it in greater detail.

Zoo-bred animals, whether first generation or not but whose ancestors are wild, still need to be tamed, as they, too, have to overcome flight reaction. However, with animals born in captivity or who have become habituated to captivity, the flight distance is somewhat less than in conditions under freedom in the wild. As we shall see, such zoo-bred animals have gone further down the line of 'domestication' than their counterparts just caught from the wild and freshly made captive in a zoo.

Conclusion

For the moment, let us sum up the main points which have been established in the light of exploring the conceptual links between the terms 'wild', 'tame', 'captive':

1. The phrase 'wild animals in captivity' is an oxymoron. This is because 'wild' as an attribute is not of the same kind as 'weighing 48 kg' in the context of captivity. For captivity to occur, it requires that the wild animal be tamed, that is, to lose its flight reaction and no longer to regard humans as the perceived predator which it would in the wild. 'Tame' is therefore the antonym of 'wild' (in one of the senses of 'wild'). While the phrase 'wild animals in captivity' amounts to an oxymoron, the phrase 'tame animals in captivity' amounts almost to a tautology, as to be a captive animal is to be a tame animal, and to be a tame animal is to be a captive animal in the context of zoos and their management. Empirically, it is also the case that without taming and training the animals, freshly caught from the wild as well as zoo-bred animals whose original ancestors are animals-in-the-wild, to respond to human ways of handling and relating to them, zoos cannot function and operate.

2. Taming amounts to a fundamental change in the behaviour of the animal taken straight from the wild or whose ancestors belonged to the wild. This change, programmed and orchestrated by its human captors and keepers, is so profound that it can be said to amount to the first but crucial stage of domestication. As a result, the gap which zoo professionals and scientists perceive to exist between zoo animals on the one hand, and ordinary domesticated animals such as horses or cows on the other, may not be as great as it is made out to be. They are at different ends of the same spectrum. However, as it is taken for granted that zoo animals are, unlike cows and horses, not domesticated animals, they are then mistakenly and wrongly called 'wild animals in captivity', implying in this context that 'wild' is synonymous with 'non-domesticated' and is the antonym of 'domesticated'. But as we have seen, at the level of surface grammar the phrase makes sense, but at the level of depth grammar it is incoherent, contradictory and hence unintelligible and senseless.[12]

4
Decontextualised and Recontextualised

The last chapter has argued that it makes no conceptual sense to talk of 'wild animals in captivity' by establishing a fundamental meaning of 'wild' in terms of its antonym 'tame'. This and the chapters following will go beyond the conceptual issue to unravel the nest of complexities behind the meaning of 'wild animal' and at the same time explore a crucial matter underlying those complexities, namely the ontological difference in status between animals-in-the-wild and their counterparts kept in zoos as captive animals. The difference may be explored in terms of the following basic aspects in the transformation of wild animals into what this book calls 'immurated' animals in zoos, that is to say, in transforming naturally occurring animals to become, to an extent (to be clarified in later chapters), biotic artefacts at the hands of *homo faber*:

1. Geographical dislocation and relocation.
2. Habitat dislocation and relocation.
3. Lifestyle dislocation and reaccommodation.
4. Suspension of natural evolution.

In other words, the animal caught from the wild, first becomes decontextualised and then recontextualised according to human design and intention. In particular, this chapter will consider the first of these two aspects. It will also begin to tease out the implications of the official definition of zoos as institutions which house collections of animal exhibits open to the public, in order to establish eventually that zoo animals are indeed the ontological foil to wild animals-in-the-wild.

Geographical dislocation and relocation

Chapter 1 has already mentioned that zoo professionals differ from lay persons in the connotation of the term 'exotic'. The latter regard an exotic animal as simply one which is unusual, strange, rare or from distant parts; the former understand the term to refer merely to animals whose natural home range is in a part of the world other than where the zoo which houses them is to be found, irrespective of whether they are charismatic or rare. In other words, in spite of the differences, the lay connotation does, to an extent, overlap with the technical understanding of the term – 'being from distant lands/parts' is perfectly compatible with 'whose natural home range is in a part of the world other than where the zoo is'. In the majority of cases, as a matter of fact, it would be true to say that exotic animals found in zoos are from distant lands/parts and, indeed, because of that fact are considered by the public to be unusual, rare, strange.[1] For instance, historically, most zoos have been found in Western and Central Europe, North America, Japan and Australasia. The *World Zoo Conservation Strategy* (1993) reckons that there are probably at least 10 000 zoos in the world, of which just 1200 could be said to be members of national or regional zoo associations – amongst this more exclusive group, North America counts 175, Europe 300, East Asia 545, Australasia 30, Latin America 125 and Africa 25.

In other words, zoos which have the most resources and are the best organised are zoos in the developed world which, by and large, are found in temperate climes, in the Northern as well as in the Southern Hemispheres, whereas the more exotic (in the lay sense of the term) and charismatic of the zoo animals, such as the lion, the rhinoceros, the chimpanzee, the elephant, the jaguar are also exotic in the technical sense. These have been transported to more temperate climes from their natural home ranges in tropical and savannah Africa (respectively the rhinoceros and the lion); in Western Africa, lying just above the equator from the coast to inland (the chimpanzee in tropical rainforests, woodlands, swamp forests and grasslands); in India (the elephant); and in the tropical forests of Central and South America (the jaguar).[2]

If it is correct that zoos deal with exotic animals, *ex hypothesi* it follows that its captive residents have been subject to geographical dislocation. The polar bear is no more a native of Australasia than the penguins are natives of Western Europe, or the elephant (whether Indian or African) of North America. As these animals have been wrenched from their natural home ranges, they have necessarily become decontextualised. Different geographical locations in the main mean different geographical features, different climates, different flora and fauna, different habitats, different environments *tout court*.[3]

Such exotic animals have to be recontextualised within a zoo setting. That is to say that zoo management has to provide them with a new setting. In the bad old days, their new setting was a cage; today's more enlightened zoo management provides them instead with an *ersatz* habitat and environment. The operative word is *ersatz*, as the most which can be offered is a naturalistic, that is, a simulated one, added to which is a programme of environmental enrichment.

While an examination of the notion of environmental enrichment will be deferred to the Appendix, that of a naturalistic environment needs to be looked at in brief details now.[4] However, before doing that, one must first focus on a fundamental point concerning the recontextualisation of such exotic animals.

Recontextualised as a collection of exhibits

In the wild, as we have seen, animals live 'for themselves' in following their own trajectories and unfolding their own *tele*. They do not live for us humans, no matter how fascinating we may find them, no matter how desirous we may be to understand more about them from the standpoint of pursuing scientific knowledge. But the moment they become exotic animals, they lose the status of existing only 'for themselves'; instead, they become incorporated into a human structure and a human project.

A zoo is a human social institution like a museum. Every museum has a collection policy regarding the type of artefacts it aims to acquire, in identifying gaps in its extant collection, and the like. While a costume gallery concentrates on acquiring garments belonging to a certain culture and a certain period in history, while a natural history museum collects stuffed birds, mammals and so on, a zoo exists to collect (a certain number of) exotic animals which are not stuffed, but alive and living. A collection in each of these three kinds of institutions just identified is necessarily a *collection of exhibits*.

According to *The World Zoo Conservation Strategy* (1993), what distinguishes zoos from other sorts of collecting institutions is precisely this: 'The fact that all zoos exhibit living specimens of wild animal species is what underscores the difference between zoos, most museums and other cultural or recreational institutions, and is what gives zoos their own unique character' (ch. 1.3).[5] It refers to 'different concepts in exhibition that play an important role in establishing the character of a zoo' (ch. 1.2).

Every zoo, therefore, necessarily possesses a number of exotic living animals which constitutes its collection of exhibits. By turning animals from the wild into exotic exhibits, zoos have robbed them of their onto- logical status as beings who live 'for themselves' and transformed them into beings who live for us humans, to serve our ends. As we shall see in greater detail later, they no longer live out their own trajectories and their own *tele*, but are constrained to exist in accordance with a plan, a purpose dictated by a power outside of themselves. This is to say that the thesis of intrinsic/immanent teleology has been displaced by the thesis of extrinsic/imposed teleology – as a collection of exhibits, they serve the ends which humans claim zoos are designed to advance, be these to educate (humans), to make them the object of scientific research and study, to save them from extinction in the wild, or provide entertainment by way of a day's outing for the family. Their value in the context of such a human project is, therefore, an instrumental one. In the wild, animals, while living 'for themselves' may, by happy coinci- dence, also have instrumental value for us humans; but in zoos, as part of a collection of exhibits, they assume (almost exclusive) instrumental value for their human captors/keepers/visitors.

A population of cheetahs or a population of antelopes in the wild is not a collection of exhibits – the population is not a collection, nor are its individual members exhibits. Similarly, a population of cheetahs and a population of antelopes coexisting in the wild are not a collection of exhibits; safari parties may set out in their specially constructed vehicles to roam their habitats with the explicit purpose of looking out for them, and if luck really holds, of even coming across a cheetah in hot pursuit of an antelope. The fact that such tourists can see them does not necessar- ily turn them into mere exhibits; in any case, these animals may even adapt their hunting habits to avoid the intrusion of such human voyeurs by no longer hunting during the day but at night, or at a time when these unwelcome lookers-on are not about. But if such animals are captured and relocated to zoos, then their *raison d'être* is precisely to be visible, to be seen by humans, at times predetermined by zoos. If certain animals do not, alas, keep zoo opening hours, they are made to do so – nocturnal ani- mals, in so-called nocturama, are made to come out 'at night', to be active, to hunt for the benefit of zoo visitors, and made to rest, to sleep during 'the day' with the help of ingenious artificial lighting.

A population of cheetahs in the wild, whose members hunt, rest, groom, mate, reproduce, look after and teach their young survival skills so that in turn they can grow up to hunt, to reproduce, to transmit knowledge to their offspring, is from the evolutionary point of view an important bearer of information, genetic, ecologic and cultural. But the living out of their existence and the transmission of such knowledge have nothing to do with human intentionality of any kind, and most certainly not with the particular human project of rendering them conveniently and suitably visible to the human gaze.

Animals-in-the-wild, as individuals or as species, are simply not exhibits. However, once made captive in a zoo, they become exhibits, much as Leonardo da Vinci's *Mona Lisa* hanging in the Louvre is an exhibit, much as the skeleton of a dinosaur hanging in a natural science museum is an exhibit. The difference between a stuffed polar bear and a polar bear in a zoo is that one is a dead organism and the other is a living organism – they both have labels, so to speak, hanging round their necks and are displayed to advantage for the benefit of those who visit natural science museums or zoos. Just as one would design special cabinets, cases, installations to show off important exhibits in an art museum, one would do the same for certain animals in zoos. The bare cage is not necessarily the only way by which one may exhibit the animals. Elaborate structures have been built to show them off 'at their best', according to zoo-speak. These clearly embody a desire not merely to put such animals under human control and management but also to incorporate them into the culture of their owners and controllers. In 1856, at a time when Europe was particularly gripped by the Pharaohs, their rediscovered tombs under the pyramids, not to mention the deciphering of their hieroglyphs, Antwerp zoo, for instance, built an Egyptian temple whose walls were decorated with frescoes and bas-reliefs depicting scenes from ancient Egyptian arts to display its elephants, giraffes and camels. Also popular were Moorish structures and Asian pagodas. Budapest zoo even went in for the Byzantine style of architecture when it built its elephant house, complete with Byzantine church interiors.[6]

One zoo expert even goes so far as to say that modern zoos present spectacles based on a combination of the techniques of museum exhibition and theatrical performance.[7]

> exhibits are designed to imitate the natural habitat of the species, which will enable the animals to express natural behaviours with the zoo environment. In some respects modern zoo design may be compared with a theatre where the animals are the actors, the exhibit design is the scenery, the theme or the story being interpreted is the play, the visitor area is the auditorium and the visitors are the audience.
>
> (Andersen, 2003, p. 76)

Of course, this insight is absolutely valid; however, it is proffered by the author with no apparent awareness that such a context is incompatible with the implicit claim that the animals under display continue to enjoy the status of 'being wild' animals expressing 'natural behaviours'.[8]

Habitat dislocation and relocation: the naturalistic environment

In the wild, as we have seen, animals live 'for themselves' at their own pace, in their own ways, under circumstances of their own choosing in habitats within which they have historically evolved and lived; in zoos, willy-nilly, they have to live for humans, no matter how enlightened the philosophy of zoo management regarding the conditions under which they are exhibited, as exhibited they must be. In other words, they have to live under conditions not of their own choosing but under those designed and chosen for them by their human keepers/carers.

Today, enlightened zoos endorse a philosophy of management which wants nothing to do with the oppressive image of the bare and barren cage. Instead, they prefer to allow the lion to move about within a naturalistic environment. What, then, is such a space? The very word 'naturalistic' says it all; it is a space which has been designed and engineered to 'look natural' within which the animal can be displayed to better advantage as an exhibit. This is its fundamental *raison d'être*; however, to say this is not to deny that other justifications have been put forward. For instance, enlightened zoo management says that it is intended to improve the quality of life of the animals, which undoubtedly it does to a limited extent, or that such a setting would provide better educational opportunities for the visitors. Whether these claims are mutually compatible remain to be seen.[9] However, as we shall see, the latter two are at best secondary justifications and cannot subvert and displace the fundamental one of maximising the conditions under which visitors could appreciate the animals which are and must be seen as exhibits.

It is instructive to quote at some length what an expert on the subject of animal exhibit design has to say which corroborates the analysis above:

> The foremost goal of most zoological institutions is to use the entertainment value of live animals and re-created foreign worlds to draw people into an educational situation, in which they will learn about

and gain respect for the animals they are observing, as well as nature in general. Encouraging visitors to lengthen their stay in the park is important to increase these educational opportunities, as well as increase the institution's income (from concessions sales, for example) so that it can continue to thrive and carry out its work.

Visitors spend the most time at an animal exhibit when the animals are close-up and active . . . Most animals however, prefer to keep their distance from their human observers, if space allows. When space doesn't allow, i.e., they are in a small enclosure with no visual cover, these animals suffer psychologically, resulting in their being inactive, or in their performing unnatural and undesirable behaviours. Unnatural behaviours detract from the educational value of the exhibit, as they do not accurately represent wild animals. The sight of an unhealthy or unhappy animal will not instil in visitors a sense of awe for the natural world and may contribute to a poor reputation of zoos.[10]

However, give animals vast, heavily planted spaces in zoos, in which they can lead a natural lifestyle in relative privacy, and visitors will not be able to benefit from observing them, as the the animals will often be out of view. This is why designing animal exhibits that are truly based on wild situations, and considering only the absolute needs of the animals, is often secondary in zoo exhibit design.

(Worstell, 2003, Introduction)

Furthermore, the provision of naturalistic environments must also be considered from another angle, namely, that it entails a severe monetary cost, which has to be taken into account, thereby restricting the construction of such designs mainly to exhibition areas while excluding off-exhibit (off-stage, behind-the-curtains) spaces and laboratory enclosures.[11] It is refreshing occasionally to find such a frank admission amongst certain zoo professionals such as the one cited below:

Animals on exhibit in zoos are presumed to benefit from the environmental complexity provided by naturalistic habitat enclosures . . . If the naturalistic exhibit also provides educational benefits for zoo visitors, then – so long as the enclosure does not compromise the residents' health – the time and effort may be justified. For laboratory enclosures or off-exhibit areas of zoos where every effort to enrich has a monetary cost, this may not be true.

(Crockett, 1998, p. 130)

The above simply confirms once again that the underlying fundamental assumption in today's enlightened zoo management philosophy is that animals in zoos are first and foremost exhibits.

Every design has to satisfy three basic requirements – comfort of the animal, the distance between the animal and the visitor, and habitat simulation – which are to be reconciled in the following way:

> Small enclosures, even when appropriate for the size and number of animals on display, will appear as a cage rather than a natural environment, unless they possess the illusion of being an undeterminable space. . . .
>
> In the design of animal exhibits, the optimal size for the animal enclosure achieves a perfect balance between animal comfort, animal proximity to visitors, and the portrayal of an essence of habitat. However, the relationship of the animal to the visitor, as well as enclosure features, influence the perceived space. The perception of the enclosure size by visitors and animals will be influenced by visual illusions and psychology.
>
> (Worstell, 2003, ch. 2).[12]

According to this way of thinking, the zoo designer tricks the human visitors into believing that the space is infinitely larger than it is in reality. Worstell (ibid.) shows a photo of the gorilla space at Burgers Zoo at Arnhem, Netherlands, which she says 'appears to be endless'. Endless to the human visitor, yes, but would it also be so to the animal which inhabits that space? Zoo professionals may be wrong in the assumption that the illusion works for both humans and animals. The former stand outside the enclosure and gawp at the animal within the enclosure. Visual illusion works better when the perceiver is at a distance and when the perceiver never gets any closer to what is perceived. Travellers in a desert report the mirage phenomenon, that they see water in the distance. However, we and they know that this is an illusion and not reality, for the simple reason (apart from the scientific one) that when the travellers get to the very spot where the mirage occurs, they will find that there is no water. One can only happily take illusion for reality if there is no other reality check available. The zoo visitors are exactly in such a position; they can live happily with the designed illusion that the space occupied by the exhibit is 'endless'.[13] The gorillas, however, are like the travellers in the desert and they would surely know, after the initial deception, that the space they occupy is not 'endless'. They have reality checks not available to the human visitor standing outside the enclosure. Their initial visual perception would be very quickly corrected once they start to stroll around the enclosure. So, who is the deceived party? It is surely not the inmate of the enclosure, but the human visitor whose perception has been deliberately manipulated by zoo designers.[14] After all, it is precisely the job of the designers to produce under conditions of simulation just such an illusion, in order to maximise the opportunities for exhibiting the captive animals – zoo visitors no longer want to see or derive maximum satisfaction from seeing animals in cages but prefer to see them in a naturalistic environment. Habitat simulation which consists of 're-creating the essence of a natural habitat' is considered to be essential to the visitor experience.[15]

To simulate, let us say, an African tropical rainforest, the natural habitat of gorillas, the presence of trees in the exhibit enclosure is essential. However, maintaining live trees may be costly; furthermore, the animals love to peel the bark and may destroy the trees. So the ideal solution – that is, from the standpoint of zoo management – is to construct artificial, virtually indestructible trees which from a distance look as convincing as the real stuff with which they may be mixed. The public who is ignorant in general of such ploys is happy with the visual effect and that is what counts. After all, '[a] visit to the zoo is primarily a visual experience' (Worstell, 2003, ch. 5). The gorillas would not be duped but that is another matter.

In temperate climes, tropical vegetation may present problems. But one can, like Munich Zoo, for instance, select evergreen grass, which is native to Spain, as the cover for the natural ground of its primate exhibits. African rainforest plants have leaves which look like those of laurel (*Prunus larocerasus*). So laurel would do, as it possesses desirable characteristics such as large, shiny, year-round foliage; however, it is not recommended for use in primate exhibits as it is toxic, although Chester Zoo has successfully used it in its chimpanzee exhibit where the chimpanzees appear to avoid eating it. Temperate needle evergreens do not remotely resemble any plant found in African tropical rainforests, but they are recommended all the same for gorilla exhibits in order to create a green landscape, especially useful during the winter months, as these animals, on the whole, do not destroy them; however, plant them right at the back of the exhibit to make them less obvious to zoo visitors, just in case they recognise that they are natives of temperate, not tropical, forests. To sum up briefly this line of thinking, according to Worstell (2003, ch. 6):

> [i]t is not important that gorilla exhibits in temperate zones cannot be faithfully planted with rainforest plants. What is important is that the essence of the gorilla's natural habitat is conveyed. This can be

done with hardy plants, as habitat replication depends more on the character, growth habit, arrangement, spacing, massing and diversity of plants, rather than the actual species.

Habitat simulation is a perfectly coherent ideal within the context of exotic animals as a collection of exhibits for exposure to the human gaze. The average zoo visitor is not a botanist with profound knowledge of different habitats and ecosystems in the world. However, it may be true that the public knows this much, namely that while many trees (deciduous) in temperate climes shed their leaves in the autumn, evergreens do not; that in tropical climes, plants on the whole do not shed their leaves in the decisive manner which deciduous trees do in temperate latitudes; that trees in a tropical forest tend to grow straight and tall, throwing up a broken canopy under which shrubs grow. It is this minimal knowledge about plants and trees in different parts of the world which the zoo designer is exploiting, as evidenced in the discussion above. The simulation reinforces this general, scanty botanical knowledge while at the same time providing a pleasing illusion to the public that they are in the presence of the natural habitat of the animal which constitutes the exhibit.

Judged by such a measure, habitat simulation, properly constructed, may be a commendable success. However, more than this is claimed for it by its advocates. They wish to argue, as we have seen, that it is the re-creation of the 'essence of a natural habitat'. We have already referred above to enough details about its conception, design and construction to make the obvious point that simulation is the essence of a naturalistic environment.[16] A simulation of something could not by any stretch of the imagination be construed as re-creating the essence of something, unless it is assumed that the 'essence' of something is nothing more than what it appears to be at a distance. This is analogous to an art historian claiming that by digitalising Michelangelo's *David* or any other aesthetic artefacts, making the images available on its web page, allowing the viewer to rotate the images so that the back of the object can also be looked at, that such an attempt amounts to 're-creating the essence of the art object'. No reputable art historian, to one's knowledge, has made such a claim, which would only make sense if visual perception of an electronic kind alone counts as knowing the object. Museums do not boast that virtual reality captures the essence of the art object; virtual reality is no substitute for the real presence.

It is true that museums, on the whole, do not allow visitors to touch their exhibits, although it may designate a very limited number for touching and holding under supervision, and it may also allow students of the subject to have greater access to them for the purpose of teaching appreciation of the artefacts along with students acquiring more profound knowledge of the objects of their study. 'Do not touch' is the general rule for the public for the simple, valid reason that art objects are rare and handling by all and sundry is a recipe for disaster and destruction, whether short or long term. It is *faute de mieux* that museum visitors are confined to only looking at the exhibits; museums do not claim that the gaze – whether by actually visiting a museum or by calling up its contents on its sophisticated web sites which permit multi-media interaction – is all, though it remains true that the gaze is better than no gaze, should one wish to appreciate the beauty and complexities of artistic artefacts. It is true that the gaze in the presence of the artefact in a museum works best for paintings. However, in the case of artefacts such as statues, pots, armour, swords, and so on, these can only be fully appreciated not simply via the visual sense, but also the kinaesthetic, and indeed sometimes even the auditory sense, as in the case of (what the West calls) true porcelain which, when 'pinged', gives a distinct sound, a sound which forms part of its 'essence'.

Visual perception is, therefore, only one mode of exploring the world and acquiring knowledge of it, whether the agent is a human animal or a non-human animal (like the exotic ones which we see in zoos). It is true that we, humans, rely more on visual perception than other animals. One ascertains what is out there in the world by looking, and not so much by touching, smelling or tasting. Humans, in comparison to other animals, have less acute auditory and olfactory senses; furthermore, in leaving the hunting and gathering mode of existence behind them, to become, so to speak, civilised, these senses have become even more diminished in the course of human evolution.

Furthermore, it is also true that in modern epistemology, the visual is privileged over all the other perceptual modes. By the time the age of modernity arrived (roughly dated to the seventeenth century, as far as Western Europe is concerned), verification *via* observation, using the eyes either directly or with the help of instruments (such as microscopes, telescopes) became the dominant mode of appropriating the world and of gaining knowledge about it. Modern science celebrates visual perceptual knowledge, as it appears to satisfy the scientific methodology of objectivity; objectivity takes the form of measurement and quantification of data. Sight in this broad sense is considered to be more reliable as a source of knowledge because one can measure objectively what one sees – one can determine the length, width, height and weight of the table we see in front of us with the help of instruments, such as a tape measure, a weighing machine. Touch is unreliable except in a situation where an instrument, such as a thermometer is available to measure the temperature of the liquid, and so on. Taste is a mere 'secondary quality', said not to reside in the object itself (unlike length, breadth, weight, shape, which are 'primary qualities', are real and reside in the object), and a singularly subjective one. As such, it falls outside the domain of science.[17]

The claim made on behalf of habitat simulation appears to carry the above reductionist epistemology to an absurd extreme. The 'essence of a natural habitat' cannot be captured, first, simply through visual perception, and second, through the visual perception of a simulated habitat. If it could be thus captured, it would follow that the exotic captives are living in their natural habitats, as opposed to being merely perceived as such by their human visitors in an *essentially human cultural/institutional* setting and context, exclusively contrived by their human captors/designers to frame them *essentially* as *exhibits*.[18]

Conclusion

This chapter has argued that ontological dislocation follows from geographical dislocation and relocation, as well as from habitat dislocation and relocation via habitat simulation, when exotic animals are displayed as collections of exhibits in zoos. They are no longer naturally occurring beings which live 'for themselves' in their natural habitats, but beings which primarily have instrumental value for their captors/keepers/carers/visitors as exhibits.

5
Lifestyle Dislocation and Relocation

The last chapter has critically examined two aspects – geographical and habitat dislocations – in the ontological transformation of naturally occurring living beings (wild animals) to become exotic captive beings, relocated to a different setting and framed within a context deliberately designed by humans for the purpose of displaying them as exhibits. This chapter continues to delineate that transformation by looking at the dramatic changes to the lifestyle of wild animals when they are made part of a collection of exhibits in zoos. Their ontological implications will be teased out under two further aspects:

1. Spatial miniaturisation in habitat simulation.
2. Hotelification.[1]

Spatial miniaturisation in habitat simulation

It is obvious that today's enlightened zoos do not, on the whole, confine and exhibit their animals in a bare, concrete cage; instead, as we have seen, they exhibit them in a naturalistic environment or simulated habitat as part of their philosophy of environment enrichment. So is it fair, then, to talk of spatial confinement when their enclosure is no longer the mere size of a cage? This section will argue that it is; the argument begins by bringing certain salient facts before the reader.

Exotic captive animals in zoos naturally vary in size, although, on the whole, it is fair to say that, as far as mammals are concerned, they are not usually as small as the size of say a rodent or a bat.[2] So let us first focus on medium and large mammals, – carnivores and herbivores – as they exist in the wild, and highlight some of their distinctive characteristics in their original natural habitats.

The cheetah (*Acinonyx jubatus*) is different from the other cats in that its sleek body has evolved for speed, although the claim that it could do 68 mph (110 kmh) remains unsubstantiated.[3] The male weighs between 50 and 72 kg and the female between 35 and 63 kg. They are found in open habitats such as grasslands and semi-desert, but never in dense forests. The Indian cheetah became extinct in the twentieth century; today, remaining significant populations are found in Central and East Africa. The home range of female cheetahs is extensive as they may roam over 800 km^2; in the Serengeti, they follow the annual migration of the Thompson's gazelles. Males travel less extensively; they form coalitions numbering two to four, marking territories about 40 km^2. However, in the Kruger Park in South Africa, there is no Serengeti-like migration; females hold territories similar in size to the males which do not form coalitions.

The gorilla (*Gorilla gorilla gorilla*) is found in Africa, from south-eastern Nigeria to western Zaire and eastern Zaire into adjacent countries. The habitat of highland gorillas is up to 3000 m (10 000 ft) above sea level; western lowland gorillas inhabit secondary forests with widely spaced, slender-trunk trees with a broken canopy, permitting sunlight to penetrate; such landscape features make it easy for the gorillas to move about on the forest floor, as they are primarily ground-dwelling animals. They move about in troupes with a dominant male, travelling extensively. The troupe's home range is between 10 and 40 km^2 (4 and 15.35 mi^2); the group travels up to 5 km per day. Its nomadic lifestyle means that it is subject to changes in its surroundings on a daily as well as on a seasonal basis.[4] The adult male weighs 135–230 kg; the females are smaller.[5]

There are three main species of elephants in the wild: one Asian (*Elephas meximus*), two African – African savannah (*Loxodonta Africana*) and African forest (*Loxodonta cyclotis*). Most zoo African elephants belong to the African savannah species. The fully grown African bull can weigh between 14 000 and 16 000 lbs (6300 and 7300 kg) and grow up to 13 ft (4 m) at the shoulder; the Asian elephant is smaller, the average weighing 5000 lbs (2300 kg) and is 9–10 ft (3 m) tall. The distances travelled by the herd depend on the availability of food – when food resources are scarce, African elephants cover vast distances of several hundred kilometres. However, the median herd home range is (was) 113.0 km^2 for Asian and 1975.7 km^2 for African elephants. The Asian herds travel daily on average 3.2 km while the African ones cover 12.0 km. (Home ranges of 10–800 km^2 have been recorded for Asian elephants and 14–5527 km^2 for the African elephants.) The male, when in must (which can be at any time during the year), roams extensively looking for receptive females.[6]

Polar bears (*Ursus maritimus*) are the largest land predators. The species is the only one out of eight others whose diet is almost entirely carnivorous. The polar bear is about four foot tall and weighs 400–1700 lbs, with the male being much larger than the female. It lives on broken ice packs off the northern continental edges near the North pole (but not generally above 82° latitude), a region called the circumpolar Artic, which covers Canada, Alaska (USA), Russia, Norway and Greenland (Denmark). In the winter, the bear dens and the mother bear produces her cubs. However, the polar bear is not in true or deep hibernation, although its heart rate slows down, its temperature falls a little, and it stops urinating and defecating. Its sense of smell is acute, enabling it to detect prey – it can smell a seal more than 32 km (20 miles) away. It is extremely well insulated, not merely with its fur but also with a layer of blubber which can be as thick as 4.5 inches; as a result, it gives off no detectable heat and, therefore, cannot be picked up by infra-red photography. It is an excellent swimmer, paddling at 6.5 mph; it can swim for several hours at a time, and some polar bears have been tracked swimming non-stop for 100 km (62 miles). Polar bears move throughout the year within their individual home ranges, which can vary in size, depending on access to food, mates and dens. A small home range may cover 50 000 to 60 000 square km (19 305/23 166 square miles); a large one may be over 350 000 square km (135 135 square miles). Polar bears can travel up to 30 km (19 miles) a day for several days – one polar bear managed 80 km (50 miles) in 24 hours and another 1119 km (695 miles) in one year. The female with cubs hunts about 19 per cent of her time in the spring and double that in the summer, while the male hunts about 25 per cent of his time in the spring and about 40 per cent during the summer. Polar bears are on the whole solitary beasts, with two main social groupings – adult females with cubs and breeding pairs.[7]

A recent study based on a comparison of the home ranges of wild mammals in their natural habitats and the enclosure sizes of the simulated habitats of their captive counterparts in UK zoological collections has come to the following significant conclusion:[8]

> This (sic) results show that the bigger the animal the bigger the difference between its enclosure size and its natural home range. ... megafauna ... are kept in UK zoological collections in enclosures an average 1000 times smaller than their minimum home range. ... The average difference for all the mammals study (*sic*) suggest (*sic*) that mammals in the UK zoological collections are kept in enclosures about 100 times smaller than their minimum home range. Considering these

values, and applying it to human beings, a person that has lived in a village of 1 km² most of his/her life would be in the same spatial situation than (sic) a captive zoo animal if this person was (sic) confined for life to live in a telephone box. . . . 10 per cent of the taxa show enclosure sizes similar than (sic) the minimum home ranges, more than a third of the taxa . . . 10 times smaller, more than a fifth 100 times smaller and 2 per cent 10 000 times smaller. There are no cases, though, where average enclosure size was bigger than minimum home range. . . . It is to notice that these results only take in consideration area, not volume. Many species live in habits where the third spatial dimension is very important (tree dwelling, for example), so although they might be living in an enclosure of roughly the same size as their Minimum Home Range (like 10 per cent of the taxa of this study), they may lack the third dimension, and therefore they may still be restricted. . . . Our results show that the actual elephant enclosure sizes in the UK, as opposed to the (sic) recommended by zoo organizations, are in fact an average of 1000 times smaller. . . .

Mammals kept in UK zoological collections during the period 2000–2001 are confined in enclosures that, as an average, have an area 100 times smaller than their minimum home range.

Mammals with a body mass bigger than 100 kg (Megafauna) kept in UK zoological collections during the period 2000–2001 are confined to enclosures that have an average area 1000 times smaller than their minimum home range.

(Casamitjana, 2003, p. 13)

Medium-size mammals, such as the cheetah and the female gorilla are unlikely to fall into the 10 per cent of the taxa whose enclosure sizes are the same as those of the minimum home ranges. Of the elephant, Clubb and Mason have this to say:

A survey of Asian elephant enclosures in 20 European and North American zoos show(s) a range of 100 to 48,562 m², or 17 to 4937 m² per elephant . . . wild elephants roam over considerable distances. Even considering minimum wild home range sizes, the outdoor enclosures recommended by both the EAZA and AZA are in the region of 60 to 100 times smaller.

(Clubb and Mason, 2002, p. 41)

These figures concern only the size of the outdoor enclosures. One must, however, bear in mind that the average elephant in northern zoos spends up to 16 hours a day during certain periods of the year in indoor enclosures which are naturally much smaller than the outdoor ones whose major, if not sole, purpose is that of exhibition. In other words, these elephants have to spend a large proportion of their time within a space of just 36 to 45 m^2. We have also seen that the Asian elephant travels on average 3.2 km and the African elephant 12 km, and can cover as much as 17.8 km a day. No data are available for the distance covered by the zoo Asian elephant; the best figure available for African elephants in zoos is an average of 3 km a day, which is very much below what the wild elephant in Africa achieves.[9]

Mason and Clubb (2003) have also done a more recent study (funded by the Universities Federation for Animal Welfare and six British zoos) based on an analysis of about 1200 articles published in learned journals during the last 40 years on observations of animals-in-the-wild and 500 zoos world-wide; in particular, they have this to say about polar bears, that the typical zoo enclosure is one millionth of the size of their natural home ranges.[10] It may be pertinent to point out, too, that Central Park Zoo, relative to some other establishments, can be said to be generous in providing room for its polar bears, as they are allowed a 5000 square-foot exhibit space; however, it is, nevertheless, pertinent to bear in mind that the average natural range of polar bears in the wild is 31 000 square miles. Polar bears on the move can travel, as we have seen, 30 km (19 miles) a day. To achieve that distance, the captive polar bear in Central Park Zoo would have to do many a turn in that 5000 square foot enclosure in order to cover 19 miles, just as a prisoner in a small cell would have to pace up and down its length and breadth endlessly in order to get a decent dose of exercise to keep fit. But as we shall see in a minute, unlike the prisoner who may have a strong motive to keep himself perambulating within the confines of his cell walls, the captive polar bear is deprived of any major motivation to roam or move about in an analogous manner. Central Park Zoo as part of its philosophy of environmental enrichment also provides a pool for the bears to submerge themselves in. At the back of the exhibit enclosure is an ice machine which produces piles of ice to imitate Arctic conditions. To an animal capable of swimming and travelling as much as 30 km (19 miles) a day, and to cover as much as 259 000 km^2 (100 000 mi^2) in a life-time, a pool ensconced within an exhibit space of 5000 square foot would be miniaturisation carried to extreme. Ironically, should the animal wish to travel the same distance per year as it would do/have done in the wild, it would have to go round and round the pool non-stop for most days in the year, a stereotypic phenomenon considered to be distinctly pathological.

Hotelification

The term 'hotelification', coined by this author in this context, means no more than what the word 'hotel' normally denotes and connotes. A hotel exists to provide two basic services to a traveller, namely accommodation and food. A zoo also provides these two kinds of services to the exotic animals which have been transported a long distance away from their natural habitats. The only difference between the human and the animal traveller is that while the former usually stays for only a short or limited period of time in a hotel, the latter and their descendants are made to stay in 'hotels' usually for the rest or the whole of their lives, with rare exceptions where a small number of them are returned to the wild.[11] 'Hotelification' denotes the procedures of incorporating exotic animals in captivity within the facilities of lodging and full board offered by zoos.

Animals-in-the-wild spend a lot of their waking time looking for food and a safe niche to bed down as well as, in the case of females, to give birth to their young. The polar bear is an excellent example of this latter need. Mating is some time between late March and mid-July. The fertilised ovum divides a few times and then implants itself in the uterus for up to six months – this is called delayed implantation. Around September, the embryo attaches itself to the uterine wall and starts developing again. The mother polar bear dens in October or November and the cubs are born in December or January while she is still hibernating. The cubs are born blind, hairless, toothless and very small, weighing only 600–700 grams (21–25 ounces). However, within several weeks, they would have grown to 10–15 kg (22–33 lbs) feeding on the extremely rich milk of the mother. In the case of the cheetah, the cubs are also born blind and toothless; it takes five weeks before their eyes open. The need to find a safe niche for them is obviously imperative.

Another imperative need for animals-in-the-wild is the search for food. As we know, cubs have to be patiently taught where to find, how to identify as well as how to obtain it. It takes a cheetah mother up to two years at least to teach her cubs to master the art of hunting/stalking before they can be left to hunt on their own. The cubs have to be taught to perfect a carefully choreographed stalking-cum-killing sequence known as chase-trip-bite. A similar period of apprenticeship is involved in the case of the polar bear cubs whose mother has to teach them what to eat, where and how to catch the prey, and in the case of females where to den, as well as to negotiate the many dangerous hazards facing them on the ice.

Medium and large mammals in the wild tend to roam varying distances each day primarily because of the imperative to seek food.[12] As they are perambulatory, they often, if not invariably, have to seek out new shelters each day. As a result, their waking moments tend to be taken up with three basic activities, namely travelling, feeding and nesting, leaving no time to play.[13] For instance, the gorilla spends about 30 per cent of its daily existence travelling, 40 per cent resting and sleeping (gorillas in the wild build their nests on the ground, building their nests afresh daily, choosing carefully a selection of plant material), and 30 per cent foraging. Gorillas are primarily vegetarian, although western lowland gorillas also eat large quantities of termites and ants. Their diet in the main consists of more than 70 different plant species which include bamboo, wild celery, vines, berries, roots and bark. A male can eat up to 75 pounds of bamboo a day and a female up to 40 pounds.[14] Elephants in the wild dine off a wide range of plants (the Asian elephant ingests up to 75 species), mostly grasses but also bark, twigs, leaves, roots, flowers, herbs and fruit, depending on the season; given the low quality of the vegetation and their sheer size, they naturally have to spend 60–80 per cent of waking hours feeding to get what is needed for their nutrition, consuming between 150–300 kg of (wet weight) forage in the case of adult elephants. The Asian elephant is very adept in using its feet to dig up roots and to scrape vegetation lying close to the ground. Elephants also use their trunks to great effect as a means of removing dirt or thorns from their food.[15] And as we have seen, the distances travelled are entirely dependent on the availability of food; the African elephant roams more extensively than its Asian counterpart because of this factor. The cheetah preys primarily on other mammals weighing less than 90 pounds, particularly impalas and Thomson's gazelles. It also goes for springbok, steenbok, duikers, hares, the young of warthogs and wildebeest, kudu, hartebeests, oryx, roans, sables and even birds.[16] The polar bear lives mainly on ringed seals and bearded seals; depending on availability, it also goes for harp and hooded seals. It would also dine on fish, birds, bird eggs, small mammals, shellfish, crabs, starfish, dead animals (like dead whales, walruses, narwals), mushrooms, grasses, berries and algae. In late May, when the seals emerge, it gorges on them; an adult needs to gain 100–200 pounds in weight before the ice breaks up and the seals leave the area. The polar bear uses a variety of methods to catch its prey, such as still hunting, stalking on land, aquatic stalking and stalking birth lairs (of seals).[17]

It is obvious that zoo diets for its captive animals are totally different from those which the animals-in-the-wild enjoy. Zoo gorillas (at least in

US zoos) are fed fruit, vegetables, monkey chow, nuts and seeds.[18] Captive elephants are fed dried forage (mainly timothy hay), some commercial concentrate feed in the form of pellets, a small amount of fruit and vegetables, vitamin and mineral supplements, and sometimes branches and leaves (browse), the last depending on availability. They have virtually no access to pasture grass. Given the higher nutritional value of the zoo diet, they would not have to eat as much in terms of volume of food compared to their counterparts in the wild.[19] In US zoos, the national zoo diet for cheetahs consists of ground horsemeat and sometimes beef (made up especially for zoo carnivores), rabbits and chicks. Polar bears are fed 'omnivore chow, four to five pounds of fish daily, and apples and carrots'.[20]

Zoo diets for captive mammals (medium and large) differ fundamentally from those which their counterparts in the wild pursue in the following ways: although by and large equivalent in nutritional value, the items of food in zoo diets bear little or no resemblance in phenomenological terms to those available in the wild.[21] This is to say that in terms of appearance, taste, texture, smell, the two sorts of foods are essentially different. Furthermore, the ways in which zoo foods are served up are necessarily not foods which the animals themselves have hunted or foraged – zoo animals enjoy 'room service' and 'table d'hôte'.

Conclusion

The spatial miniaturisation of simulated habitats and the services of hotelification may be necessary and indispensable components of zoo management; however, it remains reasonable to argue that these two managerial techniques have the effect of inducing an 'existential crisis' in captive mammals (at least of the size that this chapter focuses on). One must straightaway enter a philosophical caveat here – talk of an 'existential crisis' should not and need not be understood in the same way as such talk would in the human context. The term here is only used analogously. Humans, as we know, have a peculiar type of consciousness which is characterised by a capability for abstract thought, for symbols, for language in general. The linguistic dimension renders humans capable of deep reflection about life in general and metaphysical speculations about the meaning of life in particular. For humans, an existential crisis is usually a crisis of identity. Therefore to say that captive mammals in zoos suffer from a crisis of identity is not to imply that they do so in the same way humans suffer from it. Human captives sent to the Gulags can ponder and reflect, as well as articulate their thoughts and feelings under their changed predicament; captive mammals cannot do that, as they lack the unique type of consciousness that humans possess. However, this is not to say that they – those recently captured from the wild as opposed to those who are entirely zoo-born and zoo-bred – are incapable in some ways (short of linguistic articulation) of noting their changed predicament, of their loss of freedom in doing what they would normally do if they were not removed from the wild.[22] To travel, to roam (whether nearer or further afield in the wild) in the search for shelter, for food (and later as we shall see, also in some cases for mates) in the precise ways each species has evolved over geological times constitute the existential predicament of the individual who is a token of their type (species), and deeply informs, therefore, the identity of these individuals as tokens of their respective types. That identity is undermined and indeed, totally dissolved under captivity, where the natural modes of choice and appropriation of foods are suspended and replaced by 'room service' foods in a 'hotel' with full board and lodging under human management – such a drastic change in circumstances may no less be characterised as an 'existential crisis' or a 'crisis in identity', which constitutes a major part in the ontological transformation of their status from that of naturally occurring beings to that of exhibits in zoo collections. In other words, they are subjected to humanisation, being drawn into the orbit of human culture, cared for, controlled and manipulated by humans for human ends, be these the high-minded goals of education and research, or the perceived less high-minded one of recreation/entertainment. It is none other than to be deprived of their status as beings which live 'for themselves', to manifest their own *tele* (given to them by the fact they each are tokens of the respective species to which they belong), to become beings which live essentially for the sake of humans. It is to lose their independent value as naturally occurring beings to assume, by and large, instrumental value only for humans.[23]

6
Suspension of Natural Evolution

The last two chapters have looked at the ontological transformation of animals-in-the-wild as naturally occurring beings to become captive exotic exhibits in zoo collections under four aspects, namely, geographical and habitat dislocations, spatial miniaturisation of their simulated habitats and hotelification. This chapter will continue to explore in the same spirit, first yet another aspect, namely, medication, before it goes on to show that all these arguments, marshalled thus far, justify the claim that zoo management of animals amounts to the suspension of natural evolution, and hence that the products of such management would, therefore, no longer qualify to be naturally occurring beings, given the change in their ontological status.

Medication

It is not unknown that some animals-in-the-wild take steps to get rid of unpleasant/painful symptoms which they feel in themselves; this is even to say that sometimes they take steps to treat and cure themselves.[1] Cats are known to go for cat mint as a carminative and:

> Costa Rican capuchin monkeys rub a plant resin into their fur that repels insects; wild pigs in Asia prophylactically consume plants with anthelminthic properties; and certain birds maintain intestinal regularity by ingesting tomato-like fruits that contain a natural laxative.[2]

However, this is not to say that animals have a concept of medication which is equivalent to the one we have in (human) culture. They have no shamans, certainly no professional doctors and paramedics, no theories of illness/diseases, no hospitals, no machines whether high or low tech, no refined drugs produced by a pharmaceutical industry to sort out their ailments. Animals-in-the-wild succumb to illnesses and diseases from which they eventually die. No animal can survive should they hurt themselves in any serious manner – they have to be sound in their limbs, not to mention in all their organs, to be mobile in order to forage properly, to hunt for prey or to escape from predators. For instance, in the case of the cheetah, even slight injuries could prove disastrous and fatal.

In contrast, in captivity, animals are drawn into the orbit of human culture of which veterinary medicine is an aspect. The (human) concepts of diagnosis and treatment of illness, with their elaborate structure of health professionals, are extended to such animals for at least two reasons. First, practical/economic considerations make it imperative that animals in zoos be kept healthy; in general, given the change in climate, the diet, the habitat, and therefore, lifestyle, they are very vulnerable health-wise. To replace them on a regular basis as stock succumbs to disease and death could be expensive. Second, moral/welfare considerations dictate that zoo keepers and managers keep the animals they control as free from pain as possible. In other words, such animals (like domesticated ones) are 'humanised'; contemporary (human) culture, at least in the West, is based on the fundamental moral axiom that pain is evil and that one has a duty to ameliorate pain whenever and wherever one can do so, and to save a life whenever one can.[3] Thus, through incorporation into human culture, such animals lose their ontological status as naturally occurring beings with independent value in their own existence and their respective modes of existence; they become (through no choice of their own) dependent on humans for their very existence and for their changed mode of existence, as exhibits in zoo collections. Their claws and paws have to be examined on a regular basis, their blood and urine tested to determine whether they are well or unwell, and they have to take pharmaceutical products dished out to them in ways convenient to the task at hand. In other words, they have become (clinical) patients, under the care of registered veterinarians and carers. Some zoos may even have hospitals or sick bays to handle poorly animals.[4]

Suspension of natural evolution

According to neo-Darwinsim which, by and large, is the orthodoxy in the scientific domain of natural evolution, animals and their species have evolved through the mechanism of natural selection.[5] This scientific explanatory framework as well as world-view dispenses with the concept of design, whether divine or human. Natural selection amounts

to the following key assertion, namely that animals with certain characteristics which enable them to adapt to their habitat/environment
would survive to a reproductive age, mate and leave offspring behind.
Those with less favourable traits would either die in infancy without
reaching sexual maturity, would on reaching sexual maturity be unable
to out-compete their rivals in the mating game, would be infertile even
if successful in mating, or would have cubs which are still-born or weak
in some aspects, and therefore, themselves die before reaching sexual
maturity and in turn leaving posterity. A complex chain of causes and
effects would ensure that, on the whole, those which survive to reproduce are well adapted to the environment/habitat within which the
species has evolved. In this limited sense, and this limited sense only,
they may be said to be the outcome of 'competition' leading to 'the survival of the fittest'; they are not 'the fittest' under all circumstances, but
only under those specific ecological/climatic conditions to which their
species have become adapted over geological time. In the history of
Earth, there have been several non-anthropogenic extinctions of species
when climatic/ecological conditions drastically changed.[6]

What external conditions in the wild weaken animals and what do
they die of? Obviously hunger/thirst, extreme unexpected cold/heat,
wounds, disease, predation, and the like, adverse conditions operating
either singly or together synergistically.[7] Just to give one example of
synergistic causation at work – a hungry animal, provided it is healthy
in all other ways, may be able, nevertheless, to escape predation, while
one which is only slightly wounded but not hungry would also be able
to run away from its predator. However a prey animal which is both
hungry and even slightly wounded would be unlikely to do so, as its
speed would be thus greatly reduced, by a factor which exceeds the
mere adding up of the loss of speed incurred in the two disadvantaged
conditions acting in isolation.[8] It is within such a complex causal context that natural selection operates.

In zoos, their captive exotic animals are never left hungry or thirsty
(given that they do not have to forage or hunt for their food, as nutritionally adequate meals are served up to them daily as part of 'room
service'); they are checked and monitored by health professionals
daily/regularly for the least sign of dis-ease and/or disease (at least in the
better managed zoos which disseminate best practices throughout the
zoo world). Very significantly, they enjoy sheltered accommodation in
such a way as to remove the problem of predation altogether. In sum,
they are not allowed to go hungry and certainly not to die from
famine/thirst, to be (physically) unwell, to die from wounds or disease

(in so far as medical science and technology could ensure), to be killed by other animals for food, or through aggression whether in the mating context or otherwise. It is not a surprise, therefore, to learn that zoos pride themselves on providing a more painless as well as longer existence for its animals than their so-called counterparts in the wild.[9]

In the wild, the cheetah lives eight to ten years; in zoos, it could live up to 17 years, almost double their natural life-span. Not only is the wild cheetah cub vulnerable to lions and hyenas, the adult is preyed on by lions and leopards, predators bigger than themselves. In the Serengeti, 90 per cent of cubs die before they are 3 months old. Mothers sometimes have to leave their cubs for over 48 hours as they have to hunt everyday, and should they be unsuccessful in their hunt to build up sufficient milk for their young, they would abandon the litter and start all over again. As for the fate of adult cheetahs, lions in national parks kill as many as 50 per cent of them.[10] No definitive data exist regarding the longevity of (mountain) gorillas in the wild. However, the longest lived gorilla in captivity reached 35 years old; no gorilla in the wild has been known to look as old as that. So it is speculated that in the wild, it may live to between 25 to 30 years of age.[11] In the wild, six out of ten polar bear cubs die in the first year of their existence, either through starvation, accidents or attacks; sometimes, mothers even kill their cubs because they themselves are malnourished, and therefore have to cannibalise their offspring. Sometimes, cubs are prey to adult male polar bears, not to mention other predators such as the wolf. The male also sometimes kills its rivals in the competition for mates, and occasionally even kills females protecting their cubs. However, in general, the adult polar bear in the wild has no natural predators; the successful adult can live up to 25 years or more. On the other hand, the oldest known polar bear in zoos lived to 41 years.[12]

In the three instances of the gorilla, the cheetah and the polar bear, the statistics bear out the fact that such animals under captivity live longer than their wild relatives. However, in the case of the elephant, this appears not to be so, at least according to Clubb and Mason (2002) whose study shows that in the wild, African elephants live up to 65 years; Asian elephants in camps live just as long while individuals who live up to 79 years are not unknown. In zoos, their review of the data from the studbooks 'reveal that out of 517 Asian and 238 African zoo elephants of a known age (dead and alive), none have lived to 60 years, and the maximum recorded age is 56 in Asians and 50 in Africans' (ibid., ch. 7). The average span of survival of elephants in European zoos is 15 years while that of timber elephants is 30 years.

However, Clubb and Mason's study in this matter has since been challenged on the grounds that it is methodologically flawed and that in reality captive elephants in professionally run zoos live just as long as wild elephants – see Weisse and Willis (2004).

From the perspective of this book, the dispute just cited above is of interest for the reason that both sets of disputants, in spite of the obvious difference in their assessments, are, however, agreed on one thing, namely that it matters whether captive elephants live as long as elephants in the wild. Clubb and Mason, based on their unfavourable finding, argue that it may no longer be justified to keep them in zoos, while their critics argue that it is justified to keep them because captive elephants, after all, as a matter of fact do live as long as, even if not longer, as is the case of other animals, than their wild relatives. Both sides, therefore, agree that longevity and indeed (preferably) improving on longevity in the lives of captive animals is a justification for keeping them in zoos. Both sides appeal to the value that the prolongation of life is itself an unquestioned as well as an undiluted good. In human culture, this is no doubt the view held by a majority of people; hence it is generally considered to be the morally right thing to do for the medical profession to strive to save and to prolong life for as long and as far as its skills and technology permit.[13]

However, both sides have uncritically extrapolated this assumption from human culture to the existence of animals as a whole. As this book, so far, has shown, captive animals in zoos are indeed being 'humanised', having been incorporated into the human world of control and management together with the values which underpin such an arrangement; it would follow from such a presupposition that improving the longevity of captive animals is indeed a worthwhile and desirable goal. However, to claim that captive animals living longer lives than analogous animals-in-the-wild constitutes a relevant difference which justifies zoos, thereby also implying that wild animals are suffering unnecessarily, is philosophically flawed, as it ignores the ontological difference in status between animals-in-the-wild that are naturally occurring beings and zoo animals that are 'humanised' through human control and management as exhibits. Zoos, for their animal inmates, are what may be called 'totalising' institutions, in the way long-term prisons and long-stay hospitals/homes are for their human inmates. As freedom of movement is no longer available, naturally their every need (especially physical ones) has to be provided for, the most basic being shelter, food, medication. This amounts to saying that natural evolution working through the mechanism of natural selection as the most crucial factor determining natural evolution in any habitat has been deliberately removed. Death, arising from lack of nutrition, hunger, accidents, disease and predation no longer or hardly obtains. Indeed, as their animals live that much longer, the turn-over is that much lower; as a result, zoos often find themselves in the embarrassing position of over-producing and having to resort surreptitiously to putting down surplus stock. In contrast, death through hunger, malnutrition, accidents, disease and predation is built into the very nature of existence in the wild. However, the context of natural evolution in the wild operating through the mechanism of natural selection is replaced, under captivity, by one of human design and control where natural evolution has been suspended, even though the mechanism of natural selection itself may still be at work, as we have earlier already remarked upon. In nature, death, pain or suffering in itself is neither a good thing nor a bad thing; nature is not capable of making moral judgment. Such matters and events are simply integral to the processes of natural evolution.

As animals-in-the-wild and captive animals belong to different onto-logical categories of being, it makes no sense to claim that one can com-pare them along a commensurate dimension, namely longevity, and conclude thereby that captive animals are better off than wild ones as they live longer. To say that some being that lives longer is better off than another with a shorter existence is only intelligible if both beings are part of (human) culture. It follows that it makes sense to say that gorillas in Zoo A are better off than gorillas in another zoo because they live a few years longer than their counterparts in Zoo B, just as it makes sense to say that people in Japan are better off as they live longer than their counterparts in the United States or sub-Saharan Africa. However, it makes no sense to imply that captive animals are better off as they live longer than animals-in-the-wild. The characteristic of living longer may be a desirable one in the humanised context of zoo animals but is irrel-evant to the context of animals-in-the-wild. An analogy may make this point more clear – as a cat is not a dog, it is pointless to point out that it cannot bark like a dog or that a dog cannot meow like a cat. The ability to bark/yelp is integral to the identity of being a dog (unless that feature is excised via genetic modification), and the ability to meow is integral to the identity of being a cat. For animals-in-the-wild, to die in the way they do, at the age they do, is integral to their identity as naturally occurring beings whose coming into existence, whose continuing to exist and whose going out of existence owes nothing (in principle) to humankind – they are simply the outcome of the processes of natural evolution, which enable them to exist 'by themselves' as well as 'for themselves'. In contrast, captive zoo animals owe their identity to human design and manipulation – they are what they are and live for as long as they do because they owe their very original existence (except for those caught directly from the wild), their continuing to exist and hence, too, their going out of eventual existence (to be determined ultimately by advances in veterinary science and technology as well as by cost) to human control. Zoo animals, as we shall argue in the next chapter, are domesticated (though not in the classical sense of domestication) or immurated animals. It is this profound ontological difference between animals-in-the-wild and zoo animals which render the term 'wild animals in captivity' or 'wild animals in zoos' an oxymoron, a conceptual and ontological mistake which even the more sensitive of professionals and experts of zoos make, alas, all the time.[14]

Conclusion

If the line of argument pursued here and in the last two chapters is plausible, then the book will have made a convincing case so far for maintaining that the ontological status of wild animals is totally different from that of captive zoo animals. The difference may be traced in summary form to the fact that while the former is the outcome of the processes of natural evolution via the mechanism of natural selection, the latter is the product of human design, control and management.

7
Domestication and Immuration

This chapter will examine the claim usually made that captive zoo animals are not domesticated animals or domesticants. It will critically explore the notion of domestication to determine whether or not there is a sense of domestication which may apply to zoo animals. If such a sense exists and can be identified, then this would strengthen the view already arrived at in the last three chapters that, unlike animals-in-the-wild which are the outcome of the processes of natural evolution, zoo animals are the products of human design, management and control.[1]

Domestication: how it is normally defined

Let us first approach this issue by briefly raising the distinction between natural selection and artificial selection. We have already referred to the former in the last chapter; as we have seen, it is an explanatory mechanism accounting for natural evolution, which dispenses with the notion of design, whether divine or human.[2] In other words, (human) intention, whether direct or indirect/oblique, is irrelevant. In contrast, in artificial selection it is the human breeder who deliberately selects a trait or traits of a plant or animal deemed to be desirable with the aim of enhancing and improving upon such properties, leading eventually to the generation of new breeds; conversely, if a trait is deemed to be undesirable, the breeder will similarly try to breed it out. These techniques of artificial breeding have been used for millennia (and are still used in some parts of the world) since the dawn of agriculture and husbandry, until the second agricultural revolution in the 1930s began using the technology of double-cross hybridisation induced by the fundamental discovery of classical genetics by Mendel, the Mendelian laws having been rediscovered at the turn of the twentieth century. From the 1970s onwards, one could say

that a third agricultural revolution has arrived in the form of biotechnology based on the fundamental discovery of the structure of DNA by
Francis Crick and James Watson in 1957.[3] All three agricultural revolutions resting, first, on craft-based technology, then on Mendelian science
and the technology it generates and, third, on molecular genetics as well
as molecular biology and the technology they in turn have generated
in the last 30 years or so, enabling humankind to select, at a deeper
and deeper level of manipulation, genetic characteristics and material
containing what we, humans, consider to be desirable traits in a plant/
animal and to exclude what we consider to be undesirable traits.[4]

As the above brief account of the history of agriculture shows, breeding
of plants and animals is intrinsically tied up with the concept and practice of artificial selection which long preceded the concept of natural
selection itself. Indeed, it is said that the phenomenon of artificial breeding and selection inspired Darwin to see an analogue between it and
observations he had made in the wild, leading him to formulate the
notion of natural selection itself.[5] However, the crucial conceptual difference between the two remains – natural selection dispenses with the
notion of design altogether. This conceptual difference is bound up in
turn with the ontological difference between the outcomes of natural
selection in the context of natural evolution on the one hand, and those
of artificial selection on the other – the processes of the former involve
naturally occurring beings, while the procedures of the latter involve
beings who embody human design and intentionality. The traditional
cow who grazed the meadows before the arrival of Mendelian
science/technology, the Mendelian cow, bred using the technology of
double-cross hybridisation as well as the transgenic cow (who, for
instance, produces a human hormone in her milk) are all the results of
artificial selection; to labour an important point, they are the products of
procedures designed by humans with a particular specific end/outcome
in mind.[6]

Having clarified the distinction between natural and artificial selection, let us move on to examine the notion of domestication. Let us
begin by looking at an example, the case of elephants. We know that
there is a long history of the use of elephants as haulage animals in Asia;
from this, some authors have concluded that timber elephants are
domesticated or at least semi-domesticated. What about zoo elephants?
Are they domesticated or are they wild in all senses of the term 'wild'?
We know that they are tame; however, as Chapter 3 has shown, being
tame is a necessary but not a sufficient condition for the transformation
of an animal to a domesticant. One must, also, straightaway point out

that elephants, on any scale, have been kept in zoos for only perhaps
a tiny fraction of the time that elephants have been tamed and trained
as beasts of burdens.[7] So, clearly, the difference in time-scale would
make a difference to the respective outcomes – in the evolutionary history of organisms, the time dimension is crucial, as pointed out in
Chapter 1.[8] In the discussion which follows, one is not so much talking
about whether zoo animals, as they stand today, are fully domesticated
or semi-domesticated in the way dogs are clearly domesticants after
thousands of years of breeding, but whether the procedures put in place
by zoo keepers/managers, which transform animals caught in the wild
and transported to zoos, as well as the majority of zoo animals bred in
captivity but whose ancestors were caught in the wild, to become exotic
exhibits, as well as the processes of adaptation, which arise from such
drastically changed circumstances, have the effect of domesticating
such animals, and if so, in what sense of the term 'domestication'.[9]

The concept of domestication has been defined in numerous ways.[10]
Here is one definition:

> that process by which a population of animals becomes adapted to
> man and to the captive environment by genetic changes occurring
> over generations and environmentally induced development events
> occurring during each generation.
>
> (Price, 1984 as cited by Clubb and Mason, 2002, ch. 3)[11]

An alternative definition of a domesticated animal or domesticant is as
follows:

> one that has been bred in captivity for the purposes of economic
> profit to a human community that maintains total control over its
> breeding, organization of territory, and food supply.
>
> (Clutton-Brock, 1999, p. 32)

An influential definition is given by Bökönyi who claims that his is
close to that of Clutton-Brock:

> The essence of domestication is the capture and taming by man of
> animals of a species with particular behavioural characteristics, their
> removal from their natural living area and breeding community, and
> their maintenance under controlled breeding conditions for mutual
> benefits.
>
> (1989, p. 22)

Yet another may be found in Issac in terms of a list of characteristics which animals must possess if they are to count fully as domesticants. Animals which meet some but not all of the conditions are semi-domestic. These conditions are:

1. The animal is valued and there are clear purposes for which it is kept.
2. The animal's breeding is subject to human control.
3. The animal's survival depends, whether voluntary or not, upon man.
4. The animal's behaviour (i.e., psychology) is changed in domestication.
5. Morphological characteristics have appeared in the individuals of the domestic species which occur rarely if at all in the wild.

(1970, p. 20)

The last three accounts of domestication are broadly similar; however, Bökönyi's account holds that domestication is for the benefit of both humans and domesticants – this claim may be too strong, although mutual benefits do happen sometimes.[12] It may have been inspired by the fact that domestication does involve a complex symbiotic human–animal relationship; however, the relationship is, surely, heavily weighted in favour of the domesticator and not the animal to be domesticated.[13]

Clutton-Brock's account in terms of economic profits appears unusually restrictive, ignoring one very important non-economic motive in the history of domestication, especially in its early period, namely the religious motive.[14] Isaac's Condition 1 listed above covers both economic and non-economic motives, and therefore, is preferable to Clutton-Brock's formulation. On the whole, Isaac's account is the best articulated of the four cited; the remainder of this discussion on domestication will bear it in mind, but, however, with two small caveats.

First, it does not refer to genetic changes over generations which Price's definition does. What then could be the relationship between morphological changes (Isaac's Condition 5) and genetic changes? One could say that the latter cause the former, that is to say, all morphological (or phenotypical) changes are caused by genotypical ones; and/or that in turn, all genotypical ones are the result of mutations.[15] However, this way of looking at the relationship is open to two criticisms. First, empirical evidence is usually insufficient in historic cases (based in the main on archaeological sites) to determine that mutations were definitely the cause. Take a slightly different case, of colour variation in modern domestication, cited by Bottema, in which he argues that colour variations in captivity are 'more likely to be the result of recessive factors already present in the wild population than to mutations' (1989, p. 41).[16] Second, the explanation in terms of genetic changes in general ignores environmental causes of morphological changes. For instance, Bottema says that from the unusually large size of the udders of the modern dairy cow, one may be tempted to argue that the morphological change is caused by a genetic change but one would be wrong: 'Although the shape of the udder is hereditary, excessive size is induced by the milking regime, whether milking is done by hand or by machine' (1989, p. 31).[17] Genetic changes cannot be said to account for all the phenotypical changes observed although such changes which are brought about by domestication may be the result of mutations and/or of permitting the emergence of recessive factors under the peculiar circumstances of domestication.

Second, the reference to morphological changes may turn out to be too restrictive; it embraces primarily features of so-called infantilism usually found in domesticants. Domestication involves more than such changes; other biological phenotypical changes also occur at the level of biochemistry/physiology as well as in matters of sexuality, fertility and reproduction in general.

We have so far looked briefly at the concept of what might be called classical domestication. However, none of the definitions quoted above make any explicit reference to the notion of artificial selection.[18] What is the relation then between the two concepts, that is, of domestication and artificial selection? By clarifying the relationship between them, is it possible to identify and articulate another account of what domestication might amount to? Let us see. The notion of artificial selection is, indeed, intimately linked to the classical definition of domestication. Without the concept and the practice of artificial selection, there would be no domesticants in the classical sense. But is artificial selection pertinent only to the creation of domesticants in the classical sense? Can its use in the zoo context lead eventually to animals which can be said to be domesticated in another sense of that term?

The classical type of domestication involves, as we have seen, the deliberate selection of a trait deemed to be desirable or useful in some sense to us, the human selector. If we fancy very small dogs, we breed small dogs with fellow small dogs and after numerous generations of breeding, we end up with the Chihuahua, a dog so small that female owners have been known to pop them in their bosoms. It may be true that some zoos have also indulged in similar practices with regard to some of its animals. For example, an albino tiger may be born in a zoo. Tigers in the wild are rarely born albinos. However, zoo visitors appear to love to see freaks. Such a tiger is a crowd puller and some zoos have been known to give in to such market demands and have tried to breed albino tigers, although zoos with so-called 'best practices' would condemn such a project, on the grounds not merely that it gives in to vulgar clamour but also that it distorts the reality about tigers in the wild.[19]

Artificial selection in the context of classical domestication involves human intention which is direct/explicit, rather than indirect/oblique. Attempts to breed albino tigers – to bring about biological changes – in order to ensure that zoos would always be supplied with such freaks of nature clearly embody this kind of human intentionality. However, today's reputable professionally run zoos claim they do not engage in such scientifically disreputable practices, as we have just remarked. On the contrary, for instance, they claim that individual captive tigers (or whatever other captive animals are said to be endangered) form part of the entire population of zoo tigers, which may be spread out among several zoos throughout the world, and which may participate in their *ex situ* programme of conserving the species (*Panthera tigris*); zoo scientists do their best to replicate the populations of tigers in the wild as far as their genetic variability is concerned. Such zoos deliberately avoid artificial selection of the kind engaged in by less reputable ones when they breed freaks of nature. They imply that, therefore, they engage in no deliberate selection, and also that their populations of captive-bred tigers are not domesticants, that is, in the classical sense of the term.

However, the situation is much more complex than that. In one obvious sense, zoos which participate in specific *ex situ* conservation programmes do deliberately select, say, the individual golden tamarin for the purpose of reproduction, in order to achieve their stated goal of preserving as much, as it is possible, of the species's genetic variability found in wild populations of golden tamarins, in the population of the captive animals in zoos. If golden tamarins in zoos are allowed to reproduce on a random basis of availability and access of animals, then the captive-bred population could, in some instances, reflect the preponderance of the genetic contribution of a particularly successful male, and hence dilute the genetic variability of the population as a whole. Zoos have to intervene to prevent such an undesired outcome. A particular zoo might have to import genetic material from another zoo, which sends the animal or its sperm in order to impregnate its female(s), as it itself has no suitable male for the purpose in hand. Such deliberate selection does not lead to a deviation in genetic variability, as matters stand. For instance, the captive-bred golden tamarins, destined for introduction back to the wild, will, as we shall see in Chapter 9, have been treated in a very different way from other ordinary zoo captive-bred animals not included in the *ex situ* conservation programmes. Scientists, as we shall see, do their best to ensure that these exclusive animals are as little exposed to the processes and procedures of the normal zoo environment as the zoo context of management and control would permit. *Ex situ* conservation zoo animals fail, therefore, to satisfy the classical concept of domestication, in spite of the deliberate intention to maintain the same genetic variability as exists in a wild population, because no particular biological traits (and any underpinning genetic properties) have been directly selected or induced. It, therefore, fails Condition 5 of Isaac's list, as well as Condition 4, which states that 'the animal's behaviour (i.e., psychology) is changed'; as we shall see in Chapter 9, scientists have to take special measures to keep, as much as it is possible, the young captive-bred animals from human sight. As Chapter 9 will also argue, the goal and procedures of *ex situ* conservation do not sit well both in theory and practice with the conception of zoos as collections of animal exhibits open to the public. One may conclude that in spite of meeting the first three of Isaac's conditions, the failure to meet the other two in the special context of *ex situ* conservation, which does its best to minimise the effects of Conditions 1–3, renders that kind of conservation programme outside the confines of classical domestication.

To see if one can find a true departure from classical domestication, one should consider the vast majority of zoo animals, that is to say, those not taking part in *ex situ* conservation programmes and whose *raison d'être* is simply to be part of the zoo's permanent collection of exhibits open to the public.[20] We need to explore in this context a more complex idea based on the relationship between what is directly and what may be said to be indirectly (or obliquely) intended rather than confining oneself to the straightforward version of directly intended artificial selection which we have examined in classical domestication. One also needs, straightaway, to introduce a new concept, that of immuration.

Immuration

This concept may be readily elucidated in terms of some of the highlights in the findings of the chapters preceding this. We list the following:

1. All zoo animals must be tamed, if they are to play the role of being exhibits in a zoo collection. Unlike an animal in the wild, a tame animal permits humans to approach them without showing flight. This is a pre-condition for being a zoo inmate; the same holds true of

a household/farm domesticated (in the classical sense of the term) animal. 'Tame' is the antonym of 'wild' (in one sense of 'wild'). Taming constitutes the minimal content of training which consists of getting the animal used to the numerous ways of being handled by humans. In other words, it is part of a procedure of (human) acculturation.[21]

2. Zoo animals are, in the main, exotic animals. If they are freshly captured from the wild, they suffer geographical/climatic dislocation; and even if they are captive-bred and have known no other environment than zoos, nevertheless it remains the case that they now live in regions of the world which fall way outside the natural home range of their wild forebears.

3. Zoo animals suffer from habitat dislocation as they live in a totally man-made environment. At best, their exhibit enclosures are simulated naturalistic habitats. The operative words are 'simulated naturalistic' as they are designed primarily (though, perhaps not exclusively) to show them off at their best as exhibits. The space permitted them is miniaturised space, which bears no comparison to the size of the home range in which their wild relatives/forebears roam.

4. The loss of freedom to roam is compensated by hotelification. (However, this wrongly presupposes that the need to roam is entirely a means in order to forage/hunt for food, and that it is not intrinsic to the animal in the wild.) Full board and lodging are provided. However, instead of foods found in nature, foraged or hunted, they are given zoo meals which may be nutritionally adequate as diet, but bear no resemblance in all other significant aspects to their respective foods in the wild.

5. Medicare is available and its provision mandatory. They are checked daily for any dis-ease and disease. Should accidents happen and they get injured, they are put right immediately.

6. They are not exposed to the hazards of weather, to the perils of starvation arising from hunger/famine/thirst, to those of wounds/disease which could lead to death. More crucially, they are immune to predation, to being killed in general by other animals. In other words, as we have argued earlier, in zoos, *the forces of natural selection in the context of natural evolution in the wild are in abeyance.*[22] Humans, not nature, are in charge. Under such sheltered accommodation, it is not surprising that they live longer.

7. They are not allowed to choose with whom they mate in three aspects: (a) necessarily they cannot choose their mates from their counterparts in the wild, (b) neither do they necessarily have the choice amongst

conspecifics in the same zoo, as reproduction strategies are under the control of zoo management, and (c) in the case of male animals, they are not allowed to fight for the female they would like to mate with.

8. The features listed above render zoo animals different in ontological status from those of animals-in-the-wild. The latter are naturally occurring beings whose existence is (in principle) independent of human design and manipulation, while in the case of the latter, existence in all key aspects is dependent on humans and their goals and purposes.

The features mentioned above which are intrinsic to zoos collectively define the term *immuration* and its procedures. The term coined comes from the Latin word 'mur' which means wall. A zoological park is not a cage, it is true. The enclosure allocated to a group of animals is undoubtedly many times bigger than a cage; however, the fact remains that an enclosure has walls in the literal sense or limits/boundaries which act as walls (such as a deep ditch around the enclosure). The confines within which movement alone is permitted are designed into such enclosures – zoo animals are captive animals and the space allowed them is allocated by their human managers/keepers.[23]

Zoo professionals might well agree that all the features discussed so far are indeed deliberately designed into the very concept of the zoo. Zoo animals, *ex hypothesi*, have no freedom to roam vast distances, to engage in the lifestyle to which animals-in-the-wild have evolved to do. While admitting this, zoo professionals may well deny that it is part of their direct/explicit intention to create new breeds of animals. However, to say this is to overlook that the notion of direct intention may be understood in ways different from directly intended artificial selection under classical domestication as well as to imply, at the same time, a scientifically misleading understanding of the complex causal relationships between animals and their environments in terms of their adaptation to and their evolution within their new habitats.

We need now to turn to the long delayed task of elucidating the distinction between direct intention and indirect (oblique) intention to see if a more complex version of directly intended artificial selection may not emerge which could cover the case of immurated animals.[24] The distinction is used in everyday discourse not to mention in more specialised ones like legal discourse. The conceptual exploration which follows bears all these discourses in mind:[25]

1. All human acts have consequences which are not merely directly intended, but which can be said to be indirectly/obliquely intended. For instance, drivers directly/explicitly intend to get from A to B using their cars; however, certain consequences follow from their acts of driving; these are congestion and pollution. In other words, these consequences, generally speaking, are unwanted but which, nevertheless, follow from the means chosen to achieve an explicit goal. The goal is directly intended and so is the means chosen to achieve it, as, from the standpoint of a practical agent, whoever desires the end necessarily desires also the means to achieve the end. The car journey as the means to get from A to B is desired and directly intended, but not the congestion and pollution to which the car journey contributes.

2.   Consequences other than directly intended ones may be foreseeable or unforeseeable at the time of action. With regard to the former category, one must distinguish between what, as a matter of fact, is foreseen and what is not foreseen at the time of action. In the case of motoring mentioned above, congestion and pollution are both foreseeable and foreseen.

3.   Foreseen and unforeseen may in turn be looked at from the perspective of either the individual in question or in terms of the extant collective body of knowledge available at any one time. Congestion and pollution are foreseeable as well as foreseen, both by society in general and by individual motorists (today).

4.   Those foreseeable and foreseen consequences, other than directly intended ones, which follow from the explicit goal and the means chosen to achieve it, are often said to be indirectly intended (indeed, sometimes even said to be unintended).

5.   Such consequences may be roughly classified as merely probable, highly probable, or amounting to what lawyers call practical certainty.[26] As things stand today, congestion and pollution in the motoring example fall into the last category.

6.   On the whole, in the West, the concept of responsibility holds agents responsible only for foreseeable consequences.[27]

7.   While there is consensus that agents may be held responsible for a directly intended (bad) consequence of their act(s), there is less consensus whether agents should be held responsible to the same degree for a (bad) consequence which is said in one terminology to be indirectly intended and, in another, even to be unintended. For instance, the law of

homicide in England and Wales wavers between holding a defendant guilty of first-degree murder and holding him/her guilty only of manslaughter, when the defendant claims that s/he did not foresee the bad consequence in question, even though any reasonable person could and would have foreseen it.[28] Consider the case of a defendant who conceived the plan of driving her lover's mistress from town by posting a kerosene-soaked and lit rag through the letter-box of the lover's mistress's house in the dead of night and, as a result, a full-scale fire ensued killing the two children of the lover's mistress, who were sleeping upstairs in the house and were not rescued in time. The former verdict argues that it is not a defence to say that the defendant (being too agitated or too preoccupied with her project) did not, as a matter of fact at the time of posting the lit kerosene-soaked rag through the letter-box, foresee the highly probable consequence of destruction to life; the defendant is, after all, someone with normal intelligence and as a reasonable person, she ought to have foreseen such consequences. Therefore, as a reasonable person, she should also have taken the precaution of ascertaining that the house was empty before posting her burning rag through the letter-box. If her direct intention were simply to frighten her rival in love so that she would leave town, the defendant would and should, as a reasonable person, have taken precisely such measures. As she failed to pursue such precautions, the defence that she did not directly intend to cause death on the grounds that she did not as a matter of fact foresee death is not open to her. The law deems her to have directly intended the death and hence, under this perspective, would find her guilty of first-degree murder, not merely, of manslaughter.

Take another case: the defendant managed to carry out his scheme of getting a bag containing a time bomb into the luggage hold of a certain plane, without anyone knowing that the piece of luggage did not belong to any of the passengers checked in for the flight. The plane in mid-flight would explode at the moment pre-programmed, destroying the plane and killing all the passengers at the same time, and he would then be able to collect a vast amount of insurance money on a policy he had taken out on the plane. He could plead that he only directly intended to enrich himself with the insurance claim regarding the plane, and that he only at best indirectly intended (or even that he did not intend) to kill anyone. No court today, however, would let him off scot free or with the lesser crime of manslaughter and not first-degree murder.[29] This case differs crucially from the burning rag case in one respect – the causal link between the acts (as described by the defendants) and the consequence of death is different. In the burning rag instance, the causal link between posting a kerosene-soaked lit rag through someone's letter-box (the means chosen to drive a rival in love from town) and the ensuing death of two children is not one which amounts to practical certainty, but only to high probability. On the contrary, the causal link in the insurance case between the act of getting a bag containing a time bomb into the hold of a plane (the means chosen to obtain insurance money) and the ensuing destruction of plane and people amounts to practical certainty. In other words, the defendant had done all he could to ensure that death and destruction would occur under the circumstances he had chosen to enact – short of the device failing to trigger off as intended, or the flight being cancelled because of unexpected bad weather, the deaths were bound to occur given the laws of physics and chemistry.[30]

8. Today, jurisprudential thought in the West appears to favour the so-called objective test of liability (the reasonable man test) especially outside the law of homicide; for instance, in environmental law. A firm which pollutes a river cannot simply plead that it is not its direct/explicit intention to pollute the water but simply to get rid of its factory effluent in the most convenient and cheapest way possible; environmental legislation holds such a company liable as it could and should have foreseen (like a reasonable person) that the river would be polluted if it (and others like it) were to use the river in that way. In other words, firms as (rational) agents are held responsible for bad consequences of their acts which are not directly/explicitly intended; the defence is not open to them to plead that at best they are only indirectly intended (or even unintended), as they are readily foreseeable and ought to have been foreseen by any reasonable person. *In this area of law, as in certain types of cases in the law of homicide, the law considers the distinction between direct and so-called indirect intention to be irrelevant, collapsing the latter into the former, especially under circumstances where the causal link between intention and the bad consequences amount to practical certainty.*[31]

The examples of the motorist, the jealous lover, the insurance fraudster and the polluter cited above may also be said to raise the so-called package-deal solution to the problem of intention.[32] The analysis of intention entailed by the package-deal perspective is given in terms of four principles of practical reasoning, in the view, at least, of one philosopher:

Principle of the holistic conclusion of practical reasoning

If I know that my A-ing will result in E, and I seriously consider this fact in my deliberation about whether to A and still go on to conclude in favour of A, then if I am rational my reasoning will have issued in a conclusion in favour of an overall scenario that includes both my A-ing and my bringing about E. . . .

Principle of holistic choice

The holistic conclusion (of practical reasoning) in favor of an overall scenario is a *choice* of that scenario.

The choice-intention principle

If on the basis of practical reasoning I choose to A and to B and to . . . then I intend to A and to B and to . . .

Principle of intention division

If I intend to A and to B and to . . . and I know that A and B are each within my control, then if I am rational I will both intend to A and intend to B.

(Bratman, 1999, pp. 144–5)

However, Bratman has two objectives in mind in formulating the four principles of practical reasoning above: First, he wants to say that those who adhere to the package-deal perspective believe that by applying precisely those principles they would come to the conclusion that the motorist, the jealous lover, the insurance fraudster and the factory owner/manager respectively (directly) intend to cause congestion, to kill the children, to destroy the plane and kill the passengers, to foul up the river. Second, however, Bratman wants to say that the package-deal adherents are wrong to come to that conclusion; they are wrong, he argues, because as his critique shows, one can distinguish between choosing a scenario and intending the expected results which follow upon the choice of that scenario. He wishes, in other words, to maintain that the defendants in the four cases outlined above, as rational agents, could not be said to have (directly) intended the expected results which follow from the choice of the overall scenario. On his reasoning, those results, though expected, are unintentional and thereby implies that the agents who have chosen the scenarios in which such expected results are embedded are not responsible for those results. This seems counter-intuitive. It would be beyond the remit of this book to give either a detailed and full account of Bratman's critique of the package deal perspective or to give a detailed and thorough reply to Bratman's critique. Suffice it here to make the following brief observations regarding the latter:

1.   As his account of the four principles stands, they are unable to distinguish between say the insurance fraudster on the one hand and the jealous lover on the other. We have already observed that the casual link in the two cases differ – in the latter, the link between posting the kerosene-lit rag and the death of the two children as well as the burning down of the house is highly probable, while in the former, the link between planting the time bomb in the hold of the plane and the results of killing the passengers as well as destroying the plane amounts to practical certainty. So one might need to add another principle of practical reasoning:

Principle of causation in terms of practical certainty

If on the principle of practical reasoning I choose to A and to B when B is not merely highly probable to follow but practically certain to occur, then I intend to A and to B.

2.   His problematic distinction between choice of scenario and the denial of intention to do B may be sustainable only because in his fourth principle, he has postulated that the agent knows that A and B are each within his control. The phrase 'each within his control' is questionable especially in cases where the causal relation between A and B is not merely highly probable but practically certain to follow. Bratman cites the example of two bombers to make his case that the distinction between choice of scenario and intention is sound. Terrorist Bomber and Strategic Bomber have both chosen the scenario to bomb a munitions depot, which happens to be next door to a school with children in it, in order to promote military victory. However, only Terrorist Bomber may be said to intend killing the children, although both bombers as a matter of fact would have killed the children by choosing and then executing their choice in action; Strategic Bomber, in Bratman's view, has simply killed the children unintentionally. This is because Terrorist Bomber would behave differently from Strategic Bomber should their chosen scenario in reality start to change when they commence their bombing operations. The Enemy might have got wind of bombs dropping on the munitions depot thereby killing the children, and start to evacuate them. Terrorist Bomber, the Bad Guy, on perceiving the evacuation, nevertheless, would pursue the children and drop separate bombs on them, killing them, whereas Strategic Bomber, the Good Guy, would heave a sigh of relief at the sight of the evacuation and simply concentrate on bombing the munitions depot. The respective reactions of the two bombers to the changing scenario would vindicate attributing intention to kill the children on the part of the Bad Guy while denying such intention on the part of the Good Guy. The scenario

Bratman has chosen to illustrate the difference between the two bombers works for the simple reason that the evacuation of the children which is part of the changing scenario is part of another causal chain, initiated not by the bombers themselves but by the Enemy; furthermore, the foreign causal chain thus initiated, is such that it makes sense for the two bombers to choose to do X (to kill the children in flight) or to do not-X (not to kill the children in flight), something which is within their control.

However, the changing scenario introduced into the discussion has muddied the waters. The two bombers have been chosen especially by Bratman, it appears, to tailor his own conclusion, namely that there is something deeply flawed about the package-deal analysis of intention and that the right intuitive conclusion to draw is that Strategic Bomber does not intend to kill the children, whereas Terrorist Bomber does. Bratman's argument would fail in the case of the insurance fraudster, because while A is within his control (planting the time bomb in the plane), destroying the plane as well as killing the passengers (B and . . .) are not within his control once he has completed putting A into action. B and . . . will follow with practical certainty and there is nothing more he could do, either to hasten the expected results of A or to deflect the expected results of A.

The thought experiment Bratman asked the two bombers to perform is the wrong one to perform. Instead they should have been posed the following question at the time of their deliberation regarding their choice of scenario: Assuming that as expected, the bombs would be dropped and successfully exploded, causing the damage they were expected to cause, namely the destruction of the munitions but also of the schoolchildren, supposing that no divine intervention were to occur in the form of a very strong wind directing the bombs only to drop on the depot but waft them away from the school such that only some but not all the children would be killed or severely maimed, would you in order to achieve the military objective of defeating the Enemy still choose that scenario in question? If the answer to the question is yes, then Strategic Bomber could be said or deemed to intend A and to intend B and . . . . Similarly, the thought experiment posed to the insurance fraudster should be: Suppose that no divine intervention occurred such that the bomb destroyed only the plane but not the passengers as it exploded in mid-air, would you choose the scenario in question? If the answer to that question is yes, then Fraudster intends to A (plant the bomb to get the insurance money) and intends to B and . . . (destroy the plane and kill the passengers).

Posing the hypothetical question in the case of the Fraudster analogous to that which Bratman poses to the two bombers is not helpful: Suppose that miraculously divine intervention occurs to destroy only the plane but spare lives, would you cause another bomb to explode to make sure that the passengers would also be killed? If the answer to that question is no, then Fraudster does not intend to kill the passengers (he merely killed them unintentionally); if the answer is yes, then he does intend to kill the passengers. However, either answer makes no sense as the posing of the question in the first instance is beside the point and absurd: if the bomb (containing the right amount of dynamite calculated to do the job at hand efficiently) successfully explodes in the way planned, then the expected results, which are practically certain to ensue, are not something that are within Fraudster's power of control. Fraudster cannot control or defy the laws of nature. In the real world occupied by real agents acting in accordance with the principles of practical reasoning, no *deus ex machine* appears, no science fiction super-technology exists either; in many contexts, once the rational agent chooses a certain scenario and initiates a certain causal chain of events, the agent has no further control of them since the events unfold according to the known laws of physics, chemistry, biochemistry, biology.

In the light of the clarification above, it may be appropriate to revise Bratman's suite of principles which he attributes to the adherents of the package deal solution to the problem of intention. The recommended revision looks as follows:

> Principle of holistic conclusion of practical reasoning
> Principle of holistic choice
> The choice-intention principle
> Principle of causation in terms of practical certainty, and lack of control with regard to the relevant expected results.

The zoo context, of course, has nothing to do with either criminal or civil liability, or even with moral responsibility in general, in any obvious way; however, the revised principles of practical reasoning as set out above are, nevertheless, applicable. It is undeniably the case that human agency is at the heart of immuration which itself constitutes the essential core of zoo activities. The zoo environment/habitat (or scenario in the terminology used by Bratman above) is directly intended and designed for a specific purpose, namely to exhibit animals to the public (A) and such a (cultural) arrangement, in turn, has biological consequences (B) for which an analogous case can be made (analogous to liability in the criminal and civil law described above) for saying that they may be deemed to be directly intended, as they are foreseeable and, indeed, are foreseen to follow (in broad outlines, if not in their minute details) as a matter of practical certainty, given the extant body of scientific knowledge today.[33] A pertinent consequence of keeping animals as exhibits in an entirely human-designed and controlled environment is, in the long run, a new kind of animal, very different in behaviour and biology and, indeed, eventually in genetic variability from those of its ancestor species in the wild.

At this point of the argument, let us go back to Isaac's five conditions and test them against the claim made here that immuration, under the circumstances specified above, counts as a variant of directly intended selection. Condition 1 – the animal is valued and there are clear purposes for which it is kept – is satisfied as zoos are intended to be establishments where animals are kept as permanent collections of exhibits open to the public. Condition 2 – the animal's breeding is subject to human control – is satisfied as the animals have first been isolated and become exotic and, furthermore, are not in general allowed to breed with any other captive conspecifics but in accordance with the zoo's plans on reproduction, which could indeed even include sterilising or controlling their fertility by means of contraceptives. Condition 3 – the animal's survival depends, whether voluntarily or not, upon man – is satisfied as they are kept and fed, looked after in all ways by humans. Condition 4 – the animal's behaviour is changed in domestication – is satisfied as zoo animals are tame, that is to say, they have lost their fear of humans, their flight tendency and flight reaction, not to mention that they have acquired interest (in certain instances) with computers, pie fillings and other human cultural products and artefacts deliberately introduced into the lives of zoo animals. Condition 5 – morphological characteristics have appeared in the individuals of the domestic species which occur rarely if at all in the wild – may be said to be satisfied in spite of certain reservations.

As already argued earlier, Condition 5 should be enlarged to include other important biological changes which have been observed such as earlier sexual maturity, or more intense sexual activities as well as physiological changes which differ from their wild counterparts. Immurated animals are also heavier in weight. Admittedly, these do not constitute morphological changes of the kind normally found in domesticants in the classical sense, such as foreshortened and widened skulls, decrease in size, features which constitute overall infantilism. One should recall that morphological changes of such kinds take time to emerge even under classical domestication; immurated animals have a very short life-history compared to domesticants which are the products of classical domestication. Finally one should also bear in mind that morphological changes are the last to appear in the procedure and process of classical domestication; there is no reason to think that it would be any different in the case of immuration. Bökönyi has put the point well:

> Since domestication is a complex interaction between man and animal, its consequences are influenced by society, economy, ideology, environment, way of life, etc. Any successful definition of domestication must reflect all these possible aspects of the evolutionary process. The result of domestication is the domesticated animal that first culturally and later morphologically differs from its wild form.
>
> (1989, p. 25)

Further support may be found in Price's more recent account of domestication:

> Animal domestication is best viewed as a process, more specifically, the process by which captive animals adapt to man and the environment he provides. Since domestication implies change, it is expected that the phenotype of the domesticated animal will differ from the phenotype of its wild counterparts. Adaptation to the captive environment is achieved through genetic changes occurring over generations and environmental stimulation and experiences during an animal's lifetime . . . . In this sense, domestication can be viewed as both an evolutionary process and a developmental phenomenon.
>
> (2002, p. 10)

In other words, the cultural conditions are in place for eventual morphological/biological changes and changes in genetic patterns to take place under immuration, changes which are already underway today, permitting a more complex understanding of artificial selection.[34] Artificial selection under classical domestication involves direct selection of specific characteristics, while *artificial selection under immuration involves not so much direct selection of specific characteristics but the conscious choice of an environment (scenario), thereby wilfully permitting certain characteristics to emerge under the cultural Conditions 1–4 put in place by immuration. As the consequences of this kind of artificial selection are perfectly foreseeable and are foreseen with practical certainty to ensue upon the conscious adoption of the said scenario, by biologists in general, and zoo biologists in particular, one can argue that, as a result, zoos may be deemed directly to intend such consequences to happen at present as well as in the near or distant future.*

Such a situation and logic appear to have an analogue in the dim past of humanity, at a time even before classical domestication took place mainly in the Neolithic period, namely when animals were corralled and kept inside stockades. Bökönyi refers briefly to this practice: 'There could be primitive animal-keeping without conscious, but with unintentional, breeding selection . . . (1989, p. 26).' The analogue is far from perfect for the simple reason that a primitive corral and the methods used by our primitive forebears would not lead to an environment as tightly controlled and designed as zoos are today. *Furthermore, given our contemporary sophisticated scientific understanding of animals, one can even go so far as to say that reputable zoos today refrain deliberately and intentionally from selecting for certain characteristics, such as for the white trait in certain animals.*[35] *In other words, paradoxical though it may sound, reputable zoos today directly intend not to undertake artificial selection of traits for the vast majority of their immurated animals in the sense of classical domestication, just as they directly intend to preserve the full range of genetic variability in the small minority of animals which have been chosen for their ex situ conservation programmes and destined eventually to be introduced into the wild. However, it remains the case that they directly intend the zoo environment and all the biological changes to the animals, both in the shorter as well as in the longer terms which confinement within such an environment would inevitably induce.* It is our more profound knowledge of biology which makes it meaningful for us to say that zoos directly intend such outcomes, while exempting our dim and distant forebears from a similar attribution.

However, we shall argue in Chapter 11 that such a policy in so-called reputable zoos could well change, should they become convinced by the plausibility of the arguments mounted in this book. At the moment, they consciously and deliberately refrain from selecting for certain traits (say traits deemed to be attractive by zoo visitors) primarily because zoo experts are of the view that zoo animals are simply 'wild' animals which happen to live in captivity. Zoo theorists and zoo keepers, therefore, deliberately intend and ensure that their exhibits retain the morphological characteristics which their wild counterparts possess in order, first, to sustain the claim that they are 'wild' but happen to be exhibited to the public in zoos, second, to uphold their central justifications of zoos in terms of *ex situ* conservation and education-for-conservation.

However, we have already strenuously argued that the term 'wild animal in captivity' is, conceptually speaking, an oxymoron and, ontologically speaking, it conceals a dissonance between being a wild animal and being an immurated animal or an animal living under captivity in a zoo. We shall argue in Chapter 9 that their two central justifications are fundamentally flawed. If the ontological stance of this book is sound and the critique it mounts of zoos which follows from it is also sound, then even reputable zoos may have to alter their justificatory as well as their policy goals. Chapter 11 shows that should such a reorientation take place, then even reputable zoos may do what they today condemn as unacceptable, that is to say, to go for selection of traits deemed attractive and desirable on the part of zoo visitors, such as colour, and so on. Indeed, they might be even more enterprising than that and endorse the cloning of historically extinct animals, and to create chimeras.

If the main line of reasoning pursued so far is plausible, then one could argue that the procedures of immuration, which are directly and intentionally designed to achieve the end of keeping animals as exhibits in zoo collections, do, nevertheless, have practically certain outcomes which are foreseeable as well as foreseen, and therefore, under the principles of practical reasoning already referred to above, may be said to be directly, rather than indirectly intended (or unintended); such outcomes follow from biological processes at work which can lead to changes in the behaviour as well as the biology in general of zoo animals and their morphology, including changes in genetic variability in the long run, properties which are very different from those exhibited by their ancestors/relatives in the wild. Following this perspective, one could justifiably claim that these procedures and processes constitute domestication in a sense different from the classical account given earlier, but domestication, all the same. E. O. Price, a world-leading authority on domestication, appears to lend support to this view:

> the domestication process is difficult to avoid when animals are brought into captivity. Most captive-reared wild animals will express certain aspects of the domestic phenotype simply by being reared in captivity. The application of artificial selection together with the effects of natural selection in captivity can greatly accelerate the domestication process.
>
> (2002, p. ix)

However, this author would not wish to insist on using the term 'domestication' in case it should lead to a futile verbal controversy. As the matter raised goes beyond mere terminology to substantive issues as well as conceptual and ontological matters, it may be best to anticipate a possible red herring and use a different word altogether to characterise the complex issues discussed above, namely, the term '*immuration*'. Zoo animals, in the main, may be said to be immurated animals.

Earlier on, we remarked *en passant* that experts acknowledge that elephants in timber camps may be said to be either semi-domesticated or fully domesticated.[36] Clutton-Brock (1999) argues for the latter. She notes that although elephants in timber camps and camels are not subject to deliberate artificial selection in the way dogs and pigs have been over the millennia, nevertheless they are tamed, trained and exist in a more or less human-controlled environment as an exploited captive.[37] In this aspect, she would be in agreement with the analysis which is emerging, that there is another type of domestication, apart from classical domestication *via* artificial selection of certain traits.

If the environment and lifestyle of timber elephants differ from that of wild elephants, those of zoo elephants differ even more drastically, as we have seen, from those of their wild ancestors/relatives. It is, therefore, not at all surprising that zoo elephants show distinctly different behaviour in several key aspects. For instance, they are much heavier (the result of their zoo diet, offered 'on a plate', and their inability to roam large distances on a daily basis) – the female zoo elephants are 31 per cent to 72 per cent heavier than non-zoo elephants. They engage in sex more often than their counterparts in the wild; the females reach puberty earlier, and could start to breed as early as 11 or 12 years old while their counterparts in the wild or in timber camps reach sexual maturity at 18 years. On the other hand, a third of zoo females fail to breed at all. Furthermore, up to a quarter of calves born to Asian female elephants in captivity are stillborn; up to 18 per cent of the calves may be killed by their mothers (the latter is probably the result of the fact that the family structure of the wild elephant herd is not replicated in zoos). It is reasonable to postulate that such significant differences in behaviour are foreseeable and foreseen consequences of immuration, which are practically certain in broad outlines.[38]

Conclusion

This chapter attempts to establish that classical domestication is not the only kind of domestication at work under artificial selection by arguing that the notion of artificial selection should not be understood simplistically to exemplify only what takes place under classical domestication.

Another version of artificial selection is identified which occurs under immuration, which is just as much a form of domestication. However, in order to avoid potential needless controversies about a futile matter of terminology, this new term has been coined.

Immuration, as well as classical domestication, involves changes in the animals operating at two levels, the biological as well as the cultural levels. In the latter domain, the animals have to be explicitly incorporated into the social structures of human institutions, of control and management, not to mention ownership and exchange. Such animals have to be tamed (to a greater or lesser extent, depending on whether they are wild- or captive-born) and trained to respond to the presence and the peculiar ways and idiosyncratic demands of their human controllers. They have to learn to adapt to another entirely different set of social relationships (which includes humans as their masters/controllers, who also generally impose on them a different family/herd structure regarded as fitting or convenient for zoo policy and management), to different physical environments/habitats, to different lifestyles, including different feeding and reproductive strategies.

These cultural changes in turn set in motion biological processes of adaptation and evolution. In the case of classical domestication, artificial selection occurs for reasons, whether economic, aesthetic or cultural in the broadest sense, leading to the emergence of new breeds. In the case of immurated or zoo animals, a more complex variant of artificial selection than in the context of classical domestication takes place which has two prongs to it. It rests:

(a) On the relationship between the direct intention of immurating animals with the end of exhibiting them to the public (A) and certain results (B) expected to follow causally from (A) which, on further analysis, may be said to constitute direct intention under certain conditions. Zoo experts can foresee and do foresee that immense changes of a biological nature would take place either at present, in the near or distant future once the cultural conditions of immuration are put in place; as these consequences can be foreseen and are foreseen with practical certainty, and as there is no way of preventing them happening short of abandoning the scenario of immurating animals altogether – that is to say, of abolishing zoos as collections of exotic animals for exhibition to the public – the usual distinction between directly and indirectly intended consequences collapses in this context, and zoos may be said to directly intend all the biological consequences which result from immuration. Zoo biologists/managers, who are rational agents acting in

accordance with the principles of practical reasoning, cannot convincingly argue that although they directly intend the means to achieve the goal of exhibiting animals to the public, they only at best indirectly intend all the consequences which follow from the means adopted. (Under Bratman's account of intention, they might even claim, however wrongly, that such consequences are unintended.)

(b)   The sophisticated understanding of the biology of animals available today enables zoo experts in so-called reputable zoos to opt deliberately and intentionally not to select for certain traits deemed (for instance, aesthetically) desirable as happens under classical domestication. The omission of such experts to select for, say, albinism is itself a fully conscious act; in that sense they may be said to have directly intended not to endorse artificial selection as it operates in classical domestication; instead, they may be said to directly intend to endorse all the biological changes in the short and long run (B) ensuing from immuration (A) which they directly intend.

Immurated animals, no less than domesticants (in the classical sense) are biotic artefacts, a theme to which we turn in the next chapter.

8
Biotic Artefacts

The preceding six chapters have attempted to demonstrate that zoo animals differ from animals-in-the-wild in profoundly different ways which, in turn, underpin the difference in their respective ontological statuses. Chapter 2 has argued that, ontologically speaking, the latter are naturally occurring beings while the arguments marshalled in Chapters 3, 4, 5, 6 and 7 have as good as shown that the former may be said to be biotic artefacts. This chapter will explore more fully the concept of biotic artefacts as the ontological foil to naturally occurring organisms.

Artefacts

In this context, artefacts refer to human artefacts.[1] Chapter 2 has defined *en passent* an artefact as the material embodiment of human intentionality. Now is the time to elaborate a little on this matter.[2] Another way of making the same point is to elucidate the notion in terms of Aristotle's four causes. Take a statue as the paradigm of an artefact – its material cause is marble, its efficient cause is the sculptor, its formal cause is the blueprint either in the sculptor's head or sketched out on a piece of paper, and its final cause is the purpose for which the statue has been commissioned, such as to commemorate an event or a national/municipal celebrity.

The last three causes refer to human agency and its intentionality; the first to the material medium in which the intentionality becomes embedded. Without human agency and its intentionality, there would only be matter. With the extinction of the human species and its unique type of consciousness, the artefacts which humans have created out of matter would also disappear, leaving only matter behind. The Taj Mahal, as a mausoleum, (which Shah Jahan built in commemoration of his favourite wife, an exquisitely conceived and constructed building made of marble which the world, since its appearance, has come to admire as a great work of art) would no longer exist; only the marble (as a naturally occurring substance), from which it has been made, would continue to exist until the natural actions of wind, rain, plants and animals finally wear it down to soil. The Taj Mahal is an artefact; as such it is a human construct and construction, and, therefore, necessarily it has neither meaning nor existence in the total absence of human consciousness.[3]

Biotic artefacts

While it is evidently clear that artefacts can be made of abiotic matter (like marble) or exbiotic matter (wood), it is not so evidently clear that they can also be made of biotic matter, that humans can create artefacts out of individual organisms.[4] To see how it is conceptually possible to do so, let us further elucidate the notion of artefact, this time, not so much in terms of Aristotle's four causes, but in terms of distinguishing among three theses of teleology, namely, external teleology, intrinsic/immanent teleology and extrinsic/imposed teleology.

We have already observed that Darwin's theory of natural evolution and its mechanism of natural selection dispense with either divine or human design. Such a view amounts to the rejection of the thesis of *external teleology*; the relation between neo-Darwinism and external teleology may be spelt out as follows:

(a) The former denies what the latter asserts, namely that divine agency has created Life on Earth, either in general or in its diverse forms. Neo-Darwinism also denies, for good measure, that human agency has anything to do with the beginning of Life (since Life came into existence and evolved 3.5 billion years ago, as an earlier chapter has already observed), while strictly speaking, external teleology has nothing to say about the matter. It is said that if the age of Earth is made equivalent to a calendar year, *Homo sapiens* only appears 3.5 minutes before the year's end.

(b) External teleology implies that divine agency maintained or sustained Life in its diverse forms since its beginning on Earth while neo-Darwinism claims that neither divine nor human agency is required.

(c) External teleology holds that Life came into being and continues to exist in its diverse forms specifically to sustain human life and to serve human ends, a view which can be traced to Aristotle in his *Politics* and which has profoundly influenced philosophical as well as religious thinking for centuries in the West.[5] Neo-Darwinism denies this claim.

The relation between the thesis of external teleology and that of *intrinsic/ immanent teleology* may be spelt out as follows: the latter may be said to be an implication of denying the thesis of external teleology. In Chapter 2, we have already looked very briefly at it. Individual organisms, as we have seen, exist 'by themselves' as well as 'for themselves'. As autopoietic beings, they strive to keep alive, to reproduce, and so on, but animals-in-the-wild, however, do not do so to fulfil any end or purpose of any external agents; they do so entirely to maintain their own functioning integrity. They do what they do 'for themselves' alone.[6] In appropriating nutrients to sustain itself, the oak produces acorns in order to reproduce itself; it does not produce the acorn in order that the pig may have food. In turn, the pig eats the acorn, simply to maintain itself, and not in order that it would provide a nice dinner for the human hunter. And when the lion eats the human who has just dined off the roast pig, it is just as true to say that the human has not eaten the pig in order that he might himself, in turn, satisfy the lion's hunger. Of course, as a matter of contingency in general, organisms find certain other organisms useful in sustaining their own functioning integrity – plants that are primary producers, nevertheless find insects helpful in propagating their pollen, and certain mammals useful in propagating their seeds.

Organisms, in living 'for themselves', are realising their respective *tele* as individuals and as members of their species.[7] In so doing, they exemplify the notion of intrinsic/immanent teleology. An adult female frog will mate with her male counterpart. Her fertilised embryos will develop into tadpoles; in turn, the tadpoles will grow into adult frogs. As they are frogs, not birds or wolves, they can only live or thrive in certain habitats and not others; they prey upon some other organisms, like insects, but not others. In all ways, they behave as they do, entirely in accordance with their own *tele*, independently of human agency and its manipulation. Their trajectories, both as individuals and as species, have, in principle, nothing to do with humankind. In the absence of humans in the world, they would be there, coming and going, at their own pace and in their own ways. In other words, they follow their own trajectories.

Humans may turn naturally occurring organisms into artefacts, just as much as they may turn abiotic matter into artefacts. Such attempts exemplify the thesis of *extrinsic/imposed teleology*. As we have seen, humans have been domesticating plants and animals for a very long time. For millennia, their success rested on using what this book calls the craft-technology of selective breeding; however, in the first half of the twentieth century these traditional methods, as we have already mentioned, were radically overhauled by a new technology, which was informed by the theoretical understanding given by the science of classical Mendelian genetics. The last quarter of that century also witnessed, as we have seen, the arrival of a yet more powerful technology, called biotechnology or genetic engineering, which is informed by the theoretical understanding given by the even more basic sciences of molecular genetics and molecular biology. It is more powerful precisely because it allows humankind to cross boundaries between species and kingdoms by manipulating organisms, no longer at the level of whole organisms but at the molecular – DNA – level. For instance, one can insert into bacteria, DNA that may belong to the human genome. As we have seen, one can get cows to produce human proteins in their milk.

The examples just mentioned illustrate the process of transforming naturally occurring organisms, as in the case of the bacteria, to become biotic artefacts, or in the case of the cow, which as a domesticated animal is already a biotic artifact, to embody a deeper level of artefacticity.[8] The transgenic cow, unlike the more usual domesticated cow, has been commandeered by humans to use its autopoietic powers of self-maintenance to produce, not cow's milk, but milk which contains a human protein. In other words, biotechnology has succeeded in severing, in the clearest manner possible, what has been an inseparable link between being an organism, which exists 'by itself', and one which exists 'for itself'. Up to even 25 years ago, the distinction between 'by itself' and 'for itself' was one that could only be made intellectually but not empirically. But recently biotechnology has managed to sunder them as a matter of fact.

The transgenic cow, *par excellence*, no longer exemplifies existence 'by itself' – *ex hypothesi*, such an organism would not have existed without the direct and deliberate intervention of humans. The same is true of the transgenic bacteria. Humankind, *via* biotechnology, has captured the biological mechanisms of cows or bacteria in order to make them be what humans want them to be, and not how they themselves would be in the absence of human manipulation and control. In other words, although they may still perform biological functions such as eating, breathing, digesting, nevertheless, in carrying them out, they have been made to subvert their own respective *tele*. The cow no longer produces cow's milk, fit to nourish her own offspring, in principle at least when the milk is not whisked away for human consumption. Instead, she is made to produce milk whose constituents are not those in accordance with her *telos* as a cow. The same is true of the bacteria – their own *tele* have been subverted and made to execute a human intention and human end instead. This embodies the notion of extrinsic/imposed teleology. However, the transgenic organism, made possible by biotechnology, is but the most extreme form, to date, of a biotic artefact.

As we have shown in some detail in Chapter 2, the fact that organisms, in maintaining their own functioning integrity, exist 'for themselves' has not stood in the way of human success in transforming them to become biotic artefacts. Biotic artefacts, as much as abiotic artefacts, are not autonomous, as they are no longer *simpliciter* naturally occurring entities. However, in spite of the profound similarity of sharing the same ontological status of being artefactual entities, there is a residual difference between them. Imagine the sudden disappearance of *Homo sapiens* from the face of Earth. Empirically, over time, abiotic artefacts like houses, jewellery, computers would no longer exist; such artefacts, without constant human maintenance and repair, as we have mentioned earlier, will just disintegrate and eventually become dust and/or soil. But at the philosophical level, in the absence of humans and their type of consciousness, there would necessarily be no human artefacts, just chunks of physical matter, even before they disintegrate and decay. The other sentient beings which remain, like the leopard or even the chimpanzee, would not and could not know these things as human artefacts; only another being with a similar sort of consciousness like ours could recognise them as such.

But in the case of cows or pigs (whether transgenic or the more ordinary non-transgenic ones), in the absence of human maintenance many of them would die, or might not even succeed in reproducing themselves. But some might survive, and after many generations over evolutionary time, the humanly selected characteristics or the inserted transgenic material might become very rare in the genetic makeup. They could become naturally occurring again, like feral pigs, except for their remote genetic history. What this implies simply is that natural evolution and its processes (in the absence of human manipulation and control) have their own trajectories.

As we have argued in the last chapter, immurated animals – just as much as classically domesticated animals and transgenic animals – are the result of artificial selection, the difference lies merely in the fact that immurated animals are the outcome, in the main, of artificial selection within a context which is much less straightforward than its use in the context of classical domestication. Both forms of selection are manifestations of the thesis of extrinsic/imposed teleology which issue in animals whose main *raison d'être* is to serve human ends/purposes, rather than to follow the trajectories of their ancestors in the wild; they no longer live 'by themselves' and they no longer live 'for themselves', as they are biotic artefacts.

Type/token distinction revisited

In Chapter 2, we raised the distinction between type and token in the context of naturally occurring organisms. Here we explore two things: the distinction in the context of immurated animals as biotic artefacts and in the further context of assessing whether, from the ontological standpoint, an immurated animal can be said to be a token of a naturally occurring species in the way its wild ancestor/relative can unproblematically be said to be a token of the species.

But first, a few words about the type/token distinction in general. The actual letters of the Latin alphabet which are used here to write this book are tokens of a type, namely, the letters of the Latin alphabet as opposed to the letters of another alphabet, say, the Greek one. The individual Rolls Royce is a token of the type of car called the Rolls Royce. The word 'type' is usually used in reference to abiotic/exbiotic artefacts, or to abstract constructs such as letters of an alphabet or of musical compositions (the rendering of Beethoven's Ninth Symphony by the BBC Philharmonic Orchestra is a token of the type which once existed in Beethoven's head, in outline at least, if not in all absolute details, but which he had committed to paper using musical notations).

In the context of biological classification, the term 'type' is not usually used; instead, one uses the term 'species'.[9] More precisely, the term 'species' used in this book refers, in the main, to two different, but related understandings, namely, in terms of the biological-species concept as well as the evolutionary-species concept. As Chapter 2 has already touched on the latter in sufficient depth, no more will be said here except to remind the reader that zoos, by rendering animals exotic, have deliberately wrenched them from their evolutionary context and past; that such animals and their descendants have no choice but to live in an environment within which the processes of natural evolution and natural selection have been suspended (as shown in Chapter 6). A few brief words, however, will now be said about the former. Chapter 7 has argued that one must recognise that the human-designed habitat of the zoo does induce profound changes in the animals' behaviour and that immurated animals do adapt to such an environment and evolve within it. Such adaptation and evolution would eventually (given a long enough time-span) lead to even new species, different from the species to which their ancestors/relatives in the wild belong. This is in accordance with scientific understanding and is not at all far-fetched.[10] Furthermore, according to the biological-species concept normally used in biological classification (in spite of some obvious drawbacks), a species is distinct from another if the members of the one do not naturally meet, mate and reproduce with members of the other, even if they can do so biologically.[11] Lions and tigers belong to different species, as they do not as a matter of fact meet, mate and reproduce with each other in the wild owing to the fact that they live in very different geographical locations and habitats. However, if humans were to intervene and mate a lion with a tiger, offspring may issue – those born of a lion father and a tiger mother are called liger, while those born of a tiger father and a lion mother are called tigon. In fact and in practice, zoo animals cannot and do not meet, mate and reproduce with their counterparts in the wild, as they occupy very different geographical locations and habitats. However, should humans wish, they could capture an animal from the wild and make it meet and get it to mate with its counterpart in a zoo; alternatively, under *ex situ* conservation programmes, captive-bred animals may be released in the wild so that they can meet, mate and reproduce with their counterparts in the wild. Neither of these two possibilities is available in reality for the vast majority of zoo animals. They remain captive-born and captive-bred from generation to generation, setting in motion certain biological processes, which would eventually lead to breeds distinct from the species of their wild forebears.

The individual polar bear in the wild is a token of the naturally occurring species *Ursus maritimus*, both in the biological-species and the evolutionary-species senses of the concept of species. It mates with fellow polar bears forming membership of the polar bear population(s) in the wild; it is a descendant of ancestors through long, unbroken, evolutionary processes in natural evolution in their home range, leading back 100 000 to 250 000 years ago when the species first emerged.

On the other hand, the individual immurated polar bear in a zoo does not belong to the species *Ursus maritimus* in the sense understood by the biological-species concept; as we have seen, such individual immurated polar bears in zoos would never as a matter of fact meet and interbreed with individual polar bears-in-the-wild, except in cases of deliberate human intervention. At the same time, such immurated individual polar bears live within a cultural space designed, maintained and controlled by humans in such a way that their very existence is a fundamental rupture from that of polar bears-in-the-wild. While those in the wild are descendants of an unbroken historical evolutionary lineage of *Ursus maritimus*, immurated individuals live their lives in a context within which the processes of natural evolution are suspended. In other words, polar bears-in-the-wild and immurated polar bears belong to two crucially different categories or types, ontologically speaking – the former are naturally occurring beings while the latter are biotic artefacts. Hence, the correct inference to make is that immurated polar bears are not tokens of the naturally occurring species, *Ursus maritimus*, in either of the two senses of the concept identified, namely the biological-species concept and the evolutionary-species concept. To mark the ontological difference between them, one could propose a new system of classification and say that the individual immurated polar bear is a token of a different species to be called *I(Ursus maritimus)*, where *I* stands for 'immurated'.[12]

Conclusion

This chapter pulls together the various strands of arguments presented in the preceding chapters (2–7) which serve to establish the main thesis of this book, namely that from the ontological perspective, animals-in-the-wild and zoo animals belong to two fundamentally different types or kinds; that animals-in-the-wild and the naturally occurring species to which they belong are the ontological foil to immurated animals and their corresponding I-species; that while the former are the outcome of natural evolution, the latter are biotic artefacts, the outcome of artificial selection/breeding. They are respective tokens of their respective types. *The World Zoo Conservation Strategy* (1993), in company with most publications on the subject, is just simply wrong when it claims that 'all zoos exhibit living specimens of wild animal species . . . .'

9
Justifications Deemed Serious

Zoos started off as the private collections of kings and princes, aristo-crats and the very rich. Modern zoos began in the eighteenth century, open initially to members only, who were interested in exotic animals from the point of view of scientific research.[1] However, in the nine-teenth century they became increasingly municipalised and democ-ratised; civic pride and prestige required that every major city in the industrialising West should have a zoo, open to the public, whether free or for a relatively small entry fee. Zoos were meant in those days to be the 'green lungs' in urban settings, where nature, domesticated, was created with trees, where people could escape from the bustle and the pollution of great cities.

The aims promoted by these original zoological institutions appeared to have been more or less the same as those proclaimed by zoos today, namely, research, captive breeding and recreation. However, unlike nowadays, they do not seem to have emphasised education of the public. Today, zoos prioritise their goals as follows: on the one hand, education of the public, conservation of species and scientific research, and on the other, recreation. Unlike in the nine-teenth century, some contemporary zoos appear too embarrassed even to mention recreation as a justification, as it is considered to be, on the whole, an undignified and ignoble goal, best ignored and for-gotten. However, this chapter hopes to show that the so-called serious justifications of zoos are in fact deeply flawed; the following chapter will, in turn, argue that perhaps the so-called ignoble one in terms of providing recreation for the masses may, paradoxically, survive critical scrutiny.

Scientific research

After the French Revolution, the Jardin de Plantes in Paris started also to house animals. One of its objectives was to contribute to the advancement of science. In 1792, H. Bernardin de Sainte-Pierre argued that it was not enough, in determining their classification in terms of species and genus, only to study the skins and bones of dead animals, but that one ought also to observe the development of living animals. He also proposed to create in zoos what we today call simulated naturalistic habitats in order to make study of their behaviour all the easier. As his perspective was considered too revolutionary (in spite of the fact that Buffon had earlier advocated it), no notice was taken of his plea until the beginning of the nineteenth century, when Frédéric Cuvier, who became the keeper of the animal collection in 1804, initiated studies of the animals under his charge in terms of their behaviour, intelligence and sociability – the zoo became a kind of laboratory in which the observations made would supersede those which one might undertake in the wild. However, not even Cuvier got very far with his project; he met with stiff criticism from scholars like Etienne Geoffroy Saint-Hilaire, who argued against it on both pragmatic and theoretical grounds: that the high mortality rate of captive animals would prevent long-term observation of their behaviour, that an individual animal could not be said to represent the species, and more significantly (from the standpoint of this book) that captivity distorted behaviour.[2] In other words, we see that a debate had arisen – even at the very beginning of the modern zoo in the early nineteenth century – which continues to be relevant today.

Today's zoos are not preoccupied with the problem of classification in terms of species and genus; instead, their scientific research is based on the study of the behaviour of the animals in their collections, as well as of their physiological, metabolic processes and other related aspects, especially those pertaining to their health and well-being under captivity.[3] However, this is not to deny that it can and does address itself to other issues outside the specific concerns of veterinary science and husbandry, leading to significant discoveries about the animals being studied; for instance, in the case of the elephant, the following discoveries have recently been made by scientific zoo researchers:

> Work on zoo animals has . . . advanced our knowledge of elephant communication . . . anatomy . . . and reproductive biology . . . Zoo elephant research has also enabled the development of many techniques likely to aid in *ex situ* conservation, such as DNA extraction

from faeces and ivory, new methods for monitoring movement patterns, the assessment of reproductive state from faecal hormone metabolites, and new potential methods of contraception . . . such research opportunities would seem the greatest benefit of keeping elephants in zoos, although arguably they also could be supplied by logging camp and orphanage animals.

(Clubb and Mason, 2002, ch. 3)

While appreciating that such research in zoos may have added to the sum of human knowledge about animals, one should also be aware of certain methodological limitations and flaws inherent in such research. For a start, the scientific research appears to be conducted from the standpoint of functional biology. According to Ernst Mayr:

the functional biologist is concerned with the operation and interaction of structural elements, from molecules up to organs and whole individuals. His question is 'how?' How does something operate, how does it function? The functional biologist attempts to isolate the particular component he studies, and in any given study he deals with a single individual, a single organ, a single cell, or a single part of a cell. He attempts to eliminate, or control, all variables, and he repeats his experiments under constant or varying conditions until he believes he has clarified the function of the element he studies. . . . The chief technique of the functional biologist is the experiment, and its approach is essentially the same as that of the physicist and the chemist. Indeed by isolating the studied phenomenon sufficiently from the complexities of the organism, he may achieve the ideal of a purely physical or chemical experiment. In spite of certain limitations of this method, one must agree with the functional biologist that such a simplified approach is an absolute necessity for achieving his particular objectives. The spectacular success of biochemical and biophysical research justifies this direct, although distinctly simplistic, approach.

(Mayr, 1988, pp. 24–5)

However, in spite of its undoubted strength, such a research orientation pointedly ignores that of evolutionary biology.[4] As we have argued in Chapter 1, the zoological conception of animal presupposes the biology of time and history, of the notion of reciprocal causation in the dynamics of organism-in-the-environment, as well as of the philosophy of ecocentrism. Furthermore, the results of observation and experimentation from the perspective of functional biology conducted within zoos must remain theoretically suspect, as the doubts regarding them spring fundamentally from the fact that the animals under study are captive animals, whose very condition of captivity might have a profound effect upon their behaviour, a criticism mounted by Sainte-Hilaire, as pointed out above, two centuries ago. If so, one should hesitate to extrapolate from the results thus obtained to problems which arise outside of zoos regarding wild animals in their natural habitats. It may be sufficient to cite only two recent examples to illustrate the methodological traps lying in wait. The first concerns subjects which are laboratory rather than zoo animals; however, this difference is not germane in this context. A behavioural scientist in California has published his findings in *Nature*, showing that caged parrots show stereotypic behaviour caused by brain damage (damage to that part of the brain called the basal ganglia).[5] The brain damage, leading to stereotypies, appears to have been caused by sheer boredom in confined space. Some attempts to remove boredom through enrichment programmes are indeed successful; however, the point that this author wishes to make here is somewhat different as it concerns a matter which cannot be remedied by simple enrichment in confined space. It is to do with the confined space itself.[6] We have already seen in an earlier chapter that, in particular, medium to large mammals in zoos are routinely kept within enclosures whose severely reduced dimensions are totally out of proportion to the size of their occupants and the distance which the wild ancestors of their occupants traversed daily within their home range.

Another concern is related to the fact that the captive environment is necessarily much simplified compared with the infinitely richer, ever-changing and complex environment within which animals-in-the-wild operate. The zoo lifestyle, which bears no resemblance to the existence animals-in-the-wild lead, may induce differences in behaviour, as the next example – cited by Steven Rose – confirms. As we have earlier argued, immurated animals lack control of their lives in any meaningful way since they exist essentially as exhibits in a human purpose-designed and managed environment. The conditions under which captive animals are reared, as we have also already observed, have a significant impact on adult behaviour – for instance, captive-born animals are, from the moment they are born, exposed to the presence of humans. In other words, they are essentially born tame. According to Rose:

> [b]ack in the late 1920s, the anatomist Solly Zuckerman reported strong dominance hierarchies and high levels of 'aggression' and fighting among the large but confined hamadryas baboon colony at

London Zoo, and developed an influential theory of social behaviour based on these studies. 'Each baboon', he wrote, 'seemed to live in perpetual fear lest another animal stronger than itself will inhibit its activities.' Violence was a constantly occurring event, quarrelling frequent and widespread, and any major disturbance of the precarious equilibrium caused the social order to collapse into 'an anarchic mob, capable of orgies of wholesale carnage.' Later researchers observing baboon colonies in much larger enclosures or in the wild failed to find similar levels of fighting. Instead, the groups seemed relatively peaceful and stable. It became obvious – and with hindsight it seems scarcely surprising – that the behaviour of Zuckerman's baboon group had been dramatically modified by restricting the space within which its members had to coexist.

The constraints of Zuckerman's reductive approach had transformed the situation he wished to study and fundamentally misled him, even though his observations within that constrained situation were presumably perfectly accurate.

(Rose, 1997, pp. 28–9)

It seems fair, in the light of all the points made above, to conclude that any scientific findings about tame and immurated animals may not be applicable to animals-in-the-wild. Uncritical extrapolation from such data is hazardous, and therefore would be unwise; it would amount to bad science. To be methodologically correct and accurate, observation and study of behaviour of wild animals should be made in the wild, not in zoos.

Conservation

This aim has become dominant even if it has not been promoted to be the number one priority in the list of justifications of zoos. However, judging by some of the key documents issued in the last ten years or so on the matter, one might be forgiven for thinking that the sole *raison d'être* of zoos is conservation, from which it follows that should a zoo fail to meet it, such a zoo ought to be closed down.[7]

Two key documents have been exceptionally influential in setting conservation as a key goal, if not the only goal, of zoos. In 1993, the *World Zoo Conservation Strategy* (WZCS) was published by the World Association of Zoos and Aquariums (WAZA): 'For both ideological and practical reasoning, nature conservation must be the central theme of zoos in the future (*WZCS*, 1993, ch. 1)'. In 1999, the European Union

(EU) issued its Zoos Directive, which requires all member states to ensure that their zoos commit themselves to the goal of conservation and to conform to the conservation measures it lays down.[8] The directive is hospitable to *ex situ* as well as *in situ* conservation.[9] However, unfortunately regarding the latter, it is obvious that it is not something that zoos can, under normal circumstances, single-handedly secure, although in reality it may be true that it could contribute in some small ways *via* its research (but bear in mind the methodological limitations of zoo research raised in the preceding section), as *in situ* conservation necessarily requires action on the parts of sovereign states with the pressure, support and encouragement of concerned world bodies, such as the United Nations, to protect whole ecosystems and habitats.[10] In other words, it is expected that zoos are best equipped to contribute to *ex situ* conservation; in particular, *via* the techniques of captive breeding, which zoos themselves proudly claim to have so successfully pioneered.[11]

This book has no desire to rehearse in detail the criticisms usually mounted against the goal of *ex situ* conservation *via* captive breeding, except to remind the reader that: (a) it is very expensive; (b) it can, at best, save only a few species; (c) saving a few species in isolation from saving their habitats is mistaken in theory and in principle; (d) given these points, its critics argue that it is neither cost effective nor sound to engage in *ex situ* conservation, and that resources devoted to it should be diverted to *in situ* conservation instead.[12]

However, this chapter will develop a critique based on the following points which will in the process take up criticism (c), mentioned above, in particular:

1. The definition of zoo.
2. The exact nature of the zoo animal which *ex situ* conservation intends to save from extinction.

Imagine for one moment that the collective efforts of zoos throughout the world in their captive breeding programmes over the next 150 to 200 years were able to save all the animals on today's lists of endangered/threatened animals, and that zoos have become a true Noah's Ark. Would this mean that zoos are justified in claiming that their chief, if not only, goal is conservation? In other words, can zoos even be defined in terms of such a goal? If so, then the definition(s) of zoos as they stand today would have changed. But, how are they defined today?

According to the 1999 *EU Zoos Directive*, zoos are 'permanent establishments where animals of wild species are kept for exhibition to the public for 7 or more days a year'.[13] Section 21 of the Zoo Licensing Act 1982 (UK) says that a zoo is

> an establishment where wild animals are kept for exhibition to the public otherwise than for the purpose of a circus and otherwise than in a pet shop; and this Act applies to any zoo to which members of the public have access, with or without charge for admission on more than seven days in the period of 12 consecutive months.[14]

According to *WZCS*, institutions which call themselves zoos have two features in common:

1. Zoos possess and manage collections that primarily consist of wild (non-domesticated) animals, of one or more species, that are housed so that they are easier to see and to study than in nature.
2. Zoos display at least a portion of this collection to the public for at least a significant part of the year, if not throughout the year.

(WZCS (1993, ch. 1.3))

In other words, the three definitions cited above agree that zoos are collections of wild animals for exhibition to the public. They, thereby, differ crucially – judged from the ontological view point – from the definition pursued by this book which stipulates that zoos are collections of animals, which are not wild but immurated, for exhibition to the public.

Note, too, that all three definitions incorporate a reference to what looks at first sight a mere administrative detail, namely, that zoos must open their collections or certain parts of their collections to the public for a limited number of days or for significant parts of the year. However, the opening-time qualification in particular is paying more than attention to administrative niceties, as it permits the *WZCS* to reconcile two things which on the surface appear to conflict, namely that zoos must make their collections of exhibits open to the general public (in pursuit of the goal of educating them as part of the mission of today's zoos, not to mention the more vulgar goal of entertaining zoo visitors) and that at the same time they are also expected, in order to attain the goal of conservation, to pursue captive breeding and related scientific activities as laid down by the current agenda of the aims and objectives of zoos. Yet a moment's reflection would show that while the goal of educating the public or that of entertaining the public *ex hypothesi* renders public their collections of exhibits, the other goal of conservation requires the very opposite, that the animals be kept away not only from the presence of the general public but even of the scientists themselves, that is to say, they must be isolated from human presence as much as it is practically possible, as an earlier chapter has already pointed out. The activities of captive breeding with the aim of reintroduction to the wild are incompatible – practically, conceptually and ontologically – with those arrangements associated with exhibition. The latter requires public access, the former its denial.[15]

Clear thinking, therefore, demands that those scientific activities related to the goal of *ex situ* conservation be detached from zoos and be conducted within private space, which may officially be designated as (*ex situ*) conservation research centres, perhaps. The proponents of the *WZCS* might well retort that one might choose to define terms however one wishes, *à la* Humpty Dumpty, and that critics, like this author, are being unnecessarily prescriptive in confining the term 'zoo' only to its normally perceived function of exhibiting their collections of exotic animals to the public. However, two considerations are pertinent to restraining this kind of definitional free for all: first, differences of verbal matters should be distinguished from those of substantive matters, and second, there are very important issues of substance involved in this particular context, namely that the activities, as currently carried out by zoos under exhortation to achieve the goal of *ex situ* conservation in particular via captive breeding and preparation for introduction to the wild, conflict with those associated with the requirement of exhibition on the part of zoos. It may be true that definitions of terms, *per se*, are merely lexicographical and verbal in character; however, the choice of definitions should be guided by other relevant considerations such as the epistemological value of clarity and the ontological value of recognising fundamentally different categories of beings that the domain of activities deals with.

In light of the above, one could even be provocative and hold that a zoo is no longer a zoo when it engages in scientific research involved in the pursuit of conservation, of the *ex situ* kind, which relies, in the main, on the techniques of captive breeding. To put very different activities – different from the ontological standpoint – under one definitional roof, so to speak, would profoundly mislead and confuse not merely the lay public but also even the scientists themselves. In other words, both the *WZCS* and the *EU Zoos Directive* are fundamentally wrong-headed and mistaken in promoting (*ex situ*) conservation as the central goal of zoos; the remit of zoos cannot be stretched quite so far as to include such activities.

This fundamental misapprehension will become more obvious as we explore the next issue, namely the exact nature of the exotic animals kept in zoos which scientists are expected to save from extinction.

We have already argued in earlier chapters that animals in captivity live under conditions which are so profoundly different from those in which animals live in the wild that the individual animal belonging to the former category cannot be said ontologically to be a token of the species to which the latter individual animals belong. As animals kept in zoos are biotic artefacts living in an environment designed, controlled and managed by humans for humans, where the processes of natural evolution are suspended, they cannot be said to be wild in any sense of that term. For a start, to labour a point already tirelessly made, it is very wrong-headed, both conceptually and ontologically speaking, for the three definitions of zoos cited above to refer to zoos as places which house 'wild animals', or 'animals belonging to wild species'. This book argues that zoos are places which nurture immurated animals, an ontological foil to wild animals.[16]

We have also seen that in one sense scientists and other related professionals working in captive breeding programmes appear aware of the fact that if captive-born and captive-bred animals are to be returned to the wild, such animals must be subjected to a different regime from those captive animals which are to remain permanent residents of zoos to be transformed into becoming full-blown immurated animals. These specially selected animals must, from the moment of their birth, be protected from exposure to humans as much as possible. As we have already seen in the last chapter, zoos are very careful indeed to ensure that as many procedures of immuration as possible are suspended in the reproduction as well as in the rearing of the captive-born.[17] For instance, we know that imprinting occurs in some animals. In the captive breeding in California of condors for return to the wild, the scientists involved did not show their faces or themselves, and their hands were covered by puppet gloves in the shape of adult condors when handling the young birds. This and other similar practices are an acknowledgement by the scientific authorities that the procedures of immuration can and do have crucial impact on the behaviour of zoo animals which would render them different from those displayed by their ancestors/relatives in the wild.[18] They also appear to be aware of another fundamental fact, namely that successful reintroduction to the wild requires that the captive-born and -bred animals be initiated into the culture, so to speak, of the species to which their wild ancestors belong.[19] In other words, some conservation scientists have learned to appreciate that the animals in their charge cannot be expected to survive, to reproduce, to lead successful lives in general in the wild, if they were only exposed to the 'room service' facilities of hotelification laid on by zoos. So they do their best to make their charges acquire, as much as feasible, the skills of living in the wild before letting them go into the unknown future in an unfamiliar habitat.[20] They seem to perceive that an animal cannot be fully understood except in the context of the environment in which it is born, in which it grows up while at the same time learning how to negotiate it not simply by personal individual trial and error, but in the company of its elders, its relatives which show it the way.

Yet the grasping of this truth – that a habitat in the wild nurtures and produces a wild animal while, on the contrary, an environment which is designed, controlled and managed by humans nurtures and produces a biotic artefact – is but dim and intermittent. This charge of partial understanding may be illustrated by one example of success much acclaimed in the literature of conservation, namely the saving from near extinction of the Père David deer (*Elapharus davidianus*), although it still remains classified today as a critically endangered species. The history of this animal is pertinent to the point made here; so some details will be given. Some accounts say that the deer had become extinct in the wild about 2000 years ago and others, at least 1000 years ago.[21] It survived by the nineteenth century only in parks, in particular, the Imperial Hunting Park in Peking (Beijing). A French missionary – Père David – came to know of its existence there, and persuaded the Emperor to allow some animals to be sent to Europe. Shortly after this, in May 1865, catastrophic floods breached the walls of the Park, and starving peasants raided and killed most of the animals for food. Those few that survived were killed a little later during the Boxer Revolution. However, by pure chance, the Duke of Bedford procured a few of the arrivals in Europe for Woburn Park; these bred. More than a century later, in 1986, 39 individuals were introduced in the Dafeng Reserve in China. The species's original habitat is said to be swampy, reed-covered marshlands; the Dafeng Reserve is a coastal, seasonally flooded area.

It is interesting as well as significant to emphasise that very little is known of its original habitat and that the animal has been extinct in the wild for roughly 1500 years, if not more. Furthermore, those individual animals introduced to China since 1986 are the descendants of animals which themselves have become twice exotic (in the technical sense), residing in a park in Britain for more than 100 years. An animal with such a history could not be said to be wild in any sense of the word. Whether the park was an imperial hunting park in Peking, or whether it was Woburn in England, the environment remained a park. A royal hunting park is a human-managed and human-controlled environment in which some of the deer under its care were destined to be killed by the royal hunters, unlike Woburn Park, where the feature of hunting was indeed absent. In other words, the deer cannot be said to be a token of the species *Elapharus davidianus* in the wild, as there had not been such a species in the wild for roughly 1500 years, if not more. The species that Père David named *Elapharus davidianus* in the mid-nineteenth century was not a wild species, but an immurated species. The individual Père David deer then and now is a token of the species *I(Elapharus davidianus)*. It is, therefore, incorrect to claim that the animal had been saved from 'near extinction', first by Père David himself, and then by the Dukes of Bedford in Woburn Park, if that claim means what it says, namely 'the near extinction of the wild species'. It is simply the case that Père David and the Dukes of Bedford had/have saved the *immurated species, not the wild species, from near extinction*. The deer now in the Dafeng Reserve are, therefore, more akin to feral animals than to the long-extinct wild deer.

It is in accordance with good science to assume that animals with such a long history of immuration are bound to be different in very significant ways from their extinct wild ancestors. In general, we know that captivity brings about changes at all levels – genotypic, phenotypic, metabolic, behavioural. Genetic decline in zoos is commonly observed.[22] I will mention a few instances of other forms of decline/degradation here. As far as birds are concerned, even first-generation captives show deformities in their wings; their inability to fly means that the entire organism is affected, their skeleton, their organs, their plumage, as they fatten on a zoo diet. Captivity undermines the defence instinct; nenes or Hawaiian geese, captive-bred at Slimbridge (UK), were killed by predators when released in the wild. Ungulates in captivity have a lower adrenalin emission, a condition which would disadvantage them if they were to be released to the wild. Captive animals tend to be sexually precocious, and engage in sexual activity more often than their relatives in the wild (probably out of boredom, with nothing better to do).[23]

Conservation scientists tend to overlook or underplay the points made above because they seem to subscribe to what may be called *genetic reductionism* or *genetic determinism*. This is to conceive an animal as no more than its genetic inheritance. Put simplistically, it says that an animal is its genes. As we have already seen, (*ex situ*) conservationists are very careful to ensure that the genes of those captive-bred animals destined for introduction to the wild should reflect the range of genetic variability as found in the wild population(s).[24] This appears to be a fundamental aim and objective by which they judge success or failure in their captive-bred programme. This goal is a necessary precondition for eventual success in reintroducing such animals to the wild but it should, nevertheless, not distract them from appreciating that an animal is more than its genetic inheritance, that it is also the outcome of the complex causal interactions between those genes and their habitat/environment.[25] To put the point more neatly: the animal is not merely its germ-line but its lifeline which emphasises that the animal can only be grasped and understood properly within the interacting contexts of its evolutionary, ecological as well as genetic history.[26] The zoo environment is not the same as the habitat in the wild to which the animal would be released. Hence the captive-bred animal is not the equivalent of its relatives in the wild. If this understanding had been fully absorbed and in the forefront of their consciousness, the scientists involved, for instance, in the Californian condor reintroduction programme would have anticipated the difficulties of reintroduction, instead of being somewhat surprised when they learned about them in an *ad hoc*, trial and error manner. It should also have made conservation scientists wary of claiming that the Père David deer has been saved from 'near extinction in the wild' when that programme has done nothing of the sort.

The lesson about the limitations of genetic reductionism and a full and proper grasp of the zoological conception of an animal, despite experience, remains only partially absorbed as is evidenced by the eagerness of some conservation scientists and commentators to embrace the cloning of animals for the purpose of saving them from impending extinction in the Noah's Ark provided by (*ex situ*) conservation, or indeed, even extending the Ark in recreating long extinct mammals, such as the mammoths *via* the cloning of fossil genetic remains.[27] Recently (August 2004) Indian scientists, including those at India's Central Zoo Authority, have announced plans to save the dwindling population of 300 Asiatic lions (considered as the symbol of India) which live in the Gir Sanctuary in Gujarat and the Indian cheetah (which became extinct in the wild half a century ago) via cloning techniques. For the latter project, the scientists intend to use not a female cheetah, but a female leopard (which unlike the cheetah is not critically endangered in India) to carry the cloned embryo, as Iran appears unwilling to lend India a cheetah for the purpose. Indian scientists are also hoping to get a nucleus containing DNA from the dead skin cells of a cheetah. However, all this excitement on the biotechnological front

appears to overlook the fact that to be able to clone an animal successfully is not to ensure that it can be successfully reintroduced to the wild,
as this presupposes, amongst other things, the availability of a suitable
habitat for the animals.[28] The Wildlife Protection Society of India has
correctly pointed out: 'We are losing forests thanks to highways and
road projects and poachers are killing our tiger population. Cheetahs
need antelope to eat and space to hunt. We do not have enough of
either.'[29]

Education

Hancocks (2001) is one of those zoo professionals who have questioned
the central justification of zoos today in terms of conservation, in particular, *ex situ* conservation via captive breeding. Instead, he advocates
the goal of educating the public as their key justification. He cites
approvingly William Conway's mission statement, uttered more than
40 years ago in 1961:

> Zoo visitors should have the opportunity to learn something about
> each animal's environment through natural habitat displays, to
> explore the mysteries of wild animal behavior, [and] to be informed
> by special displays . . . The justification for removing an animal from
> the wild for exhibition must be judged by the value of that exhibi-
> tion in terms of human education and appreciation . . .
>
> (Hancocks, 2001, p. 110)

However, in spite of Hancocks's unfavourable assessment of zoos in terms
of their central goal of conservation, in particular, of *ex situ* conservation,
it is not obvious that he disagrees in reality too profoundly with those he
criticises, as is evidenced by his endorsement of Conway's position cited
above. Today's zoo professionals and scientists see educating the public as
a key goal, especially when that goal is inextricably entwined with the
central one of conservation. Some zoo advocates go so far as to say that

> zoological institutions only have a right to hold animals in captivity
> if they satisfy one (or both) of two criteria. 1. The animals must be
> part of a managed population with the eventual aim of a 'release-to-
> the-wild' programme in mind. Such a programme may be five years
> or five generations away but it should be our goal . . . All zoos should
> strive to link *ex situ* conservation with *in situ* conservation pro-
> grammes . . . 2. The animals in zoos should form part of a *structured*

conservation-education programme. . . . In the words of Eudley (1995) 'The major function of zoos in protecting biodiversity may prove to be conducting education programs designed to raise the public's ecological awareness.'

(Stevens and McAlister, 2003, p. 99)

Hancocks himself appears to endorse such a stance:

It is within a wider definition of education that the best and most viable reason for the continuing existence of zoos can be found. They have enormous potential to shape public opinion, to encourage sympathetic attitudes toward wildlife, and to educate the public about ecology, evolution, and wild animals. Zoos can open windows to a world of Nature that people could otherwise experience only via technology . . . It is essential to reach the point where the only zoos allowed by law are those that aim to create respect for wildlife and a desire to save wildlife habitat, by making animal welfare their first priority, by adopting conservation strategies as a central tenet of their operational, budgeting and marketing decisions, and by injecting passion and daring into their interaction with visitors . . . If these changes do not occur, then zoos must surely become increasingly meaningless . . . Zoos have the potential to present holistic philosophies with greater veracity and impact than any other type of natural history institution because they can present and interpret all parts of the story . . . More zoos are becoming habitat based, explaining ecosystems rather than only reciting facts about animals.

(2001, pp. xviii–xix)[30]

If education as understood above is the key, if not the sole justification, of zoos, then we must evaluate it in terms of both its practice and its theory.

The practice, *prima facie*, is immensely successful. Of the 10 000 or so zoos estimated to exist in the world (although, however, they do differ from one another in more ways than one), the World Association of Zoos and Aquaria (WAZA) numbers roughly 2000 as federated zoos, which it considers to form the core of the zoo world, and to which the *WZCS* (1999) is primarily directed.[31] These are also the zoos which are the most likely to have implemented the philosophy of the zoo which Conway, Hancocks and others cited above advocate. Many such zoos (especially those in North America, Europe and Australasia) are rich enough to afford to construct naturalistic enclosures for exhibiting their animals, which is the way recommended by zoo professionals to be the most effective, if not indispensable, means of educating the public about the vital need of protecting biodiversity and the various and different ecological habitats in which wild life exists.[32] As zoos remain hugely popular, their numerous visitors must, therefore, have been exposed to such 'structured conservation-education' programmes. It follows, too, that the visitors should and would have absorbed the information as well as the lessons purveyed by these programmes.[33] But have they, as a matter of fact?

Alas, it appears not to be the case. For a start, visitors in general do not stay for a long enough period of time to absorb all the relevant information; 50 per cent of the visits, it is said, even to Woodland Park (in the United States), last no longer than two hours. According to one set of figures for the Reptile House at the National Zoo in Washington DC, the average time taken to cover the entire house was 9.7 minutes and the average time spent in front of each exhibit was 0.44 minutes.[34] Stephen Kellert, in a US nation-wide survey of 3000 American zoo visitors, found that they failed to identify the basic groups of animals, that they had one of the most limited levels of knowledge and understanding about ecological issues, they scored no higher than non-zoo goers in their knowledge of such matters. Indeed, amongst those with the lowest score are zoo goers – 'Animal-related activity groups with relatively low knowledge scores included zoological park visitors . . .' (Kellert, 1981, p. 14). They appeared to be more concerned with the welfare of the animals they saw in the zoos – that the animals be kept in sanitary and comfortable conditions – than in issues concerning biodiversity, conservation and ecology. 'The zoo-goers group revealed considerable affection and concern for wild animals, but had one of the most limited levels of knowledge and understanding about ecological issues' (Hancocks, 2001, p. 156).[35] The European experience does not seem to be any different. Baratay and Hardouin-Fugier note that

> [s]urveys from Frankfurt to Mulhouse have shown that 80 per cent of visitors questioned claimed to have learned something, but a 1979 investigation noted that they were less sensitive to the need to respect nature than hikers, even after their visits.
>
> (2002, p. 236)

Let us pause at this point to perform a thought experiment. Imagine that the average zoo visitor has all the time available at her/his disposal and is a conscientious seeker of knowledge. What would such an ideal visitor have learnt at the end of a long day at the zoo? It is true that s/he would have read/heard/seen and hopefully understood and digested most, if not all, the material amply available in a variety of forms – whether these be general or more specific information sheets/brochures distributed at the main gate or at the entrance of each enclosure, audio-visual shows in the multi-media centre, special lectures with regard to some aspect of the animals in the zoos, or conducted tours of selected exhibits, and so on. As a result of this kind of educational exposure, the visitor would no doubt be better informed than some other citizen who has not been similarly exposed. However, even under such ideal cir-cumstances, can we simply infer that the combined exposure to the information available and the presence of the animals themselves is either uniquely efficacious, or more efficacious than other methods, for instance, than just simply exposing the person to the information together with the showing of some wildlife films? In so far as evidence appears to be available, such evidence seems to indicate that the zoo experience is neither uniquely efficacious nor more efficacious than other forms of experience and exposure to facts about nature. Zoos are not the only source of information regarding biodiversity and ecology available to the average citizen; it is accessible in numerous other ways. What is unique is the availability of such information presented in a milieu where the animals can be seen in the flesh. Yet there appears to be nothing uniquely efficacious about such a combination of exposure, as the data already cited seem to show.

If education is the key justification of zoos, then zoos appear to be a miserable failure; they have not unquestioningly got their educational message across to the public. Furthermore, let us carry on with our thought experiment and take it a stage further. Imagine that the ideal visitor is not merely a conscientious seeker after zoological knowledge, but also someone who is reflective about such knowledge garnered within the zoo experience. What conclusions would such a thoughtful visitor come to? We have already seen that what is unique about the zoo experience is exposure to the relevant biological/ecological information in the presence of the exotic animals themselves. It is to this unique juxtaposition that the reflective visitor would, therefore, turn her/his thoughts. It would then strike such a person that a severe dissonance – both at the ontological as well as the conceptual level – exists between the message embedded in the explicit texts and that conveyed by the sub-text behind the animal exhibits themselves. The former is about conserving biodiversity and ecosystems/habitats in the wild; the latter is about the presentation of exotic animals as exhibits within a context which is human-designed and human-controlled to promote human ends, albeit including, that of educating humankind in the need for conserving wild species in their wild habitats. The earlier chapters of this book have laboured long over the point that while wild animals in their wild habitats live 'by themselves' as well as 'for themselves', and are the outcomes of the processes of natural evolution and natural selection, captive animals are immurated animals which no longer live 'for themselves', and are biotic artefacts. In other words, what jars is precisely what the *WZCS* considers to be the trump card of zoos, namely exposing the visitor to the 'living animal':

> The most basic level of zoo education is simply the display of living animals. This is the only means by which countless people will ever come into contact with living, wild animals and in a compelling manner become acquainted with elements pertaining to nature conservation.
>
> (*WZCS*, 1999, ch. 4.6)

The animals are indeed living organisms, but they are not wild, in any sense of the word. The thoughtful visitor becomes disoriented on two fronts, the ontological and the conceptual.

Upon further reflection, it might also occur to such a person that zoos themselves have failed to grasp that the dissonance noted exists because they subscribe to a view of animals which may be called *morphological reductionism*. By this is meant that an animal is no more than what it looks – in terms of its size, its anatomy, its shape, its colouring and other visible marks (such as stripes), or whatever other special features are best captured by a painting or a snapshot of the animal. However, zoos claim, of course, that seeing the animal in its living flesh and bones in a zoo is not the same as seeing a painting of the animal or indeed even seeing a stuffed specimen in a natural history science museum. According to zoos, then, what is peculiarly transformative about the zoo experience is to see that animal breathing, moving about or indeed even sleeping, and to be able to straightaway recognise it as a zebra or a lion if one has already seen paintings or snapshots of zebras or lions. However, advocates of the educational merit of zoos appear not to have paused to ask themselves the searching question why this dimension of seeing a live exotic animal should have such peculiar transformative powers.[36] They have just assumed that they exist.[37]

Perhaps the blunt fact of the matter is that it has no such transformative powers, given that the available evidence appears not to support the claim. Its failure as a hoped-for potent educational tool lies precisely in the fact that it rests on nothing more than morphological reductionism, as the sub-text of the exhibits themselves appears to confirm. That sub-text appears to say to the visitor that the exotic animal is 'wild' just because in terms of its outward appearance, it resembles, on the whole, the individual wild animal out there in the wild, be it a zebra or a lion; at the same time, it implies that the habitat of which the animal is an integral part is of no significance.[38] According to this perspective of morphological reductionism, the fact that the captive exotic zebra is kept in a naturalistic, scaled-down, human-constructed caricature of the wild habitat in which the wild zebra lives, that its lifestyle bears no remote resemblance to the life led by the wild zebra in the wild are of no relevance, meaning or importance in identifying it as a 'living wild zebra' as long as it resembles roughly in its outward appearance the individual wild zebra in the wild. In other words, the sub-text confirms that morphological reductionism is what determines that an x is a 'wild x'. This runs totally against the zoological conception of an animal – as purveyed by the explicit texts on biology and *in situ* conservation – according to which the individual animal is an integral part of its habitat, its existence led within its home range as a member of its species is itself the outcome of a long evolutionary history based on the processes of natural evolution and natural selection. It is precisely the simplistic perspective of morphological reductionism embedded in a context of exotic animals as exhibits which stands in the way – theoretically, psychologically/practically – of the zoo experience acting as a transformative educational tool with regard to conservation issues.

A very obvious instance of this dissonance would confront visitors to wildfowl and wetlands reserves such as Martin Mere or Slimbridge (both in the UK). Apart from acting as a valuable point of rest for migratory birds, they also house birds which are not wild but tame, many of them belonging to endangered species.[39] The tameness of mammals under captivity which distorts the reality of mammalian existence in the wild is not as obvious as that of birds. While the uninformed lay visitor may have to glean the fact from the information sheets and other graphics displayed at enclosures to learn about the size of the home range and other details of any particular mammalian lifestyle in the wild, no adult visitor at least, no matter how uninformed, would need to read such brochures to know that it is in the nature of birds to fly, and to fly relatively small or large distances, depending on the species. However, what they see before their very eyes is that the birds do not fly at all, in any real sense of that term. Take the pelican with its enormous wingspans which, if opened in a strong gale, can lift it off the ground, launching it into flight. To render it incapable of flying, at least one of its wings has got to be pinioned, preferably as a chick, to make the operation easier. However, when the bird stretches out its wings to sun itself, for example, it would be obvious to the visitor that one wing has been anatomically tampered with.[40] The taming, in the case of birds, involves in most cases mutilation, which amounts analogously in the case of mammals to breaking one of their legs to make them lame, in order to render them tame and captive.

One could put the point made above in another way. Should the zoo experience have the transformative efficacy attributed to it by its advocates, then the result would be ironical and paradoxical: the transformative efficacy comes from grasping that there is dissonance between the explicit text, on the one hand, about the urgency of the conservation of biodiversity and habitats in the wild and, on the other, the hidden sub-text of simplistic morphological reductionism inherent in the idea of captive exotic animals as exhibits.[41] Furthermore, grasping that dissonance would logically lead to the realisation that the justification of zoos in terms of their educational goal has no basis either in practice or in principle.[42]

Once attuned to the realisation that dissonance is inherent in the zoo experience, the thoughtful visitor would begin to find evidence of it in more ways than one. It suffices just to list a few examples, which are ultimately also derived from the central one already elaborated upon. Zoos are known to pair or group animals in what may be called an anomalous manner. For instance, most species of small felids are solitary in the wild; zoos, however, routinely exhibit them in pairs or trios for a variety of reasons, one of which is to provide the cat with company, an arrangement endorsed as part of the zoos' programme of environmental enrichment.[43] Instead of pairing animals or forming groups out of animals belonging to the same species (conspecific pairing), zoos sometimes form pairs or groups from different species (xenospecific pairing).[44] Zoos again justify this anomaly in the name of enrichment. The complex elephant social systems in the wild are never replicated in zoos, at least in the European zoos studied and cited below. Instead of the extensive family-based structure of wild herds, most of the elephants in zoos are held in small groups of unrelated elephants, consisting mostly of adults with a few infants or juveniles. In the wild:

> [t]he matriarch leads the herd, her dependent offspring, and adult daughters with their immature offspring including pre-pubescent

males – family units on average have between six and eight individuals (range 2–40) amongst Asian elephants and four to 12 (range 2–29) amongst African savannahs. These family units associate in 'kin' groups, consisting of up to five families and sometimes in turn form part of a clan consisting of several hundreds of individuals. Males either roam around alone or associate together in bull groups when not sexually active. In zoos, males and females are housed apart; in extensive systems in Asia, elephants are kept in mixed groups of between 51 to almost 3000, which approximate more to conditions in the wild. Female elephants in European zoos are therefore kept in far smaller groups compared to the wild and extensive systems. A solitary state is far more common in wild adult males, but they are also often observed in small groups when not sexually active, a situation not replicated in zoos.

(Clubb and Mason, 2002, ch. 5)

In the wild, the ratio of male to female is 1:1; in zoos there are more females than males, the ratio is 1:4.5 for Asian elephants and 1:3.4 for African elephants. In the wild populations, infants (under 5 years of age) were by far the most common age class; in (European) zoos, the most common age class does not comprise infants, but adults aged 30 to 34 years in the Asian populations and 15–19 in African populations.[45]

These anomalies distort the reality out there in the wild.[46] The mismatch, when noticed by the conscientious and reflective zoo visitor, would only serve to disenchant the observer rather than enhance the zoo experience from the point of view of its expected transformative powers as enunciated by zoo advocates whose primary mission is education.

It is worth labouring the crucial point yet again that zoos, far from showing respect for animals-in-the-wild and their behaviour in their wild habitats, necessarily present their captive exotic animals as mere exhibits, whose behaviour has nothing to do with behaviour in the wild. The behaviour of immurated animals as exhibits implies that behaviour in the wild is considered as an irrelevance to a proper understanding of what constitutes the wildness of animals, or that it is meant to be taken seriously as a way of understanding animal wildness; in reality animal behaviour in zoos is, at best, nothing more than an obscene caricature of behaviour in the wild. Hancocks has, indeed, put his finger on the point when he writes:

Visitors often do not see zoo animals as wild animals because they are not presented as wild animals. They play with beer barrels, bowling balls, and plastic milk bottles. They feed from piles of chopped carrots, heaps of premade chows, beef patties in stainless steel dishes.

(Hancocks, 2001, p. 249)[47]

Zoos proudly present such innovations in animal activity (playing with plastic toys) as laudable and successful attempts to enrich their lives; as captive residents, they would otherwise die of boredom and engage in stereotypic behaviour, upsetting their visitors into the bargain.

Shepherdson, a leading expert and exponent of enrichment programmes for captive animals writes:

> Chimps at zoos are furnished with artificial termite mounds. Instead of actual termites, the chimps plunge their sticks into gobs of treats such as mustard, pie fillings, ketchup, apple sauce and other goodies they love . . . They are doing what they do in the wild, only the taste results are different . . . Keepers try to stimulate animals' intellects, making them want to constantly explore their cages for potential food and fun with: Smells, such as spices and animal scents, are placed on logs or rocks in exhibits. Large, tough polyethylene balls go into the cages of our tigers. They like to bang them around or pounce on them as if they were prey. Our tigers . . . hunt for fish and dry meat that keepers hide in logs, giving them a chance to smell and forage for their food as they would in the wild. Bears also get the 'log' treatment, hunting in them for bone, raisins and nuts . . . Occasionally the polar bears receive lumps of ice containing frozen food items. Polar bears are also given plastic balls and tubs to manipulate, play with and ultimately destroy.
>
> `   (Shepherdson, 2005)

Hancocks (cited above earlier) is quite right. Imagine what children taken on an educational trip to a zoo would take away as lasting impressions of what such 'wild' animals do in their spare time! Children exposed to such scenes would be highly unlikely to make the intellectual as well as psychological leaps from the evidence before their very eyes to the story revealed in the literature about truly wild animals and their behaviour in the wild and thereby benefit from the postulated transformative powers inherent in the zoo experience. The transformative powers could not be realised as a result of the dissonance between what they see and what they are told to see/read.[48]

The postulated transformative powers can never be realised, no matter how much effort zoos may make to reduce the dissonance noted. In the 1970s, Woodland Park Zoo (United States) began feeding whole carcasses of sheep and goats to the big cats and of rabbits and chickens to the small cats. Outraged citizens protested and the zoo stopped the innovation at once.[49] Danish zoos have spent a lot of time and energy in preparing their visitors over the years to accept more naturalistic diets; for instance, they have been able to feed horse or cow carcasses to their lions without the Danish citizenry being outraged. Whenever the zoos get a fresh carcass, the visitors are treated to a talk by a keeper and an educational officer to explain what is happening. When presented with a whole large carcass, the lions would behave more like lions in the wild – for instance, the male lion is the first to go up to it to feed, then followed by the cubs and the female. Furthermore, unlike the normal zoo food, such a carcass takes more time to consume which makes it approximate more to consuming prey in the wild.[50] However, there is a limit to such 'enlightened acceptance'. No federated zoos at least in the Western (economically) developed world would dare to allow a large predator, such as a lion, or a tiger hunt a live prey such as a deer, even if this were practical, although some zoos may feed their large carnivores with small prey, such as a live chicken, behind the curtains, so to speak, in non-exhibit enclosures.[51] This limitation is but a reflection of the essential fact that a zoo is a human cultural space, where human values prevail and in which captive animals must be treated in ways consonant with these values. As the thrust of this book has argued, zoo values are just not compatible or consonant with the reality of life as led by wild animals-in-the-wild. There is, therefore, bound to be dissonance between the two worlds, that of the zoo and that of the wild, and into the gap falls irretrievably the so-called potential transformative powers of the zoo experience.

Conclusion

This chapter has argued that none of the three justifications of zoos deemed to be serious can stand up to critical scrutiny. The flaw lies crucially in the inevitable dissonance – fundamentally at the ontological level – between, on the one hand, wild animals, their behaviour in their wild habitats within their own home range and, on the other, immurated animals and their behaviour as exhibits in zoos which are human-designed and human-controlled institutions serving human ends, whether these be furthering scientific knowledge, conservation or education.

In the domain of scientific research, it would be bad science, into the bargain, to extrapolate uncritically from data achieved with immurated animals to very dissimilar contexts elsewhere. In the domain of conservation, we have seen that captive breeding as part of *ex situ* conservation is incompatible – ontologically and practically – with the conception of zoos as collections of exhibits open to the public; as such, these activities are best hived off to other institutions which may be called captive breeding conservation centres. The definition of 'zoo' cannot be stretched indefinitely to cover two fundamentally diverse forms of activities, without creating confusion in thought and in policy. In the domain of education, there is little or no evidence to back the claim that zoos are uniquely efficacious in driving home ultimately to the public the need for the conservation of wild animals in their wild habitats. Furthermore, the unique potential transformative power of the zoo experience, as claimed by those zoo advocates who see zoos as the crucial tool in their mission to educate the public in matters of biodiversity and ecology, cannot and can never be realised, given the distinct mismatch between what zoo visitors see with their very own eyes, which is at best a caricature of wild animals and their behaviour, and at worst is a denial of the relevance and significance of wild animals and their behaviour in the wild. Immurated exotic animals and their behaviour exhibited in zoos are the ontological foil to wild animals-in-the-wild and their behaviour in their natural habitats within their normal home range.

10
Justification Deemed Frivolous

The last chapter has critically challenged three justifications of zoos deemed serious and found each wanting. We turn now to the fourth justification, usually considered today to be less respectable, and therefore sometimes not even deemed worthy of mention by serious-minded zoo professionals and advocates. It is recreation.

Why frivolous?

It is not easy to comprehend why the recreational justification of zoos is held in somewhat slight regard in certain quarters, when it is obvious that zoos play a vital social role in providing recreation and entertainment, and that zoo-going is one form of leisure activity competing against others for the attention and attendance of the public, such as watching television, playing computer games, going to sporting/musical/theatrical events.[1] According to the statistics, zoos appear to be doing exceptionally well in attracting punters, so to speak, to their gates. We have seen in the last chapter that there are more than 10 000 zoos world-wide. However, as figures of attendance for all of them are hard to come by, it would be best to confine the numbers only to the 1200 or so federated and accredited zoos which are said to constitute the core of the zoo-world today; nevertheless, they still come to the staggering total of at least 600 million, about a tenth of the entire population in the world per year.[2]

Zoos are big draws and big business. Why? *Ex hypothesi*, it is not because zoos are institutions which carry out scientific research or captive breeding conservation programmes, for the simple reason that these two forms of activities are not open to public visitors. So that leaves education; however, in the last chapter, we have cited statistics to show that the average time spent before enclosures and the attention-span on the part of visitors do not, as a matter of fact, permit them to pause to read, absorb and digest the detailed material made available to them about matters of biodiversity, ecology and conservation. It then seems reasonable to infer that zoos are big draws precisely because the public, flocking to visit them, do so for recreational/entertainment reasons, and not primarily for the so-called noble motive of educating themselves about the wild and the urgent need to save wild species and their habitats from extinction.

Zoo-going is, undeniably, for a substantial part of the world's population, a congenial and agreeable form of family outing, rather like a picnic in some park or beauty spot in the country. It is, indeed, superior to these other examples of outdoor recreation, because when the weather becomes inclement, one can take refuge in the indoor enclosures. It is true museums provide shelter, too, when it rains or hails. But museums, on the whole, do not welcome children, and even when they do, the children, like those in Victorian times, are expected only to be seen but not heard. Museums are hushed places; for parents to keep their children under severe control so as not to disturb the quiet reverential ambiance, is so stressful that museums are best avoided. In a zoo, children are welcome; they can run around, and fellow visitors appear on the whole not to mind when children behave as children tend to do, as zoos, in general, unlike museums, are not places which expect visitors to keep a respectful and reverential silence in front of the objects they have come to see. Furthermore, children do get excited when they see a charismatic animal such as a tiger or a chimpanzee, whereas it would be a very odd and unusual child indeed to get excited in an analogous way and to the same extent when s/he is taken to stand in front of the *Mona Lisa*. So parents, being sensible, opt to take their families to zoos instead.

Shared presuppositions

Zoo-visiting as a form of recreational activity demonstrates that zoo visitors act upon the understanding that zoos are essentially collections of exhibits consisting of live exotic animals, open to the public. For them, the crucial aspect of the exhibits is that the animals are alive, not dead and stuffed, as are the exhibits they would find should they visit a natural history of science museum. In this, they are at one with zoo professionals, who also claim that the unique attraction of zoos over these other establishments exhibiting animals is that their exhibits are indeed live animals.

As long as the animals ostensibly breathe, move about, flick their tails or their trunks, make other movements such as climbing poles or frames, or make vocalisations such as growling or screeching, these appear to be sufficient to please the punters. At the same time, the animals must look like those they have seen pictures or other images of, so that they can immediately recognise the exhibit as a lion or an elephant, and so on. In other words, they subscribe implicitly, as zoos do, to the view referred to, in the last chapter, as morphological reductionism – basically, what an animal is, is what it looks like.

Zoo visitors, on the whole, appear indifferent to the ontological status of the exotic live animals before their very eyes; in other words, they are not particularly worried or concerned whether such animals are wild as zoos claim they are, or whether they are truly wild. They appear to side with the zoo interpretation of 'wildness', namely an animal is 'wild' if it looks like the wild animal in the wild existing in its own home range. They are complicit with zoos in accepting for their purpose that the animal exhibits are 'wild'. Those members of the public with ontological scruples of the kind advanced by this book are few and far between; they either keep away from zoos altogether or if they do visit, they do so not on the understanding that they would encounter truly wild animals. As such, they are not likely to ask for their money back on the grounds that zoos have violated the Trade Descriptions Act; nor, indeed, would they necessarily succeed were they to sue, because of their prior understanding that zoos do not in reality exhibit wild animals or animals-in-the-wild.[3] Zoos do not need to lose sleep over such potential litigation; the majority of their visitors are happy and satisfied customers.

As already remarked upon in the last chapter, their visitors would indeed be unhappy or dissatisfied should the animals appear to them to be mangy, undernourished, obviously unwell or living in unsanitary conditions. So long as the animals look well-fed, not ill or diseased, the visitors would have nothing to complain about. In this respect, the (environmental) enrichment programme that today's zoos are committed to fits in absolutely well with the expectations of zoo visitors. As one such expert has correctly remarked, animals in captivity easily succumb to boredom, as they have nothing really any better to do to keep themselves occupied than to engage in stereotypic activities, such as nibbling their tails, shaking their heads endlessly, pacing for no obvious good reason, or throwing their faeces about.[4] These would distress the visitors; zoos rightly attempt to eliminate them via their enrichment programmes, for the sake of their animals and of their customers as well as of their own financial interests.

We have already observed that the majority of visitors are not particularly concerned either about the ontological status of the animals they see or the explicit mission of zoos to educate them in matters of zoology, ecology or biodiversity, and as their aim is to get some entertainment looking at how live animals behave in zoos, it would not matter to them if the animals behave in ways which bear no resemblance to what real wild animals actually do in the wild. They are not perturbed that zoo animals cannot hunt or forage, or roam large distances. On the contrary, they can be mightily amused by the sight of the polar bear trying to submerge itself in the tiny pond of ice-cold water provided in its enclosure, or to catch with its paw the fish in it, or to find a chocolate encased in a lump of ice, or to play with plastic balls which it then crushes with its paws. They can be agog with delight to see a chimpanzee using a stick to poke into the artificial termite mound, not to fish out termites as their relatives in the wild would as their proper food, but supermarket treats such as mustard, pie fillings, ketchup, apple sauce and other processed foods that they themselves and their children might well be munching at that very moment as they stand in front of the chimpanzee trying to get hold of similar goodies. They are not remotely concerned with the fact that in the wild, the chimpanzee and the polar bear would not be eating the kind of food or acting in the kind of way they are constrained to in their enclosures. Indeed, the greater would their delight be, should the experts on enrichment come out with new ways of making the lives of the animals less boring and more entrancing for them. For instance, chimpanzees can be trained to use computers; they appear to enjoy themselves tremendously in watching computer games.[5] It is obvious to the visitors that chimpanzees in the wild would show no interest in such gadgets or in games and would be preoccupied with a very different set of activities altogether; however, this realisation would not stand in the way of their enjoyment in seeing a chimpanzee in a zoo doing precisely that.

Visitors to some zoos would find it just as amazing and amusing to buy a special ticket to be allowed into a special enclosure to take tea with some chimpanzees or have a so-called 'wild breakfast' in the company of orang-utans.[6] They know, of course, that chimpanzees in the wild would not be sitting down with them at a table while they drink tea or have breakfast; but the zoo is not wilderness and the chimpanzee or orang-utan hosts are not truly wild. So it is fun. The chimpanzees and the orang-utans appear to be enjoying themselves; it is far better that they sit at a table with humans than that they tear out their own hair in frustration. Some zoos also offer facilities for nuptial and other receptions, using their animals as backdrop to enhance the pleasure of such occasions. One could choose one's favourite animal amongst those on offer.[7]

The logic of recreation may be seen in its most extreme as well as its purest form in the kind of entertainment which has been honed to perfection and which pulls in the crowds at Sea World, where Shamu, the whale, does amazing turns under the command and control of its trainer.[8] Those who want to see what whales in the wild actually do are not the people to turn up at Sea World; however, the spectators of the show know full well that whales in the ocean do not behave in that fashion, obeying human commands. What thrills them is precisely the way the whale reacts to such commands; what they marvel at is also the skill with which the trainer has got the whale at his/her beck and call.

Zoo visitors equally enjoy the novelty, charm and thrill of seeing zoo keepers exercising/walking a cheetah or a tiger, an experience unique to zoos. Naturally, such a sight would never be found in the wild; nor would one find it in one's suburb. The cheetah or tiger is definitely not a dog, yet its keeper is taking it for a walk in the same way as they themselves, in their role of suburban dog owner, walk their own dog. It is tame, yet it looks wild. Herein lies the peculiar fascination which comes from the conjuncture of frisson and amusement.

Conclusion

In other words, zoo visitors are tacitly aware that a zoo is a human cultural space and that the animals exhibited within it are expected to conform to human practices, norms and values. They know, then (even if they do not articulate the thought in quite this manner), that the exhibits are immurated animals and are biotic artefacts, responding to and interacting with the stimuli orchestrated by their human controllers and managers. They, therefore, implicitly know that the animal in front of them is not a token of the wild species. The individual zoo orang-utan is not a member of the species (*Pongo pygmaeus*), but of the species *I*(*Pongo pygmaeus*).

This is to say that they know what the score really is and with that tacit awareness, they can sit back, relax and enjoy their time at the zoo in the presence of the exhibits as a form of recreation and entertainment. For those who care about the welfare of animals, they may even prefer to visit a zoo rather than a circus, as zoos, on the whole, are more successful than circuses in convincing them that the animals under their care do not suffer, but are happy and comfortable in their enclosures.

Their satisfaction and fascination, as we have already observed *en passant*, comes from the tension they perceive between the fact, on the one hand, that the exhibits look wild, as they look just like the animal-in-the-wild (their tacit commitment to morphological reductionism), and on the other, their tacit appreciation of the fact that they are not wild. This adds to their thrill of being in the presence of something which could be potentially dangerous and threatening, such as a truly wild tiger or lion while knowing that the exhibits are not truly wild, that the animals would not be able to cause harm to the visitors, since they know that the zoo is a human space made safe, but interesting and entertaining for humans.

It may well turn out that ordinary zoo visitors are more insightful about the true nature or 'essence' of zoos than the noble-minded advocates of such institutions in the name of education-cum-conservation. It is no wonder that such missionary zeal, on the whole, leaves them unaffected and passes them by – rightly, they concentrate on what they have come to zoos for, namely, to see and be amused/entertained by live exotic animal exhibits in pleasant naturalistic simulated enclosures. Good wholesome recreation and fun to be had by one and all in the family. This is not to be sneered at, especially when the satisfied customers appear not to have been taken in by the zoos' own mistaken and misleading spiel that the animals they exhibit are wild.

11
Philosophy and Policy

In Chapter 1, we identified five conceptions of what counts as an animal, namely the lay conception, the zoo conception, those presupposed respectively by Singer and Regan in their philosophy of animal welfare/animal rights, and the zoological conception. The subsequent chapters investigated the fundamental issue concerning the different ontological status of zoo (captive) animals on the one hand and that of animals-in-the-wild on the other. It is now time to look more closely at this fundamental issue in the context of the five conceptions raised in Chapter 1. There are some surprising conclusions which emerge from this examination.

The zoological conception and its ontological presuppositions

This conception of what counts as an animal and what is a wild animal need not delay us for long; it is this conception which has inspired precisely the exploration of the difference in the respective ontological status of exotic zoo animals as exhibits and animals-in-the-wild. The ontological perspective of this book, therefore, coincides with those implicit in the zoological conception concerning animals-in-the-wild. A quick summary should suffice:

- Animals include all animals – as identified and classified by the current consensus of the scientific community – whether vertebrates or invertebrates, of which there are over a million known species at the least.
- Animals-in-the-wild can only be properly grasped and understood when they are seen as the outcome of the processes of natural evolution and the mechanism of natural selection within their natural

home ranges, interacting in a causally complex manner with their habitats and ecosystems of which they are an integral part. Their morphology, their genetic inheritance, their behaviour are the outcome of such processes of reciprocal causation.

- Animals-in-the-wild live 'by themselves' and 'for themselves', pursuing their own trajectories, independently, in principle, of human design, manipulation and control.

The zoo conception and its ontological presuppositions

These constitute the *ontological foil* to the zoological conception of animals-in-the-wild:

- Zoos claim to keep animals, but in reality, they house only a tiny portion of all the species of animals known to science; these are usually confined to the vertebrates, especially large- and median-size mammals. This extreme selectivity in their reference implicitly denies the status of being animals to those hundreds of thousands of species excluded.
- Zoo animals are necessarily exotic in both the technical as well as lay senses of the term – they are animals which have been removed from their natural home ranges, transported to other climes in other parts of the world, and they are, on the whole, considered to be charismatic, rare, especially cuddly or appealing to humans because of their looks or behaviour.
- Zoos are defined as collections of animal exhibits open to the public; this definition encapsulates in a nutshell the difference in ontological status between zoo animals and animals-in-the-wild.
- As exotic exhibits, the animals necessarily lead lives in environments which are totally different from those led by animals-in-the-wild, because a zoo is a human cultural space within which the animals under captivity are expected to conform to the purpose and norms of humans who control and manage them.
- As immurated animals, they are no longer subject to the processes of natural evolution; they are, therefore, biotic artefacts. As such they are the *ontological foil* to the wild animals living in the wild.
- As immurated beings, they do not live 'by themselves' nor do they live 'for themselves' since they live to fulfil the destiny that humans have ordained for them.
- Zoos maintain that the animals under their care and control are 'wild'; yet that claim cannot survive critical scrutiny. It is, therefore, both a conceptual and an ontological mistake for zoos to hold that immurated

animals are wild animals; they are not wild and are not tokens of any wild species; they are tokens of what may be called immurated species.

The lay conception and its ontological presuppositions

According to the deconstruction pursued in the last chapter, the zoo-going public appears to have an intuitive grasp of the ontological issues embedded in the zoo conception of animals as exhibits. Surprisingly, one is able to infer from its attitude certain (implicit) tenets and construct a more or less coherent and, indeed, even a sophisticated conception of zoos and their justification.

* The majority of zoo visitors regard zoos as places of recreation rather than serious education.
* The zoo-going public agrees wholeheartedly with zoo management/philosophy that zoos are essentially collections of animal exhibits.
* The animal exhibits constitute the focal point and source of their entertainment. As such, visitors expect zoos – and zoos happily oblige – to exhibit, by and large, charismatic, rare animals which, on the whole, tend to be vertebrates, especially large- or median-size mammals. These exhibits are exotic in the lay sense of the term.
* The zoo-going public intuitively grasps that, although the animals may look like animals-in-the-wild (morphological reductionism), they are exotic (in the technical sense of the term) and are not really wild, as they live in environments, geographically far from their home ranges, which are totally different from habitats in the wild.
* Visitors do not, therefore, expect animals in captivity to behave in the same way as animals-in-the-wild.
* The zoo-going public, nevertheless, finds the behaviour of animal exhibits within their human-designed and human-controlled space fascinating, amusing and pleasing, provided it is convinced that the animals themselves do not suffer pain in being encouraged to display such modes of behaviour while passing their time in confined space. It implicitly and happily accepts that the programme of enrichment is a good thing, as enrichment makes the animals less bored and, therefore, also less boring to watch.
* *Au fond*, the zoo-going public appears to agree with the implicit ontological stance behind the zoological conception of animal which this book makes explicit, namely that zoo animals are immurated animals, and therefore are biotic artefacts, and that they are not tokens of any species in the wild.

Philosophy of animal welfare

This philosophical perspective is neutral as to whether the animals are truly wild or tame, domesticated or undomesticated, as its remit is confined solely to sentience and the prevention of suffering whenever possible. As such, it is agnostic with regard to the ontological status of zoo animals. The morality of sentience demands that all sentient animals – in particular, mammals, in which zoos take an inordinate interest – ought to be treated in ways which do not cause them pain (or cause them less pain), especially physical pain. Such a philosophy chimes in admirably with the philosophy of enrichment which today's progressive zoos practice, as such a programme does have at its centre the welfare of the animal. That is why while animal welfare activists picket animal laboratories (which breed animals for use in scientific experiments which often, if not invariably, involve pain), they do not stand outside the gates of well-run zoos, which are perceived to do all that can be done to reduce, if not eliminate, suffering in animals under their care and control. However, organisations such as the Royal Society for the Prevention of Cruelty to Animals (RSPCA) do protest and demonstrate against poorly-run zoos which ignore the health and well-being of the animals in their charge.

Philosophy of animal rights

Unlike the philosophy of animal welfare, the relationship between this philosophical stance and that pursued in this book regarding the ontological status between zoo animals and animals-in-the-wild, is somewhat more complicated. One needs to distinguish between two contexts. The first involves the commonly perceived situation regarding zoos, that they house wild animals, so to speak. It is safe to assume that those today who subscribe to the morality of animal rights, also share this perception; it, therefore, follows that their philosophical view about animal rights is distinctly incompatible with the philosophy of zoos and zoo management including that of enrichment, as zoo practices and procedures make life difficult for animals which are 'subjects of a life'. The morality of animal rights sees the denial of the freedom to roam, to lead lives as animals do in the wild as amounting to slavery. On this view, slaves are slaves, whether they are slaves suffering and miserable in chains, or contented slaves living in safe, sheltered and luxurious accommodation – the equivalent of the Ritz Hotel – and are cared for and looked after by the best medical consultants in the world and their team of health workers. The Born Free Foundation is not the RSPCA; when pressed, it would surely argue that freedom to lead independent lives which bring inevitably in their wake, hunger, cold, thirst, disease and early death is better than living in security from famine, disease and early death but at the price of being unfree. A gilded cage is still a cage. Zoos, in principle, are morally dubious institutions and, therefore, unacceptable.[1] However, the second context is a potential one which involves the rejection of the commonly held perception that zoo animals are wild animals in the light of the ontological arguments set out in this book – those who uphold the morality of animal rights may come to agree that zoo animals, being immurated animals, are a special type of domesticant, especially those animals which are the offspring of immurated ancestors, and therefore are not proper subjects for liberation from zoos, any more than cats and dogs are proper subjects for emancipation. Those who react in this way may then be said to belong to the moderate wing of the animal rights movement; those who consider domesticants of all varieties, whether classical or the newly identified immurated type, to be the proper target for liberation, would not, naturally, be impressed by the ontological standpoint advocated by this book.

Policy conclusions

From the ontological vantage point of this book, it transpires that zoos, in principle, are not necessarily unacceptable institutions; indeed, they are a positive cultural asset provided certain important caveats are entered:

- First of all, one needs to draw attention to the fact that the ontological standpoint of this book is in a sense, on the surface, compatible with the philosophy of animal rights with regard to zoo animals. It, too, deplores the fact that animals directly captured from the wild are denied freedoms; however, it regards animals born and bred under captivity in a totally different light, as their degree of artefacticity is much greater than that of the former category of animals, which have not been transformed so thoroughly into biotic artefacts.[2] In other words, the ontological perspective belongs to a totally different type of discourse from that of the philosophy of animal rights – the central value of the latter rests on a particular normative ethical theory, that of rights, while the central value of the former is derived from the premise that animals-in-the-wild have come into existence, continue to exist and go out of existence, in principle, independent of human design and control. These are different in ontological status, and therefore, possess a value different from animals whose existence is shaped and controlled by human intentions and goals. The wildness of animals-in-the wild is an intrinsic value which has nothing to do with the goals and purposes of humans; in contrast, the value of immurated zoo animals, which are biotic artefacts, is primarily instrumental, to serve human intentions and purposes. As a result, unlike the morality of animal rights which condemns all zoos in principle, the ontological stance is not *per se* inhospitable to the idea of well-run zoos from the point of view of animal well-being. In this aspect, it is compatible with the philosophy of animal welfare; zoos which are human cultural spaces are, therefore, quite morally acceptable if they are run on the 'humanitarian' lines endorsed by a philosophy based on sentience. It is precisely the emphasis of this study on the ontological difference in status between animals-in-the-wild and animals in zoos which enables it, as a result, to adopt a much more nuanced attitude to zoos.

- It also follows from the above that zoos must not, however, claim that animals under captivity, living at best in simulated naturalistic environments, are wild, and recognise that they are not tokens of species in the wild.
- Once these crucial concessions are made, there is no ontological objection to zoos, provided that zoos are prepared to acknowledge, in consequence, that immurated animals are in fact domesticated animals. One objects neither to immurated animals nor to traditional domesticants; one could only legitimately complain should their ontological status be misrepresented by the claim that the former are 'wild animals in captivity' when, in reality, they are biotic artefacts.
- Hence, the ontological stance endorsed by this book is sympathetic to the attitude of zoo visitors who, by and large, are very comfortable with the reality they are confronted with, namely, that they are looking at, and being entertained by a unique kind of animal, that is to say, immurated animals which are found only in zoos, and that such animals do behave very differently from animals-in-the-wild. For them, this uniqueness is what precisely constitutes the zoo experience.
- Zoos are mistaken in believing that their true goal is not recreation, but some other more high-minded ones such as education about animals-in-the-wild for the purpose of their conservation, as well as *ex situ* conservation which zoos undertake through introducing certain of their captive-bred animals to the wild. Ontologically speaking, zoos are simply not the right space for the pursuit of such aims, noble though the goal of educating the public about biodiversity and ecosystems in the wild undoubtedly is.[3] *The World Zoo Conservation Strategy* (1993) and the European *Union Zoos Directive* (1999) are essentially misguided in the goals and the priority they have set out for zoos.[4]

• Zoos, in the long run, appear to be ideally positioned to create new immurated species in two ways: (a) out of extant wild species through the sheer fact of sustaining generations of captive-bred animals; and (b) more radically through modern biotechnological techniques, of either re-creating extinct species such as woolly mammoths via cloning, relying on fossil DNA, or entirely new chimeric animals/ species. Chimeric technology has already permitted the creation of the geep or shoat (a chimera with a sheep and a goat as genetic parents).[5] Such novelties are bound to thrill zoo visitors of the future and would be sure-fire draws, should zoos choose to embrace this route.

The usual arguments which are justly mounted against using such techniques as part of *ex situ* conservation would no longer apply, as the products of these procedures are intended to be immurated animals which would be tokens of new immurated species. For instance, one argument against cloning extinct animals in the context of conservation is that such clones would have no suitable habitat to be released into the wild, and even if such a habitat could be found, they would not necessarily know how to survive, as we have earlier seen in the analogous case of the Californian condors. However, as immurated animals, they would never leave the zoo environment and never be expected to fend for themselves. Nor would they be expected to found (genetically) sustainable populations in the wild.

To clone an extinct animal would require using an extant female animal belonging to a different species to provide the egg as well as to act as surrogate womb – this means that the clone, in reality, is a hybrid, as the mitrochondrial DNA remains in the egg in spite of having its nucleus removed. The contribution of mtDNA to the development of the embryo and later the animal could not be ruled out; nor could the contribution of the surrogate womb be ruled out on scientific grounds. However, in the future zoo context of creating artefactual biodiversity, none of these objections would be valid.[6]

• Zoos, once emancipated from the myth that their collections of exhibits are 'wild animals' may also choose to embrace the strategy of artificial selection of certain features of animals in their keeping, deemed to be attractive to zoo visitors. This would, indeed, enhance their recreational value. For instance, zoo visitors find albino tigers fascinating. At the moment, reproduction of such animals in zoos is frowned upon because they are said to be 'freaks' in nature, that is to say, they occur very rarely; therefore, it would be wrong for zoos to reproduce them because this would mislead zoo visitors about the true state of affairs which obtain in the wild.

Conclusion

Before setting out a few succinct conclusions to be drawn from the philosophical/ontological exploration of zoos, it may be useful to remind the reader of the details embedded in the distinction between animals-in-the-wild on the one hand and immurated/zoo animals on the other.

'Wild' versus 'tame/immurated'

Wild	Immurated
Not habituated to human presence; flight tendency and distance deeply engrained	Tame: Habituated to human presence; flight tendency overcome and flight distance overcome/reduced through training procedures
Roaming within ancestral/natural home range	Exotic: removed from home range, transported to different geographical/climatic locations
Existence led in natural habitats	Existence lived in miniaturised, simulated naturalistic environments in exhibit enclosures during zoo opening hours but otherwise often in cages in non-exhibit areas for the rest of the day/night
Daily life dictated by need to get food/shelter, to run away from danger	Hotelification: 'bed and full board' with 'room services'
Diet: hunting other animals in the case of carnivores; foraging for plants in the case of herbivores	Diet: scientifically/nutritionally prepared food pellets, chopped up meat from the carcasses of domesticants such as horse, cattle, supplemented by some foraging in the case of herbivores

(Continued)

Wild	Immurated
Seasonal activity: males actively seeking and fighting for suitable females to mate	Exaggerated sexual and mating activity; mating partners determined by zoo managing policies which include techniques such as *in vitro* fertilisation
Vulnerable to hunger/thirst, wounds/injuries, disease leading to early death	Totally protected from hunger/thirst, wounds/injuries, disease leading to early death
Vulnerable as prey to dangers including wounds/injuries and often death inflicted by predators	Neither prey nor predator and the 'ills' thereof, including early death, exist
Self-medication in certain cases is known to science	Full panoply of veterinary/medical care and services to prevent disease and early death
Life of (usually) females taken up largely with looking after their cubs, teaching and initiating them to become independent mature adults	Nurturing tasks taken over in the main by zoo carers and keepers; teaching tasks are redundant given hotelification and medication
Subject to the processes of natural evolution and the mechanism of natural selection	Suspension of natural selection: human design, control and management act as substitute, involving a form of artificial selection, leading to domestication (though not in the classical sense of the term)
Naturally occurring beings; beings which live 'by themselves' and 'for themselves'	Biotic artefacts which exist neither 'by themselves' nor 'for themselves'
Pursue their own trajectory, fending for themselves	Existence ordained by zoo managers; looked after in all ways by zoo carers and keepers
Exemplify the thesis of intrinsic/immanent teleology	Exemplify the thesis of extrinsic/imposed teleology
Individual animals are tokens of a wild species	Individual animals are not tokens of any wild species, but of novel immurated species
Individual animals form part of a natural population within a certain habitat	Individual animals are exhibits and form part of a collection of animal exhibits open to the public
'Wild animal in their natural habitat': antonym and ontological foil of 'tame'/'domesticated'/'immurated' animal	'Tame', 'immurated': antonym and ontological foil of 'wild animal in their natural habitat'

Summary of ontological implications for policy making

- Zoos do not house naturally occurring wild animals; nor does it make sense to say that they house 'wild animals in captivity'. Instead, zoos house tame, immurated animals. The latter, as biotic artefacts, are the *ontological foil* to wild animals in their natural habitats.
- Zoos should not, therefore, be expected to take on tasks for which they are ill suited, such as *ex situ* conservation as well as education for conservation.
- Zoos through their collections of animal exhibits are uniquely placed, as a human cultural space, to provide recreation and entertainment to their visitors.
- Zoos, far from protecting threatened wild species from extinction as a modern-day scientific Noah's Ark, are well placed to be the creator of new species of animals in the long run; however, these are immurated species. They add to biodiversity; but the biodiversity is not that of extant naturally occurring, wild species but novel artefactual species. (For details of the differences between natural and artefactaul biodiversity, see Lee (2004).)
- In the light of the above, this book throws out a challenge to the World Association of Zoos and Aquariums (WAZA) and the EU to reconsider their justifications for zoos in terms of *ex situ* conservation and education-for-conservation.

Appendix: Environmental Enrichment or Enrichment

This short appendix is not meant to be a comprehensive account of the zoo management technique called either environmental enrichment or, more simply, enrichment. Instead, it aims only at assessing it from the ontological standpoint adopted in this book, namely, that zoo animals are immurated animals, a new type of domesticants in the making.

Let us begin by stating what environmental enrichment is through the words of one of the world's leading experts on the subject:

> 'Environmental enrichment is a process for improving or enhancing zoo animal environments and care within the context of their inhabitants' behavioural biology and natural history. It is a dynamic process in which changes to structures and husbandry practices are made with the goal of increasing the behavioural choices available to animals and drawing out their species-appropriate behaviours and abilities, thus enhancing their welfare. As the term implies enrichment typically involves the identification and subsequent addition to the zoo environment of a specific stimulus or characteristic that the occupant/s needs but which was not previously present. (American Zoo & Aquarium Association, 1999)'. In practice, this definition covers a multitude of innovative, imaginative and ingenious techniques, devices and practices aimed at providing adequate social interaction, keeping animals occupied showing an increased range and diversity of behavioural opportunities, and providing more stimulating and responsive environments. Examples range from naturalistic foraging tasks, such as the ubiquitous artificial termite mound, puzzle feeders constructed from PVC pipes, finely chopped and scattered food, novel foods and carcasses, to objects that are introduced for manipulation, play and exploration, novelty and sensory stimulation. Appropriate social stimulation, both within and between species, and even training can be considered as enrichment. On a larger scale, the renovation of an old and sterile concrete exhibit to provide a greater variety of natural substrates and vegetation, or the design of a new exhibit that maximizes the opportunities for natural behaviours, are also considered as environmental enrichment.
>
> (Shepherdson, 2003, p. 119)

From Shepherdson's account above, one may tease out the following theses:

1. It is regarded by its practitioners to be in accordance with the philosophy of promoting animal welfare.[1]
2. It tacitly acknowledges that the zoo environment is very different from the wild environment; we have seen that in the latter, animals have to fend, hunt/forage for themselves while in the former, all services are laid on as far as food, shelter and protection from danger go.

3. It recognises that the animals are bored by the zoo environment, and hence the need to provide for opportunities to stimulate the bored animals.[2]
4. It implies that it is Janus-faced, that is to say, it looks back to the animals's 'behavioural biology and natural history' as well as to the present and the future with regard to their behaviour in their new exotic environment.

In other words, it is at the very centre of the procedures and processes of transforming the inmates to becoming immurated animals, to becoming domesticants. While recognising that the animals carry with them a biology and a natural history peculiar to the species which have evolved naturally in the wild, it also recognises that zoos cannot satisfy the needs which arise from that biology and that natural history in the same way as their ancestors or their wild counterparts can satisfy them in the wild. It cannot do so for two reasons: (a) the simple contingent one that the zoo environment is utterly different from their natural habitat; and (b) the theoretical and logical one that zoos cannot meet the needs of the animals in the way their ancestors meet theirs without abolishing zoos themselves, and returning the animals to the wild. Hence zoos offer naturalistic environments, naturalistic foraging and hunting instead.

As this book has demonstrated, zoo experts and managers hold that zoo animals are 'wild animals in captivity', but such a claim is conceptually speaking an oxymoron, and ontologically speaking wrong-headed and totally misleading because wild animals are naturally occurring beings, whereas zoo animals are biotic artefacts, the ontological foil to the latter. These deep-seated flaws are reflected in the programme and techniques of environmental enrichment which, as we have just remarked, are Janus-faced. In turn, these techniques for ameliorating boredom and thereby improving the welfare of the animals serve to hasten the procedures and processes of transforming them into biotic artefacts, to being domesticants. Over time, as the animals adapt to and evolve within the context of selection for captivity, zoo experts, too, would evolve their techniques of enrichment such that they will increasingly be less and less naturalistic and more and more oriented to human culture rather than to animal culture as found in nature amongst animals living in the wild.

According to Shepherdson (2001) and others, an enriched environment should allow the animals to perform 'natural behaviours' implying that the behaviour of animals in zoos are 'unnatural' or 'abnormal'. But what counts as 'unnatural or abnormal behaviours'? In the widest sense of 'abnormal' or 'unnatural', all behaviours of zoo animals may be said to be abnormal or unnatural – eating zoo pellets as much as playing with plastic toys are abnormal or unnatural, as such behaviours are not found amongst animals-in-the-wild. The kinds of 'abnormal/unnatural behaviour' which zoo experts are keen to eliminate are those which appear to indicate that the animals under their charge are unhappy, such as stereotypic behaviours. That is why one clear goal of environmental enrichment is to improve their psychological well-being as well as physical welfare.[3] Take the case of chimpanzees and the oft-quoted termite mound. In the wild, chimpanzees have been observed to fish termites out of a termite mound with a stick. So, zoos provide the chimpanzees with an artificial termite mound, not, however, containing termites which the chimpanzees can fish out with a stick, but pie fillings, mustard and other Macgoodies instead. This enrichment, undoubtedly, serves to amuse and distract the chimpanzees, thus preventing them from performing stereotypic behaviours out of boredom. It does not, however, instantiate the natural behaviour of chimpanzees in the wild; these actually use a stick to fish out live termites from the termite mound which they have found for themselves, not out of boredom, but as part of their search for food in order to survive. To say the least, pie fillings and mustard are not part of the diet of chimpanzees in the wild. The mound is artificial, without the smell, the texture of a real termite mound in the wild; the Macfoods are mere snacks for which the chimpanzees have, unfortunately, developed a taste. Macfoods are known not to do humans any good; it is also unlikely to do good to the captive chimpanzees, whereas the termites, which chimpanzees in the wild adore, are of significant nutritional value to them. The two contexts are so fundamentally and utterly different from each other that it would be grossly misleading and grotesque to say that fishing for Macfoods from an artificial termite mound in a naturalistic enclosure with human visitors gawping at them is an instance of 'natural behaviour' as performed by chimpanzees in the wild.

It would be better explicitly to admit the ontological gulf between the two contexts and simply adopt whatever techniques zoo managers and experts can devise in order to keep boredom at bay as far as their charges are concerned. So why not, then, opt not for 'natural behaviours' and naturalistic analogues but simply for the most effective way(s), which in certain contexts can be the most charged with human culture? Let's go back to the chimpanzees. Why not allow them to play with computers, even to play with specially devised computer games, if these can be taught to them? Immurated chimpanzees, after all, are part of human culture; it would be in keeping with that logic to let them be exposed to the latest developments in hi-tech should such exposure achieve at least two of the three desired goals of the enrichment programme as outlined by Shepherdson (2001), namely to improve the psychological (and physical) well-being of zoo animals as well as to render them more interesting to zoo visitors.[4] The same goes for elephants. Elephants in the wild do no painting. But elephants in zoos, who are bored out of their minds, should be given access to paint pots and a large canvass, so that they can have a whale of a time tossing the pots and the paint at the canvass with their trunks, thereby creating at the same time works of elephant 'art', which the zoos can auction to the highest bidder – surely, a win-win situation for all the parties concerned, the elephants, the visitors and the owners of the elephants. Of course, it is true that in tossing the pots the elephants are doing something 'natural', so it counts to that very limited extent as 'natural behaviour', since it is with the same trunk and the same action of tossing that the elephants use to toss the pots as their wild counterparts would to toss an offending human against a tree trunk in the wild. But again, the two contexts are so different in ontological character that it would be highly misleading simply to say that the one is as 'natural' a behaviour as the other.

The logic of the above perspective leads to a conclusion which would make zoos more like circuses. Of course, circuses are often condemned for the cruelty of the methods used in training animals for their acts; reputable zoos, undoubtedly, would be against cruelty in any of the methods and devices they use in their enrichment programmes. Furthermore, it offends our human sensibility to see animals in circus acts dressed up in human clothes; reputable zoos, today, do not impose human accoutrements on their animals. However, these important differences apart, the similarities between them remain striking. The performing whales, each indifferently called Shamu, are a case in point. If one must keep whales in aquaria, then one, at least, ought to make sure that they are not bored out of their minds; what better means of making their lives more meaningful than to train them to enter into a relationship with their keepers/trainers who, seemingly without cruelty, manage to get them to perform according to their orders, by making use of movements they would 'naturally' use if they were living wild lives in the wild? Looked at from such an angle, and bearing in mind that these animals are not 'wild animals in captivity' but immurated animals which are on the way to becoming domesticants, there is nothing odd or uncomfortable about their trained behaviour. Their trained behaviour is only odd and uncomfortable to behold if one mistakenly holds them to be wild animals. Similarly, the orang-utans who have been trained to slide down the vine to greet the human guests gathered in their enclosure for breakfast are not all that different from cats and dogs which go into the kitchen to greet their owners when they saunter down for their breakfast. The orang-utans do not appear to be unhappy doing their act; on the contrary, the act is part of the enrichment programme designed to make life less boring for them. Zoos may come to displace circuses as the more acceptable face of training animals to perform acts which appear both to amuse the animals as well as their human spectators. Now, of course, circus trainers have long maintained that their training also plays such dual roles; however, circuses carry a burdensome past which makes it difficult for them to convince modern sensibilities that their methods are those that the philosophy of animal welfare could endorse with a clear conscience, whereas zoos are in a better position towards to do this PR job.

The direction towards which the analysis points reflects the shift in terminology from 'environmental enrichment' to 'enrichment'. The former conjures up artificial trees and branches in enclosures from which monkeys could swing; it attempts to provide a naturalistic environment for the animals and thus to lessen their boredom by making it possible for them to swing from branch to branch. The latter conjures up the chimpanzee amusing itself with a computer, which focuses on the animal and attempts to lessen its boredom by providing whatever devices or situations that happen to work. The simpler term looks forward to active training – usually involving a close and intimate relationship between the human trainer and the animal – as a means to improve the psychological well-being of immurated animals, while the more restrictive term appears to focus more on improving the physical welfare as well as the psychological well-being of such animals by making their physical environments more naturalistic.

Conclusion

The concept of environmental enrichment or enrichment tacitly acknowledges that zoos are exotic environments for exotic animals. It is at the centre of those procedures and processes under immuration to transform animals to becoming new kinds of domesticants. The logic of environmental enrichment leads to the logic of enrichment. Under that latter logic, the distinction in theory and principle between circuses and zoos would be difficult to make, as both may claim that they are enriching the existence of the animals under their charge as well as enriching us human spectators who watch the animals performing their trained acts.

Notes

Introduction

1. Ontology is that part of philosophy which deals with being, with different kinds of being in the universe. For instance, God, the devil, angels and demons are supernatural, transcendent beings; Hamlet and Anna Karenina are fictional beings; Michael Jackson and Prince Charles, on the other hand, are flesh and blood individuals whom, in principle, one can meet face to face and whose hands one can shake. The latter are material beings, with space-time co-ordinates, even though they may not be wholly physical beings. In contrast, fictional and supernatural beings are, *ex hypothesi*, non-material, non-physical beings. If someone literally claims to have met Anna Karenina in Russia and kissed her hands, that person runs the risk of being incarcerated in what was once called a lunatic asylum. Of course, some people in history have also seriously claimed that they have seen God, that they have spoken to God, or that God has spoken to them – atheists regard such people to be equally suffering from delusions, but religious believers consider them to be very special individuals whom they call mystics.
2. There is more than one sense of tame. 'Immurated' is a term coined to refer to zoo animals, the detailed meaning of which will be explored, in due course, in the book.
3. Note that the operative phrase here is: 'the suspension of the mechanism of natural selection within the context of natural evolution'. We shall see in later chapters that natural selection does operate in contexts outside of natural evolution, such as in the context of zoo management and domestication. The crucial difference between them is that while the former leads to the emergence of naturally occurring living beings, that is, wild animals in the wild, the latter leads to the emergence of biotic artefacts, the ontological foil to wild animals in the wild.

1 What does the public find in Zoos?

1. Some of the issues raised in this chapter may be found in an earlier article – see Lee, 1997b.
2. The ontological significance of this will be explored in Chapter 4.
3. In contemporary literature in ethics, there are three main types of ethical approaches:

   (a) Consequentialist: The most often invoked of which is utilitarianism. Consequentialism considers an act to be right only if its good consequences turn out overall to outweigh its bad consequences. In other words, the notion of good logically precedes that of right.

(b) Deontological: The Kantian variety has been long dominant in modern Western moral philosophy. Deontology considers an act to be right irrespective of its consequences, emphasising the motive of the act. In other words, the notion of right is central to ethical discourse, and not that of good.

(c) Virtue ethics: The Aristotelian variety is most often invoked; it considers character of the agent to be at the heart of ethics, not so much the agent's acts.

For details, see Baron, Pettit and Slote (1997).

4. The term 'morally considerable' primarily refers to beings which we humans deem to have moral needs or to be the bearer of moral rights in virtue of the fact that they possess certain relevant empirical characteristics, such as sentience or mental life.

5. See also Rachels (1991).

6. The term 'human chauvinism' was coined by Richard Routley (Richard Sylvan) (1973) to draw attention to and critically question the anthropocentrism (human-centredness) deeply embedded in Western moral philosophy. Human chauvinism considers human beings alone to be morally considerable.

7. Of course, there are a few zoos which specialise in exhibiting domesticated animals.

8. According to *The World Zoo Conservation Strategy* (1993): 'It is difficult to estimate the total number of animals and species in zoos (5.1).' It only gives the world total of zoo vertebrates as 1 million animals.

9. I owe this point to Mary Midgley.

10. For an account of classical and molecular genetics, see Lee (2005a).

11. As we shall see in a minute, Darwinian evolution and natural selection must not be understood in terms of what Karl Popper has called 'passive Darwinism', that is to say, as if it implies that 'organisms are the mere playthings of fate, sandwiched as it were between their genetic endowment and an environment over which they have no control' (Rose, 1997, 140).

12. On the themes of time and space in biology, apart from Rose (1997), see also Mayr (1982, 71).

13. On reciprocal causation, see Dickens and Flynn (2001); on its equivalent, non-linear causality, see Lee (1989).

14. In Chapter 9, we critically look at genetic reductionism and show that it is methodologically wrong-headed in the context of *ex situ* conservation of endangered species, a flaw which undermines a key role, if not the key role, assigned to zoos by bodies like the World Zoos Conservtion Strategy (1993) and the European Union Zoos Directive (1999).

15. On ecocentrism, see Rolston (1988). In contrast, the philosophy of animal welfare as well as of animal rights are explicitly biocentric in outlook, that is to say that the focus is on the individual animal which is capable of suffering or which actually suffers pain on the one hand, or which is the subject of a mental life on the other. Such a fundamental difference in theory/philosophy between ecocentrism and biocentrism is bound to have implications for policy-making. For instance, the former may not object to letting animals die of hunger under unusually harsh weather conditions, provided they are not

anthropogenic in origin (not caused by humans), whereas the latter would be in favour of saving those individual suffering animals.

Natural biodiversity should be distinguished from another kind which is artefactual, not natural, in character. (See Lee, 2005a and 2004.) One of the main burdens of this book is to argue that zoos are well placed to create arte-factual biodiversity and that they are ontologically misguided in claiming that one of their important aims, if not the most important, is to save extant threatened natural biodiversity through their programme of captive-breeding and *ex situ* conservation – see Chapters 9, 11 and Conclusion.

2   Animals in the wild

1. Note, however, that this historical fact of evolution hides two very different types of phenomena which ought to be distinguished – vertical evolution where there is change but without speciation and evolution which involves speciation. According to E. O. Wilson (1994), Darwin was primarily concerned with the for-mer, not the latter – for instance, a genetic mutation in a population of white moths which happens to bestow survival advantage could end up by being one with predominantly black moths. There has been change but no speciation; you start and end with one species. However, Darwin's account of finches in the Galapagos is an instance of vertical evolution with speciation.
2. This is in contrast to the theory of creationism (popular amongst certain fun-damentalist Christians) which holds that there is intelligent design in life forms, present and historic, and that God is that intelligent designer and cre-ator.
3. According to Mayr (1982), this is the kernel of truth behind population think-ing in biology – that individual (sexually reproducing) organisms are unique, that there is no 'typical' individual, and that mean values are abstractions. Natural selection works on such unique biological individuals.

   The notion of ecosystems is central to ecological thinking. Very briefly, an ecosystem is that ensemble of biotic and abiotic components which interact in a causally reciprocal manner, of which, of course, the animals form an inte-gral part. The boundaries of ecosystems may not be easy to delineate in all cases, but by and large, there is consensus where the limits may be usefully drawn. One should also bear in mind that ecosystems are dynamic, not static in nature. See, for example, Botkin (1990); Botkin and Keller (1995).
4. Apparently, the age of *Homo sapiens* could then be followed by the age of rodents, as these animals can take refuge underground during a nuclear holo-caust; moreover, they are better able to withstand radiation than humans and other mammals, should they be exposed to it.
5. For a full exposition, see Lee (1999).
6. See Maturana and Varela (1980) for the introduction of the term 'autopoeisis'.
7. Note that this sense of autonomy has nothing to do with the Kantian sense of autonomy of the (human) will.
8. However, in Chapter 8, we will argue that the intimate ontological link in an organism between existing 'by itself' and 'for itself' has finally been dramati-cally ruptured by biotechnology – transgenic organisms exist neither 'for them-selves' nor 'by themselves'.

9. For example, a mammalian species on average lasts a million years; this kind of extinction is entirely natural, non-anthropogenic, that is to say it is not caused by humans and their activities.

10. There are obviously other senses of the term 'nature' which will not be considered here. For a thorough and detailed clarification of the different senses, see Lee (1999) or Lee (2005b) for a briefer account. Lee (1999) also argues that the distinction between the natural on the one hand, and the artefactual on the other is fundamental given that there are entities in the universe (or more narrowly construed in our solar system) which are totally independent of humans and others, which are the direct products of human intention, ingenuity and manipulation. However, the distinction is meant, in philosophical terms, not as a dualism (in the Cartesian sense) but as a dyadism. Examples of dualisms are: mind/body, male/female, human/ non-human, where the first term mentioned in each set is considered to be superior or to belong to the master class, while the second term refers to an inferior or slave class. The dyadism – naturally occurring/(human) artefactual – which this book is basically concerned with, has no such hierarchical connotations; the dyadic distinction is simply necessary to an understanding of the history of Earth and of life in general on it on the one hand, and of the role played by *Homo sapiens* on the other. Humans belong to a species which happens to possess a unique kind of consciousness, enabling it to develop not only language but also very powerful technologies for transforming the natural to become the artefactual. In other words, human technology makes it possible for humankind to manipulate nature in order to make it embody human intentions and ends. The dyadism in question therefore has ontological import (but has no hierarchical import in terms of superiority or inferiority). The categories of the naturally occurring on the one hand, and the artefactual on the other are distinctly different, ontologically speaking – the former has existed and (in principle) continues to exist and will eventually go out of existence in the absence of humankind, while the latter exists, continues to exist and will exist only as long as humankind itself exists.

11. Chapter 7 will look at the issue whether animals in zoos could be said to be domesticated animals and, if so, in what sense of that term; Chapter 8 will explore the notion of zoo animals as biotic artefacts.

12. For details on the science, philosophy and technology of genetics, see Lee (2005a).

13. Chapter 8 will look more closely at the notions raised here.

14. See Irwin (2001).

15. Three theses of teleology (of which intrinsic/immanent teleology is one) will be distinguished and characterised in greater detail in Chapter 8.

16. The emperor penguins (*Aptenodytes fosteri*) in Antarctica breed during the Southern winter. After a few weeks of courtship, the female lays an egg and then sets off to the sea to feed herself (travelling up to 50 miles or 80 kms), leaving the male, famished for as long as 65 days in the hostile Antarctic environment, to guard and keep the egg warm, and eventually to hatch it. At the end of that long period, the female returns, 'miraculously' recognises her family and immediately starts to feed the recently hatched chick by regurgitating food from her stomach; whereupon her mate leaves straightaway, on his equally long journey, to replenish and refuel himself at sea. See Rockliffe and Robertson (2004).

17. To make empirical and conceptual sense of this kind of phenomena requires the so-called biological-species concept which may briefly be defined as follows: 'a species is a population whose members are able to interbreed freely under natural conditions' (Wilson, 1994, 36). However, this is not to say that the concept is without difficulties. For instance, it is not applicable to organisms (mainly plants) which reproduce asexually.

18. This refers to the deep themes set out in Chapter 1 of the biology of time and history as well as of the biology and philosophy of reciprocal causation of organisms-in-the-environment.

19. See http://www.polarbearsinternational.org [01/12/04]

20. The evolutionary-species concept is different from the biological-species concept which will be raised again in Chapter 8. In other words, there is no one single meaning/definition of species which can do justice to the notion in all the contexts in which it is invoked. One needs to distinguish, at least, between these two different, though related, understandings of the notion. For a fuller discussion of this and related matters, see Mayr (1982, pp. 256–75, 286–7, 295–6). See also Ereshefsky (1998).

21. The token/type distinction will be raised again in Chapter 8, but in the context also of biotic artefacts, as well as assessing whether a biotic artefact such as a captive animal in a zoo could be a said to be a token of a naturally occurring species in the same way a wild animal in the wild may be said to be a token.

22. The variations are both phenotypic and genotypic. Today, with regard to the latter, scientists, in studying the genomes of various species, can ascertain with precision what constitutes the genetic variations between individuals of the species – in the case of the human genome, they have identified what are called SNPs/snips, that is, 'single nucleotide polymorphism'. A SNP represents a DNA sequence variation amongst individuals of a population. SNPs promise to be a money-spinner as they can be used to identify individuals who could be vulnerable to diseases like cancer.

23. After all, as we have earlier remarked, a peacock and a peahen hardly look alike; nor does a caterpillar look like the adult butterfly.

24. There are in fact three species in the wild today: one Asian (*Elephas meximus*), two African – African savannah (*Loxodonta Africana*) and African forest (*Loxodonta cyclotis*).

3 'Wild animals in captivity': is this an oxymoron?

1. A more recent volume is Kleiman, Allen, Thompson and Lumpkin's (1996) *Wild Animals in Captivity*.

2. Increasingly, professionally accredited zoos make it their official policy to use as exhibits zoo-bred animals and would only sanction in exceptional cases the import of animals freshly caught from the wild. In such reputable zoos, only the last two categories of animals would, presumably, form part of their respective collections of exhibits.

3. Of course, 40 years after the event, should the man still be incarcerated, most people in society would say that he should be released especially when his character might even have changed during those long years of imprisonment and he is no longer that dangerously violent man in prey of women. However, the matter is then one of justice and morals, not of conceptual

sense, provided one changes the tense by referring to him as that man who was once dangerously violent denied his freedom and under captivity. Analogously, one could say intelligibly, using the past tense, that this animal under captivity in a zoo was once a wild animal; however, what one cannot intelligibly say, as the arguments in this chapter show, is that such an animal is a wild animal in captivity. The latter is precisely what zoos want to say.

4. As we shall see, animals under long-term captivity could weigh much more than their counterparts in the wild.

5. From now on, whenever appropriate, the term 'animal-in-the-wild' rather than 'animal in the wild' will be used in order to emphasise that the property of wildness in animals can neither be understood apart nor is it detachable from the animal's existence in the wild, as well as to highlight the conceptual incoherence of the term 'wild animals in captivity'.

6. Here is a brief biological summary of tameness and taming, according to one leading zoologist on the subject of domestication:

> Reduced flight distance in the presence of people is one of the most obvious behavioral changes accompanying the domestication process. ... The degree of tameness attained is heavily influenced by the animal's experience with people. Tameness is facilitated when people become associated with positive reinforcers such as food or pleasurable tactile contact. ... they are stressed less by interactions with people and may experience greater reproductive success and productivity. Researchers selecting for tameness in wild silver foxes ... have found predictable changes in the activity of the serotonergic and catecholamine systems of the brain. ... Overall, tameness is becoming one of the better understood behaviours associated with the domestic phenotype.
>
> (Price, 2002, p. 129)

Price also points out the tameness is the single most important effect of domestication of behaviour:

> the single most important effect of domestication on behaviour is reduced emotional reactivity or responsiveness to fear-evoking stimuli (i.e. environmental change). This characteristic is observed in virtually all populations of domestic animals and pervades a wide variety of behavioral responses to both the social and physical environments (e.g. intraspecific social interactions, reactions to the presence of humans, responses to novel objects and places). Reduced responsiveness to fear-evoking stimuli is seen as an adaptation to living in a biologically 'safe' predator-free environment with: (i) limited opportunities for perceptual and locomotor stimulation; (ii) frequent invasions of personal space, with little opportunity to escape from dominant conspecifics; and (iii) frequent association with humans, who are prone to cull untamed and intractable individuals. Available information supports the hypothesis that individuals less reactive to fear-evoking stimuli experience reduced levels of stress in captivity, greater reproductive success, greater productivity (e.g. growth rate, animal products) and are handled by humans with greater ease. ... It is not surprising that one biological trait can be so important to the domestication process. Consider the importance of emotional reactivity to the fitness of animals living in nature.
>
> (Ibid., p. 180)

7. The dodo, because of its evolutionary history, had not encountered enemies which it had to fear, least of all humans; it did not develop flight reaction, nor escape distance. As a result, it became extinct when humans arrived in their habitat, killing them with ease.

8. Hediger (1968, p. 43) also points out that the animal-in-the-wild is capable of assessing very finely and precisely the situation it finds itself in the context of exercising its flight reaction. For instance, an antelope would not necessarily display such a reaction every time it meets a lion; if it judges that its normal predator has already just dined handsomely, it would nonchalantly ignore its presence.

9. Furthermore, Hediger goes on to note perceptively that '(m)an is moreover the only creature able to free himself from the elementary function of escape. By this self-release, man clearly stands apart from the rest of creation, and, as the arch-enemy, is the focus of all animal escape reactions' (1968, p. 49).

10. In recent zoo literature, the term 'tame'/'taming' seems to have dropped out of usage. Instead, it talks of 'desensitization', which in part, if not completely, refers to the same thing. See the following example:

> The first step in teaching an animal to allow husbandry procedures to be performed consists of desensitizing the animal to human touch . . . Another important aspect . . . involves using desensitization techniques to reduce an animal's fear responses to unfamiliar objects and uncomfortable procedures.
> (Kuczai II et al., 1998, p. 319)

11. Obviously, domestication which leads to domesticants such as cats and cows involves more elements than taming. The point made here is simply that taming is an essential first step in ultimately producing domesticants or domesticated animals.

It may be worth drawing the reader's attention to another related matter. Reindeer and yaks are tame, but they are not domesticated animals or domesticants. Some writers have called such animals 'domesticated animals' but not 'domesticants'. This author chooses to use the terms 'domesticated animals' and 'domesticants' interchangeably and would just simply refer to reindeer and yaks as 'tame' which is the antonym of 'wild' in one sense of 'wild'; it follows that reindeer roaming in the Artic north are wild in other senses of 'wild', which is made clear in this book. In other words, 'domesticated animals/domesticants' mean more than just 'tame'; 'wild' means more than just 'untamed', though for an animal to become tame is an essential first stage – a necessary though not sufficient condition – in its transformation to a domesticated animal; for an animal to become tame is to lose one aspect of being wild, though not all.

12. Philosophers talk about the distinction between surface and depth grammar in the following way:

A1. Sunday of by the way great.
A2. Sunday is the Muslim holy day.
A3. Sunday is the Christian holy day.
A4. Sunday is large in girth.

A1 is unintelligible at the level of surface grammar as it is not a properly constructed sentence in English. In contrast, A2, A3 and A4 are intelligible at the level of surface grammar because each is a properly constructed sentence in English. A2, though intelligible, happens to be false (if the sentence were to be uttered) while A3 happens to be true. However, it would be inappropriate to say of A4 that it is either true or false, as it is unintelligible at the level of depth grammar – it just makes no sense whatsoever to apply the attribute 'large in girth' literally to Sunday, when Sunday is the name of a day in the week. (Of course it would make sense if 'Sunday' refers to a child, the son of Joe Blogg.)

4 Decontextualised and recontextualised

1. However, it remains true that the public, by and large, are mainly interested in charismatic animals like the lion, the tiger, or cuddly ones like the panda.
2. At least, that is, one of the three species of elephants found in zoos comes from India, the other two from Africa.

    It is true that, historically, the jaguar's home range extended to some of what we today call the southern states of the USA, such as Arizona.
3. Historically, capturing wild animals from the wild to make them residents of zoos also went hand in hand with breeding such animals in captivity; of course, the earlier historical motive for doing the latter is different from the contemporary one which, as Chapter 9 will critically examine, heroically focuses on the goal of saving endangered animals-in-the-wild from extinction. The zoo venture in both aspects began in the nineteenth century in several European zoos – London, Antwerp, Marseilles, Turin – during a period when European imperial power was at its height, making it possible in the first instance for agents from such countries to capture animals-in-the-wild from various parts of their respective colonies, thereby rendering the animals exotic. This set of dislocations involves two things: acclimatisation or naturalisation as well as domestication, which would enable zoos to produce beasts that could be put to work, or crossed with indigenous ones to produce larger and more vigorous versions. (See Baratay and Hardouin-Fugier, 2002.)

    For a discussion of climate on captive animals, see Price (2002, ch. 16).
4. See Appendix for a brief critical exploration of the notion of environmental enrichment.
5. It would be boring and tiresome here to point out once again the unintelligibility of the phrase 'all zoos exhibit living specimens of wild animal species'. The reader should take it as read that, as far as this book is concerned, phrases such as that or similar about the so-called 'natural behaviour' of zoo animals in zoo environments which may occur as part of quotations from zoo literature, are all objectionable.
6. See Baratay and Hardouin-Fugier (2002).
7. This dramatic, theatrical style of exhibiting animals was pioneered by Hagenbeck:

    The panoramas, for which Hagenbeck had received a patent in 1896, were made up of a series of enclosures, laid out like theatre stages, each one behind and slightly higher than the other and separated by hidden moats. Artificial rockwork and plantings concealed the holding quarters and service ways. . . . The obscured moats, dramatic rockscapes, and numerous ponds and lakes created scenes of expanding vistas in the most audacious zoo development to that time. The African panorama was the first to generate the illusion of an open savanna, populated with gazelles, flamingos, storks, cranes, antelopes, zebras, lions, and in the distance, ibexes and wild sheep on rocky outcrops.

(Hancocks, 2001, pp. 66–7)

An earlier but no less colourful way of presenting an animal as an exhibit may be found:

[w]hen (Charles X of France) received a giraffe as a gift from . . . the Ottoman viceroy of Egypt, in the summer of 1827, he arranged for her to wear a cape embroidered with the French fleur-de-lis and the Egyptian crescent on her walk from the docks in Marseilles to Paris. . . . the giraffe's winter quarters were quite elegant, with parquet flooring and the walls insulated with an 'elegant mosaic' of straw matting: 'truly the boudoir of a little lady,' wrote Geoffroy Saint-Hilaire.

(Ibid., pp. 33–4)

8. At first sight, this charge appears unwarranted. For instance, he has a point when he says:

Observing animals in a zoo is often closer to reality when compared with other media, for example film. In a zoo, a pride of Lions . . . rests and sleeps for most of the day, just as it would on the African savannah, whereas in a 50-minute-long television programme the Lions are often shown being active for the majority of the time. Therefore, a visit to the zoo provides a more realistic representation of the daily life of a Lion than an edited film programme.

(Andersen, 2003, p. 77)

However, in a later chapter, it will be argued that such a view is at best super-ficially correct.

9. Chapter 9 will examine these claims from the viewpoint of their mutual compatibility in the light of the ontological exploration pursued by this and following chapters.

10. This argument will be examined in detail in a later chapter.

11. As a result, most animals spend most of the time, throughout the year and especially all nights in barren cages, where they have to put up with a great deal of noise produced by the clanging of steel panels reverberating through the walls – see Hancocks (2001, p. 141).

The cost of creating Jungle World at Bronx Zoo in 1985 was $9.5 million while that of creating Penguin Encounter at Sea World in San Diego in 1983 was $7 million.

12. One such naturalistic exhibit, commonly acclaimed to be the best of its kind, is Jungle World at Bronx Zoo, New York, which opened in 1985. Below is an account of what it really is:

Its real architecture is not the building but 'the design and construction of space . . . found inside, and what is inside is a representation of the rainforest, the mangrove swamp and the scrub forest of Asia. . . .' Obviously, this is not an Asian forest, as Asian forests do not grow in New York. Yet in this world full of trees and dense foliage punctuated by the colour of bright tropical flowers through which birds fly and other animals move, and with the rich smell of vegetation and the sounds of a busy jungle, it is almost impossible to remember that one is in fact within a building. Yet it is a man-made forest not just in the sense that trees and plant have been put in a particular location by man, but in the more profound sense that many of the trees are actually manufactured by man. The huge tree which dominates one of the areas is actually made out of steel tubing over which there is metal cloth which is itself covered by an epoxy resin textured and painted so carefully that most people would never guess that it is fake. But the vines which climb around it are real vines. Some vines however are *not*, and those which are provided for the gibbons to swing on are fibre-glass. The mist which envelops the tree tops is real mist but it is produced not by natural conditions, but by the sort of machines used in commercial citrus groves. The rockwork (except for the small pebbles) is artificial but it is a base on which real peat moss and algae grow. Although it might seem that one is in the midst of an undivided tract of forest this is not so; the rocks help form barriers to separate species which may not mix in these conditions. Here one can see animals which actually do live in Asian forests, but what one does not see is the animals living as they would do in that forest. The sound of the cooing of the forest dove is real but it was recorded in Thailand.

(Mullan and Marvin, 1999, pp. 53–4)

13. Apparently, so cleverly done are some of these simulated/naturalistic habitats that not simply are lay visitors taken in by them, but also field biology students. One particular firm of zoo habitat designers, Jones and Jones, has created so realistically the gorilla exhibit in Woodland Park (Seattle) that photographs of it sent to Dian Fossey have fooled her field biology students into believing that the gorillas are wild and in their natural habitat. The National Geographic has also published, in one of its publications on Africa, a photograph of the patas monkey exhibit in that same zoo, in the mistaken belief that it is a piece of photography of wild things in the wild. (See Hancocks, 2001, p. 139.) Such incidents only show how easily humans can be visually misled about what reality is.

14. This reasoning is based, however, not on direct empirical studies of the matter as the author is not aware that any such work has been done. However, it is not implausible to assume that the animal is intelligent enough, given all its faculties and senses with which evolution has endowed it for survival, to work out that there is a discrepancy between immediate perception and reality.

15. See Worstell (2003, ch. 6).

It is not obvious how habitat simulation can re-create the 'essence of a natural habitat'. But this point will be looked at in what follows in the remainder of this section and also in Chapter 5.

16. For some brief details about the technology behind constructing such enclosures, including biodomes, see, for instance, http://www2.ville.montreal.qc.ca/biodome/e1-intro/ef1_rens.htm [15/02/05].

17. That is, until of late, as the food industry becomes interested in the project of manufacturing tastes and flavours.

18. As Mullan and Marvin point out, it is a pity that zoo professionals, on the whole, have not taken to heart what Hediger has said about the limitations of creating naturalistic exhibits:

> The best guarantee of complete naturalness is assumed to be a faithful copy of a piece of natural scenery. This apparently logical conclusion is based on a false ecological estimate that may have serious results. Even an untouched section of the natural ground, enclosed within six sides (i.e. the closest possible imitation of a section of a biotype) is likely to be unnatural ... Mistakes of this kind, resulting in a pseudo-natural arrangement of space, are due to ignorance of the following elementary fact: a cross-section of nature is not an equivalent part of the whole, but merely a piece which, on being completely isolated, alters its quality. In other words: nature means more than the sum of an infinite number of containers of space (cages) however natural.
>
> (Hediger in *Wild Animals in Captivity* as cited by Mullan and Marvin, 1999, p. 77)

It appears that David Hancocks, too, has missed Hediger's point: 'in the animal exhibit areas there must be one constant and inherent design philosophy: Nature is the norm' (Hancocks, 2001, 145). One should not confuse nature with a simulation of nature. If nature were truly the norm, there would be no simulated habitats; indeed, quite simply, there would be no zoos.

5   Lifestyle dislocation and relocation

1. This term does not sound elegant but has been coined by the author; its sense will soon be made clear in the chapter.

2. Of 24 exhibits of mice and rats in the UK, 17 are at London Zoo. Zoo collections, as we have already seen, are not representative of the animal kingdom; if they were, a quarter of their mammals would be bats and a third would be rodents. See http://www.goodzoos.com/Animals/small.htm [12/11/04].

3. See http://members.aol.com/cattrust/cheetah.htm [12/11/04].

4. See Worstell (2003, ch 1).

5. Although the average male qualifies to be megafauna, the female does not. (Any animal weighing more than 100 kg counts as megafauna.)

6. See Clubb and Mason (2002).

7. For further details, see http://www.nwf.org/wildlife/polarbear/;http://www.seaworld.org/infobooks/PolarBears/home.html [13/11/04].

8. For a more detailed discussion of the biological effects of living in miniaturised space under captivity, see Price (2002, ch 17).

9. See Clubb and Mason (2002, 53).

10. See Clubb and Mason (2003). The study makes similar findings with regard to lions, tigers, cheetahs – animals which roam over large areas in the wild – namely, that these are kept in analogously reduced spaces in zoo enclosures, conditions which impinge on the welfare of such zoo animals.

11. A few very rich eccentrics may choose to have no home of their own except for their permanently retained suites at the Ritz, the George V or the Hilton.

12. From this, one is not entitled to draw the conclusion that the need to roam is entirely a parasitic one, a mere side-effect of the need to look for food. In zoos, where the latter need is superseded by hotelification, it would then follow that the need to roam also becomes redundant. However, a recent study shows, for instance, that polar bears in zoos are distressed not simply because they are not allowed to hunt, but also to roam – see Clubb and Mason (2003); see also http://www.admin.ox.ac.uk/po/031001.shtml [01/10/03].

   The idea that the need to roam in the case of mammals or to fly in the case of birds is at best a parasitic one has been very influential and was first made clear by Hediger, who seemed to think that to argue otherwise is merely to be anthropomorphic. He held that it is not a physiologically necessary activity for birds of prey to fly, ignoring the basic understanding that the origin of such a bird's anatomy and physiology stems from its very ability to fly, and that the very organism has been shaped and has evolved within such a context. Hediger was in turn influenced by German philosophy in the 1930s, especially that of Martin Heidegger who gave courses on animals and animality at the University of Freiburg im Bresgau in 1929–30. Heidegger argued that freedom is impossible for animals roaming in the wild as they are consumed with trying to satisfy the essential biological needs of finding food, water, shelter, etc. On the contrary, they are only truly free in zoos as zoos by laying on 'hotel services' relieve them of the need to satisfy such basic functions; consequently, the zoo as such satisfactorily replaces the animal's home range and territory. Such an influential view did not get challenged until the later appearance of animal ethologists such as Lorenz, Tinbergen and others. (On these points above, see Baratay and Hardouin-Fugier (2002, pp. 262–3).

13. The young does so occasionally, but the adult hardly ever.
14. See http://www.worldwildlife.org/gorillas/ecology.cfm [06/01/05].
15. See Clubb and Mason (2002, ch. 4).
16. See http://www.hlla.com/reference/anafr-cheetahs.html [17/12/04].
17. See http://www.nwf.org/wildlife/polarbear/behavior.cfm [18/12/04]. http://www.seaworld.org/infobooks/PolarBears/pbdiet.html [18/12/04].
18. *See* http://www.denverzoo.org/animalsplants/mammal01.htm [10/01/05].
19. See Clubb and Mason (2002, ch. 4).
20. http://www.denverzoo.org/animalsplants/mammal01.htm [10/01/05].
21. The nutritional equivalence is only approximate – according to Clubb and Mason (2002, ch. 4), the zoo diet for Asian elephants contains more fat than the wild diet.
22. It would be too philosophically exhausting as well as unnecessary in this context to defend this thesis in great detail here and now. Suffice it to say that such an assumption is implied by the contemporary programme of (environment) enrichment which enlightened zoo management goes out of its way to emphasise as part of its philosophy. However, the concept of enrichment will be examined in the Appendix.
23. Note that independent value and instrumental value (for humans) are mutually exclusive although the exclusion could be on a continuum – less of one and more of the other; however, this does not mean that a being which has lost its independent value and acquired instrumental value is not a morally considerable being; for example, from the point of view of its ability to suffer pain.

6   Suspension of natural evolution

1. So impressed have the pharmaceutical sciences and industry been by this observation that even a new branch of pharmacology has been established.
2. Carrico (2001); see Plotkin (2000); see also Kuroda (1997) for more examples at http://www.shc.usp.ac.jp/kuroda/medicinalplants.html [11/01/05].
3. This is utilitarian ethics to which reference has already been made in Chapter 1. Although it is not the only normative system, it is, nevertheless, an exceptionally powerful one since the nineteenth century. As we shall see again in a later chapter, the philosophy of animal welfare rests on this axiom: if it is morally good, and therefore, morally obligatory to ameliorate pain in humans who are sentient, it is equally morally good and obligatory to ameliorate pain in all other sentient beings. Some exponents are even keen to extend the reach beyond domesticated and zoo animals to animals-in-the-wild – the logical conclusion to which this perspective can be pushed is a pain-free world where carnivores have been genetically modified to become herbivores, where the lion would literally lie down beside the lamb. For exposition, see Easterbrook (1996); for a critique, see Lee (1999).

   The imperative to save a life (amongst animals) with the same devotion and resources as any human life in peril may be seen in the following account: In 2001, a keeper at Bristol Zoo hand-reared a baby gorilla (Djengi). His mother died soon after his birth. (It is unclear from the account given whether the baby gorilla was found in the wild or that it was captive-born. The title of the article refers to it as 'a wild animal'; but as the word 'wild' is also used when speaking of zoo animals, its reference in this context is none too clear.) The keeper stayed with the infant in the spare bedroom at the house of the keeper of the primate section. The keeper had to feed Djengi every two to three hours in the night; he slept in the keeper's bed for the first few weeks. At about four months, the animal was transferred to a cage put in the living room. His bottles had to be sterilised till he was about 7 months old. He had to be winded until he got big enough to bring up his own wind. The gorilla, like human infants, wore nappies (as it might otherwise foul up the living room). He was bottle-fed Baby formula till 7 months old; then he was fed Complan at bedtime as well. He also had puréed fruit. The keeper sat and watched TV with him; brightly coloured things apparently caught his attention and he liked to play with the remote control. He was also clothed in jumpers. At 9 months, he left Bristol for Stuttgart Zoo, which has been running an orphanage for gorillas for 20 years where the keepers there would continue to look after him until 4 years old, an age when young gorillas in the wild would have become independent of their mothers. Djengi got the same care and attention as a human infant and lived the life of a human infant while at Bristol Zoo. See Wright (2001).

   Caesarean operations are sometimes performed on zoo animals. Jones (2000) reports that The Jersey Zoological Society and Trust (set up by Gerald Durrell) delivered through a caesarean operation of a lioness some lion cubs; the cubs were then bottle-fed.
4. At Emmen Zoo in northern Netherlands, it has recently been reported that two rhinos are being given sun-bed treatment during the winter months in custom-built 4-metre-long sun-beds. As rhinos get older, their skin gets flakier.

Exposing them, especially in the winter, to infra-red sessions of up to 20 minutes and to shorter bouts of ultraviolet rays would improve their skin–blood circulation, as well as give them vitamins. See *The Guardian* (12/02/05, p. 18).

5. In the context of natural evolution (in the wild), natural selection and natural evolution inextricably go hand in hand. Without the former, the latter would not have come about. However, the mechanism of natural selection, nevertheless, does operate in a context other than natural evolution (in the wild) – it may and does operate even in the context of artificial selection, under human-administered selection, for certain characteristics such as colour:

> man does not necessarily select those individuals with the greatest fitness (for captivity) as breeding stock for his selection programs. That is the role of natural selection. Differential mortality and reproduction, including reproductive failure, among artificially selected populations is one way that natural selection in captivity is manifested. . . . [One] focuses on the various ways that natural selection is expressed, namely, mortality and reproductive failure, and changes in these parameters over generations in captivity. . . . [N]atural selection does not cease once a population of animals is brought into captivity, but rather continues to operate regardless of whether or not artificial selection is applied.
>
> (Price, 2002, p. 51)

We should bear the above in mind in Chapter 7 which looks at the notion of artificial selection and domestication.

6. Non-anthropogenic, for the simple reason that they occurred before the evolution of *Homo sapiens sapiens*.
7. Not all animals would be subject to thirst deprivation; cheetahs, for example, do not need to drink water as they get their liquid from ingesting their prey.
8. For a more detailed exploration of synergistic causation, see Lee (1989).
9. The operative phrase here is: 'so-called counterparts in the wild'. This wording leaves it open at this stage of the discussion whether animals-in-the-wild are the true counterparts of zoo animals, a thesis which will be looked at critically in later chapters.
10. See http://members.aol.com/cattrust/cheetah.htm [11/01/05].
11. See http://www.cotf.edu/ete/modules/mgorilla/mgbiology.html [11/01/05].
12. See http://www.seaworld.org/infobooks/PolarBears/pblongevity.html [05/01/05].
13. It is obvious that in such a culture, euthanasia, even with all the practical safeguards against abuse in place, is morally problematic.
14. One such is Hancocks. He talks of a good zoo as one which gives the visitors 'lessons about life' and at the same time 'provide wild animals with safe and contented lives' (Hancocks, 2001, p. 206). *Ex hypothesi*, animals-in-the-wild can never lead safe lives; only animals in zoos can lead lives free from predation, starvation, disease, etc. It makes no sense to say that animals-in-the-wild in their natural habitats can lead contented lives as the lives they lead, with all their hazards and privations, are the only lives they can ever know and can lead. It is possible to say that some zoo animals lead more contented lives than others; those which do not mutilate themselves, do not eat

their own faeces are clearly happier than those which do. However, they engage in such activities only because they have no choice but to live in zoos.

7   Domestication and immuration

1. For the most comprehensive recent account of domestication in all its biological aspects, see Price (2002).
2. As in this context divine design is not pertinent, the discussion will confine itself in this chapter and in the rest of the book to human design only.
3. For details of these historical developments as well as the philosophy of science and technology that they embody, see Lee (2005a).
4. In the case of biotechnology, it may not be so appropriate to talk of breeding new breeds with or minus certain traits, as its techniques permit scientists to insert directly into the genome of another organism or excise from the genome of an organism a DNA sequence that is said to account for the desirable or undesirable trait in the phenotype, especially in the case of single-gene characteristics. While traditional craft-based technology and even Mendelian technology in the case of animals depends on mating, biotechnology by-passes mating altogether; furthermore, while the first two agricultural revolutions lead to the creation and generation of new breeds by mating individuals belonging to different varieties of the same species, biotechnology transcends species and, indeed, even kingdom barriers. (See Lee, 2005a, for details.)
5. See Lee (1969).
6. Note that this author deliberately uses two different words in the two different contexts – 'processes' in the case of naturally occurring events and 'procedures' in the case of technological interventions. Nature involves processes, but uses no procedures. In this usage, 'procedures' imply design and deliberate structuring which have specific outcomes in the mind of the designer.
7. Clutton-Brock (1999, pp. 144, 148) says that the use of elephants in warfare, circuses and zoos and as beasts of burden has a history of at least 3000 if not more than 4000 years.
8. According to one authority (Bökönyi, 1989), one of the constituents of domestication – morphological changes – would have taken up to 30 generations to manifest themselves, at least during the early periods of domestication. Other experts contend that evidence in modern times shows that such changes can take place within a much shorter generational span. See, for example, Bottema and his observations of greylag geese:

> It is well known that after a few generations of domestication greylag geese (*Anser anser*) become fatter and heavier, losing the power of flight . . . Besides, after some time, early maturing and loss of the permanent and monogamous pair-bond occurs. Colour variations such as white, piebald, and buff appear, and feet turn orange whereas they were originally pink. The fact that greylag geese become heavier after a few generations is not a genetic change, but a result of feeding. Next to this process, a selection on weight took place, resulting in various extraordinarily heavy breeds.
>
> (1989, p. 32)

9. The adaptation on the part of an animal to its simulated habitat/environment in part involves processes which are natural, and in part procedures which zoo management deliberately imposes on the animals as part of its new existence and lifestyle. An example of the former would be the use of a stick on the part of the chimpanzee to fish for pie fillings, in this case apple sauce, from an artificial termite mound instead of using it to hook up termites from a real termite mound as its relative in the wild would do. This would be an instance of a natural adaptation on the part of the chimpanzee to its (cultural) zoo environment. An example of natural biological processes at work would be the animal becoming heavier than its wild counterpart as a result of the zoo diet, which itself is an instance of zoo policy and zoo procedures. Another instance of the latter would be the obvious fact that freedom to roam is no longer permitted or that a zoo diet is what the animals would get instead of foraging/hunting for their own food.

   Clutton-Brock has made the point more generally as follows:

   > I believe that domestication is both a cultural and a biological process and that it can only take place when tamed animals are incorporated into the social structure of the human group and become objects of ownership. The morphological changes that are produced in the animal follow after this initial integration. The biological process of domestication may be seen as a form of evolution in which a breeding group of animals has been separated from its wild conspecifics by taming. These animals constitute a founder population that is changed over successive generations by both natural and artificial selection, and is in reproductive isolation.
   >
   > (1989, pp. 7–8)

10. There are numerous competing definitions of the term 'domestication'; for a quick discussion see Bökönyi (1989) and Ducos (1989).

11. Note that the passage cited uses the term 'process'; the preferred term, for this author, would be 'procedure'.

12. The pigeon, for instance, is a 'classic case of the exploitation of a symbiotic tendency, for it is essentially a self-domesticating bird which seeks out human fields and settlements' (Issac, 1970, p. 112).

   On the subject of domestication as a symbiotic relationship, see also Budiansky (1994).

13. One way of combating this criticism is to say that the domesticated animal enjoys benefits which its wild ancestor/counterpart would not enjoy, namely that it would have a more secure food supply, be relatively better protected from certain predators and dangers. This, though, might not have been true in all instances especially during the early history of domestication. However, the important point is surely that humans would not have invested consistently over time so much effort, energy and resources to the enterprise unless they believed that domestication would benefit them greatly, irrespective of whether it would also benefit the animals in any way at all. The aim of the exercise is simply to make the animals serve their purposes and goals.

14. One example is the pigeon; see Isaac (1970, p. 112). Another motive which may not necessarily also be economic is ornamentation – some breeds of

dogs, birds and fish fall into this category. There is today another category, animals specially bred or genetically engineered as laboratory animals destined for scientific experiments.

15. Morphological changes are one kind of phenotypical change underpinned in many instances by changes in genetic patterns.

16. He writes:

> Such rapid appearance of deviating colours in many species (of ducks) cannot be explained by mutation during domestication, but it may be due to recessive factors in the wild population. The colour of wild duck species is generally dominant over other colours. The wild-colour pattern is caused by many genes responsible for the various components or for the distribution of the colours. If a mutant factor is present in a duck in heterozygous form, it will not show up in the appearance of the bird, because of the dominance of the wild-colour factors. In practice, the chances of a duck meeting a partner with the same recessive factor are limited: offspring in which combinations of the factor have occurred, e.g. white in homozygous form, will therefore be very rare. Besides there is strong selective pressure against these white mutants, as predators can see them from a great distance, For the same reason, a dominant white mutant will have little chance of surviving. On the other hand, a recessive factor, if present in heterozygous form, is not visible, cannot be eliminated by selection, and thus survives to produce a colour variant only if the owner meets a partner with the same genetic combination. The white trait, a clear negative property in the wild, can be positively valued in captivity by man. As this is a recessive trait, it will be very easy to develop a pure breeding stock of white ducks.
>
> (Bottema, 1989, p. 41)

17. An example from the distant history of domestication, concerning cattle, reinforces the point:

> In cattle, for example, a foreshortened and widened skull, decrease in the dimensions of eye and ear openings, shortness of backbone, decrease in size – in short, overall infantilism – distinguishes domestic from wild varieties. Some of the changes in the soft parts are reflected in skeletal remains. Muscular development or atrophy and changes in brain volume due to environmental modifications, such as differences in food supplied by man or the specialized physiological performance required of domestic animals, mark the skeleton and lead to the development of characteristic crests or ridges.
>
> (Isaac, 1970, p. 21)

18. However, Price (2002) discusses it at great length, although the definition quoted does not.

19. In Chapter 11, we shall be returning to this issue to assess its validity.

20. There are at least four reasons to account for why zoo animals are increasingly captive-bred and zoo-resident for their entire lives. First, as Chapter 9 will argue in detail, zoos exist primarily to exhibit exotic animals to the public; engaging in *ex situ* conservation is a parasitic activity which sits ill with the zoo's main business. Second, not all the animals exhibited in zoos are endangered. Third,

the cost of *ex situ* conservation is prohibitively high. Fourth, current thinking discourages, if not forbids, the replacement of zoo stock by capturing animals-in-the-wild. As a result, world-class zoos engage in captive breeding as a normal method of replenishment; in their bid to be conservation-minded, they hold that the default position must always be no replenishment of stock from the wild. Any deviation from this fundamental prescription will only be permitted under the most stringent conditions and would only be justified under the aegis of *ex situ* conservation. This does not mean, however, that zoos world-wide do not, as a matter of fact, buy animals captured from the wild to replenish their stock – between rhetoric and practice, there is a gulf in many cases.

21. One aspect of acculturation involves responding to the sounds of certain human commands in appropriate ways. One fairly amusing example of this phenomenon has occurred recently when Paris Zoo donated 19 of its surplus zoo-bred baboons to the zoo at Hythe in Kent. The keepers, to their amusement, discovered that these baby baboons only respond to French sounds/words such as 'dejeuner' and 'bonjour' but not to their English equivalents. As a result, the keepers had to buy a French phrase-book to communicate with these animals, who would, probably, remain 'French speaking' for the rest of their lives. (See Ward, 2005.)

22. Recall that in Chapter 6, we were careful in distinguishing between natural selection in the context of natural evolution in the wild on the one hand, and that of natural selection and artificial selection in captive environments on the other. At this stage of presenting the arguments in favour of zoos being an instance of artificial selection in captive environments, it would be appropriate to remind the reader that natural selection can and does occur in captivity.

23. This point is effectively illustrated by the following quotation from a novel by Alexander Dumas which describes the captive (future short-lived) Napoleon II who spent his childhood at the imperial Schönbrunn in Austria. One day, he managed to escape from his allotted quarters in the château to the park surrounding it. The child commented: 'I am as much captive in my room, only that instead of my prison being twenty paces in diameter, it is 3 leagues in circumference. It is no longer my window which is barred; it is that my horizon has a wall' (*The Mohicans of Paris*). (The passage is loosely translated by this author.)

24. The discussion to come and the issue it is meant to elucidate should not be confused with another different and separate issue, namely that deliberately intended artificial selection under domestication may have results which are inadvertent. For example, by consciously keeping animals in captivity, one inadvertently selects for tameness. Conscious selecting for early spawning in hatchery-raised salmon broodstock produces young, which are larger at traditional release times or can be released earlier in the spring. Conscious selecting for the Rex hair colour in rabbits has inadvertently produced certain metabolic and endocrine disturbances which increase mortality and render the animals susceptible to diseases. On these points, see Price (2002, pp. 43–44, 55).

Furthermore, as already noted earlier, the mechanism of natural selection may not entirely be displaced in captivity. It is relevant to cite Price again:

> In general, natural selection in captivity is most intense during the first few generations following the transition from field to captive environments.

Evolution and adaptation to the captive environment occur rapidly during this time because of the change in direction and intensity of natural selection on so many different traits and the relatively large number of correlated characteristics affected. . . . The degree of adaptation to the captive environment will increase as the frequencies of 'favorable' genes increase in response to selective pressure. An improvement in reproduction (i.e. fitness) over the initial generations in captivity can reflect the climb to a new adaptive peak as individuals become increasingly well-adapted to the captive environment over successive generations . . .

(2002, p. 56)

25. The analysis seems to follow roughly the so-called desire-belief model – see Bratman (1999, pp. 6–15).
26. Absolute certainty is 100 per cent; practical or near certainty would be less than 100 per cent though more than what in lay terms may be called high probability. In real life, where diverse and complex variables are at work, the concept of absolute certainty is not appropriate; instead, one operates with the notion of practical certainty.
27. There are exceptions which are covered by what is called strict liability, under which one could be held liable for even unforeseeable consequences, provided one has done something illegal and such consequences flowed causally from that illegal act.
28. For an account of the criminal law (in England and Wales), see Smith and Hogan (1996).
29. Direct intention to kill is first-degree murder; indirect intention to kill attracts a slightly lower category of crime, manslaughter. In jurisdictions where capital punishment obtains, first-degree murder means death by electric shock, chemical means or traditional hanging, while manslaughter means either life imprisonment (with no reprieve) or a fixed sentence (subject to review). However, in jurisdictions which have abolished capital punishment, the distinction between first-degree murder and manslaughter may be more or less academic as both would attract only imprisonment.
30. If such eventualities of failure were to transpire, and if caught, the defendant would be tried for attempted murder, which would merit a verdict of manslaughter.
31. Bratman (1999, pp. 139–42) argues that so-called indirect or oblique intention does not count as intention at all; in his terminology, it would follow that indirect or oblique intention would be said to be 'unintentional'. Bratman's analysis of the concept of intention leads to counter-intuitive results; on his reasoning, the court would have to acquit the woman who posted a kerosene-lit rag through the letter-box of the house of her lover's mistress of either first-degree murder, or indeed, even of manslaughter, as the defendant would, according to Bratman, have acted unintentionally. In the tradition of Anglo-Saxon jurisprudence, at least, no defendant could be found guilty of either first-degree murder or manslaughter if the defendant had acted unintentionally.
32. See Bratman (1999, pp. 143–5).
33. The case of immuration is, therefore, analogous to that of the plane cited earlier. The causal link between direct intention/end (to put animals in captivity

in order to present them as a collection of exhibits to the public on the one hand, to plant a time bomb in the luggage hold in order to claim on the insurance policy on the other) and consequences (bringing about morphological and other biological changes on the one hand, killing the passengers and destroying the plane on the other) is in either case such a strong one that it goes beyond high probability to practical certainty. It is also analogous to the situation in environmental law where the causal link between the polluter's action and his contribution to the pollution amounts to practical certainty.

34. For instance, in ungulates under captivity, emission of adrenaline is low, which in the wild would have hampered the animals from escaping successfully from predators as such low emission reduces the muscular power required for flight. See Baratay and Hardouin-Fugier (2002, p. 273). Genetic decline is also commonly observed in zoos – see Crandall (1964, p. 377); Blomquist (1995, pp. 178–85).

In general, it:

> is a reasonably safe assumption that some relaxed selection will accompany the transition from field to captive environments . . . Certain behaviors important for survival in nature (e.g. food finding, predator avoidance) lose much of the adaptive significance in captivity. Hence, one would expect natural selection in captivity on such behaviors to lose its intensity. As a result, changes in the gene pool of the population are likely to occur and genetic and phenotypic variability for many traits are likely to increase. For example, behaviors of free-living prey species toward predators may be changed after relatively long periods of freedom from predators . . . Caution in accepting novel foods may decline over time in captivity. Free-living herbivores are sometimes exposed to toxic plants . . . In contrast, captive animals are generally protected from toxic food items. Hence, it seems reasonable to expect relaxed selection for food neophobia in captive animal populations . . . . In nature, locating sources of food, water and shelter, mating activities and avoiding predators can require relatively high levels of physical fitness. Physical stamina and agility are less important in captivity due to the absence of predators and provisioning of basic necessities of life by man. . . . There is also reason to suspect that natural selection for cognitive abilities may be relaxed in captive populations . . . In nature, fitness is enhanced by the ability of individuals to quickly learn the consequences of their behavior or the behavior of other animals. In captivity, humans typically provide animals with the basic necessities for survival and may buffer the negative consequences of their mistakes. Opportunities to exercise cognitive abilities are reduced when the animals' environment limits physical activity and social interactions.
>
> (Price, 2002, pp. 63–5)

35. An omission to do x can be a deliberate act, an act which is directly intended.
36. For the former, see Weilenmann and Isenbrugel (1992).
37. There is no need on the whole for deliberate artificial selection in the breeding of these animals because they are already perfectly well adapted for the task in hand within the environment they are expected to work – see Clutton-Brock (1999, p. 130).

38. Another instance of pertinent evidence concerns the case of hand-reared sloth bears; these 'showed significantly higher frequencies of stereotypic and self-directed behaviours such as masturbation, self-stimulation, and pacing as compared with mother-reared individuals . . .'(Kreger et al., 1998, p. 71) Recently, the newspapers reported a case at the Yangon Zoological Gardens in Burma in which a woman offered to breastfeed two Bengal tiger cubs, four times a day. These cubs had been removed from their mother who had killed the third in her litter. A veterinarian at the Zoological Society of London was reported to have made the following two comments: first, that human milk may lack sufficient fat and protein for a fast-growing tiger cub which can put on as much as 1 kg a day; second, that breastfeeding can cause changes in the animal's behaviour later in life, rendering it a social misfit. See Sample (2005). The latter point is pertinent to our concern here.

8 Biotic artefacts

1. However, this is not to deny that non-human animals make artefacts. We know, for instance, that the beaver makes dams. However, this sort of observation is not germane to the preoccupation of this book which sets out to examine how humans create zoos as an artefact and, in so doing, have also made artefacts of the animals kept and controlled within zoos as exhibits.
2. A more technical and formal definition may be given as follows:

   By an 'artifact' I mean here an object which has been intentionally made or produced for a certain purpose. According to this characterisation, an artifact necessarily has a maker or an author, or several authors, who are responsible for its existence. . . . Artifacts are products of *intentional making*. Human activities produce innumerable new objects which are entirely unintentional (or unintended); such objects and materials are not artifacts in the strict sense of the word. When a person intends to make an object, the content of the intention is not the object itself, but rather some description of an object; the agent intends to make an object of a certain kind or type. Thus what I want to suggest is that artifacts in the strict sense can be distinguished from other products of human activity in the same way as acts are distinguished from other movements of the body; a movement is an action only if it is intentional under some description . . . , and I take an object to be an artifact in the strict sense of the word only if it is intentionally produced by an agent under some description of the object. The intention 'ties' to the object a number of concepts or predicates which define its intended properties. These properties constitute the *intended character* of the object. I shall denote the intended character of an object o by '$IC(o)$'.

   Thus an object o is a proper artifact only if it satisfies the following *Dependence Condition*:

   . . . The existence and some of the properties of o depend on an agent's (or author's) intention to make an object of kind $IC(o)$.

   (Hilpinen, 1995, pp. 138–9)

Note, however, that Hilpinen's definition of 'artefact' is much wider than that used by this book which stipulates that the human intentionality be embodied in a material medium. On this account, unlike Hilpinen's, belief systems and concepts are not artefacts. For instance, the concept *per se* of the division of labour is not an artefact; however, that concept could be applied in practice to design/create, say, the Ford assembly production line which, is, of course, itself an artefact. (For a more thorough philosophical examination of the concept of artefact, see Lee (2005a, ch. 1.)

3. For details of this point, see Lee (2000).

4. Aristotle, in talking about the four causes, has invoked abiotic/exbiotic artefacts to illustrate them; this seems to have influenced unduly how theorists/philosophers have looked at the matter ever since.

5. This view in environmental philosophy is referred to as anthropocentrism, namely that only humans are morally considerable (or intrinsically valuable), and that all other natural things and non-human beings have only instrumental value for humans. We have seen in an earlier chapter that another term for anthropocentrism is human chauvinism.

6. However, no volition should be read into this locution in the case of plants or the lower animals.

7. In this context, *telos* or *tele* (in the plural) is used to refer to the developmental programme, which inheres in every individual organism as a naturally occurring being. For example, an acorn, in accordance with its *telos*, would become an oak sapling, which would grow eventually to be a mature oak tree, producing in turn its own acorns.

8. For degrees of artefacticity, see Lee (1999).

9. However, the term 'species' is not exclusively confined to discourse about biological matters. Historically, it has also been used (although today it appears to have an old-fashioned ring about it) to naturally occurring abiotic matter, such as different natural kinds of minerals. See Wilkerson (1995); see also Laporte (2004).

10. The time-scale is crucial. Those who advocate *ex situ* conservation as a means of saving certain species from extinction are aware of this; hence they talk of a time-span of a 100 years, at most of 200 years, if the captive-bred animals (even in the presence of precautions taken to fend off some of most obvious consequences of immuration) are to remain members of the same species as those individuals which live in the wild.

11. More formally it may be defined as follows: 'a group of actually or potentially interbreeding natural populations that is genetically isolated from other such groups as a result of physiological or behavioural barriers' (Clutton-Brock, 1999, pp. 41–2). Indeed, certain populations of bats may occupy the same space but nevertheless constitute different species, as they do not interbreed because their respective mating calls operate on slightly different frequencies. 'Genetic isolation' is itself a complex notion and may cover numerous aspects of which the bats cited exemplify but one. Another refers to the fact that even if two individuals succeed in mating, it fails to lead to reproduction, and that even should there be successful reproduction, hybrids are born which are sterile, and therefore, in turn, cannot reproduce themselves. For the purpose of this book, the most important aspect is that zoo-born and zoo-bred animals and their counterparts in the wild remain 'genetically isolated'

because they do not and cannot meet and mate, as they occupy different locations/spaces and habitats; they can only meet and mate when humans permit them to do so. (For details, see Lee (1997b).)

12. This author agrees with the view of Mullan and Marvin (1999, p. 12) that zoo animals form a new and distinct species; however, these two authors argue simply on the grounds that zoo animals appear neither to be wild nor domesticated animals. This book argues that they are domesticated, though not in the classical understanding of the term, and that therefore they are immurated animals.

In the literature about classical domestication, variations in the taxonomic designation of the domesticated and the wild are found; for instance, wild and domestic forms of the pig are often given as *Sus scrofa*, on the grounds that although the wild boar and domestic pig are typically found in different environments, their phenotypic differences and their habitat choices are not as dramatically different as between wolves and dogs and that, furthermore, domestic pigs can successfully become feral if given the chance. However, other writers list the domestic pig as *Sus domesticus* – see Clutton-Brock (1999). In general, different taxonomic names commonly mark the distinction between wild and domestic forms. For instance, the domestic chicken is either *Gallus domesticus* or *Gallus gallus domesticus* whereas its wild ancestor, the jungle fowl is called *Gallus gallus*. The controversy concerns the issue whether the domesticated represents a separate species or subspecies. Those arguing for separate status are impressed by the fact that wild and domestic forms are morphologically, behaviourally and/or ecologically distinct. Those arguing against rely on the theoretical possibility of their interbreeding, and therefore on their respective genetic distinctiveness as a subspecies. For further details on this debate and a solution to the problem raised, see Price (2002, pp. 3–4).

9   Justifications deemed serious

1. Vienna was the first to get a (modern) zoo in 1752, followed by Paris in 1793 in the wake of the French Revolution, and then London in 1826.
2. On these points, see Baratay and Hardouin-Fugier (2002).
3. The Department for Environment, Food and Rural Affairs (Defra), UK, has issued a document in connection with the European Union Zoos Directive, 1999. That document contains 7 examples of research projects which zoos could undertake; with the possible exception of one, the rest are concerned solely with issues and problems arising from zoo management and husbandry. See http://www.defra.gov.uk/wildlife-countryside/gwd/zoosforum/handbook/2. pdf, 30–31. [03/02/05].
4. Mayr has written:

> Every organism, whether an individual or a species, is time-bound and space-bound. There is hardly any structure or function in an organism that can be fully understood unless it is studied against this historical background. To find the causes for the existing characteristics, and particularly adaptations, of organisms is the main preoccupation of the evolutionary

biologist. He is impressed by the enormous diversity as well as the pathway by which it has been achieved. He studies the forces that bring about changes in faunas and floras and the steps by which have evolved the miraculous adaptations characteristic of every aspect of the organic world. In evolutionary biology almost all phenomena and processes are explained through inferences based on comparative studies. These in turn, are made possible by very careful and detailed descriptive studies. The evolutionary biologist is interested in the why question.

(Mayr, 1982, p. 71)

5. See Meek (2001).
6. The point made is therefore different to the one made below:

> **Research**: The options available for off-exhibit research animals can actually be more diverse and cost-effective because the emphasis can be purely on functional rather than on aesthetic considerations . . . Public perceptions are not as critical when animals are designated for research purposes and the research facility is not on public view. Environmental enrichment is still important, however, because atypical behaviour and associated physiological stress can add unwanted variation to the experimental design, thereby confounding the results and jeopardizing the validity of the study . . .
>
> (Kreger et al., 1998, pp. 64–5)

While recognising that off-exhibit space for research animals is much smaller and barer than exhibit space and that enrichment is still appropriate, nevertheless these authors have failed to see that the off-exhibit space designated for research animals (which is already smaller than exhibit space) may itself be a cause in bringing about physiological and behavioural changes, and indeed may even lead to brain damage induced by confinement within a very limited space. In other words, enrichment may not be always or entirely successful in counteracting the effects of the variable of severely confined space. More systematic research should be conducted to clarify matters; until that happens, scepticism regarding the validity of results conducted on research animals within confined space, whether in zoos or in laboratories, is justified.

Some zoo researchers and professionals have acknowledged the limitations of enrichment for the psychological well-being of captive animals, such as the macaque monkeys which are the subject of one of these studies:

> Our own studies added substantially to the emerging evidence that modest variations in cage size have little measurable effect on the psychological well-being of monkeys . . . Neither urinary cortisol, appetite suppression, nor abnormal behaviour varied significantly as a function of cage size . . . I want to dispel the illusions that increasing cage size, within the range likely to be possible in a research lab or behind-the-scenes zoo setting, will provide meaningful enrichment to macaques . . . Novelty can stimulate exploratory behaviour but can also elicit fear and disturbance.
>
> (Crockett, 1998, pp. 133–5)

Another set of zoo writers have also admitted the same point made above:

> increasing cage size fails to result in any measurable changes in behaviour . . ., even if that increase is more than 600 times the standard size. The data obtained from these empirical assessments suggest that increasing cage size as a means by which to enrich and enhance an animal habitat may not be worth the cost, at least under conditions in which the size of the cage is the only aspect that is altered.
>
> (Morgan et al., 1998, 160)

7. This seems to be the fate which has overtaken Glasgow Zoo which closed down at the end of September 2003. Admittedly, financial debt was the obvious cause; however, behind that truth is also that the zoo, already in a precarious financial situation, would not be able to meet the requisite demand that it could demonstrate continuous participation in, and contribution to, the goal of conservation.

8. According to the document, all zoos must implement the following measures:

- participating in research from which conservation benefits accrue to the species, and/or training in relevant conservation skills, and/or the exchange of information relating to species conservation and/or where appropriate, captive breeding, repopulation or reintroduction of species into the wild,
- promoting public education and awareness in relation to the conservation of biodiversity, particularly by providing information about the species exhibited and their natural habitats.

9. *In situ* conservation may be defined as:

> The conservation of ecosystems and natural habitats and the maintenance and recovery of viable populations of species in their natural surroundings . . .

Ex situ conservation may be defined as:

> The conservation of components of biological diversity outside their natural habitat.

These definitions are taken from the Convention of Biological Diversity and as cited in Defra's document on the *EU Zoos Directive* (1999, p. 2). http://www.defra.gov.uk/wildlife-countryside/gwd/zoosforum/handbook/2.pdf [03/02/05].

Another similar definition of *ex situ* conservation may be found in *WZCS* (1999). It refers to:

> the maintenance of wild animals in stable populations outside their original biotrope. Being out of their original habitat means that the animals were separated from the other components of their natural community, and are kept in zoos, other types of scientific institutions, breeding centres, or in semi reserves.
>
> (ch. 6.1)

10. Neither do zoos contribute anything significant financially to *in situ* conservation programmes in general, although in 1999 zoos supported more than 650 such projects – see Olney and Fisken (2003).

11. One matter which may be worth pointing out to readers is that the two key bodies (cited in this book) exhorting zoos to embrace conservation as a central justification for their existence differ in their view as to what constitutes breeding stock under *ex situ* conservation. While the *WZCS* stresses that the animals must be captive-born and -bred unless there are exceptional circumstances to justify the capture of an animal from the wild, the Defra gloss on the *EU Zoos Directive*, appears to put the emphasis somewhat differently. It says:

> Stock should only be taken from the wild, regardless of whether it is to be part of a managed programme, if there is evidence to show that collection will not have a detrimental effect on the population, species as a whole or its habitat . . . Collection from the wild is not always detrimental.
>
> (1999, p. 13)

12. A recent telling critique from a zoo professional is that of Hancocks. He says that fewer than five species have been saved from extinction – see (2001, p. xvii). Furthermore, he points out that the 'most optimistic projections state that if all the world's professionally operated zoos, in concert and under perfect conditions, devoted a full half of their facilities to breeding endangered animals they could perhaps manage about eight hundred of them in viable breeding populations' (ibid., p. 152)'. However, of vertebrate species alone, there are about 46,000 in existence.

13. See Article 2 at http://europa.eu.int/eur-lex/pri/en/oj/dat/1999/l_094/l_09419990409en00240026.pdf [03/02/05].

14. As cited by Clubb and Mason (2002, p. 11).

15. Note that captive breeding with the self-contained aim of replenishing zoo stocks is not subject to such a restriction.

16. Of course, domesticants like cats and dogs, the products of what this book has called classical domestication, are another kind of ontological foil to naturally occurring wild animals.

17. As we have seen, which animals get to reproduce is guided by the explicit goal of maintaining genetic variability.

18. There is one outstanding instance of a (private) zoo which runs an *ex situ* conservation programme based on lines which are the exact opposite of what is endorsed by the scientific consensus. This is John Aspinall's Howlett's Animal Park which claims that its success in captive breeding depends exactly on 'being friends' with the animals, encouraging keepers to have intimate physical contact with the animals, romping with them, which makes the animals happy and contented. Now this may be so, as far as the reproductive rate of captive breeding is concerned and as far as animal welfare itself is concerned. However, the point missed by such a perspective is precisely that such intimate contact and relationships with humans render the animals tame – they are happy, contented tame/immurated animals, not wild animals. See http://www.guardian.co.uk/print/0percent2C3858percent2C3962804-103390percent2C00.html (13 February 2000); http://www.totallywild.net/howletts.php?page=howletts. [23/03/05].

19. Just to cite one example regarding the Californian condor (*Gymnogyps californianus*) captive breeding programme initiated by the US Fish and Wildlife Service in 1984, but run by the San Diego Wild Animal Park and Los Angeles Zoo in conjunction with other interested bodies. The 27 last remaining condors (of a reproductive age) were captured to form the core of the breeding programme; from these, the zoos successfully bred more than 200. By 2001, half of them have been returned to the wild and some of these have bred in turn. As mentioned earlier, the chicks were reared by scientists wearing condor-like puppets etc. The first attempt at releasing two of the birds in 1992 was not a success; one of them died when it swallowed antifreeze and the other had to be recaptured when it kept landing on power lines and pylons. The second attempt later that same year was not successful either; three died when they collided with power lines and the other three also had to be recaptured because of their 'fondness' of landing on power lines. The next year, more were released but this time, far from the pylons and electricity lines; however, there was no improvement in the success rate. The scientists finally drew a lesson from these failures – they began to teach their birds to avoid pylons and such dangerous things, by setting up two electricity poles in the enclosures which gave the birds a mild electric shock whenever they approached them. This tactic seemed to have worked as none of the birds in the next batch released died from electrocution or collision with electric cables. But that alone did not ensure long-term survival, as these birds lacked the appropriate knowledge and the relevant skills of how really to survive in the wild, a deprivation brought on by the fact that they were raised by humans disguised as condors within a zoo environment. Eventually, the scientists resorted to giving them in their enclosure what may be called a mentor, an older bird which had been captured from the wild and had known existence in the wild. When these pupils were released, it was found that they behaved more like adult condors. In May 2002, the scientists went even further and released an older bird, which had once lived in the wild and was now well past the age of reproduction, together with the captive-bred juveniles, hoping that she would remember the roosting sites and the watering holes she must have visited 14 years ago. The experiment was acclaimed a success, although problems still remained. The captive-bred birds failed to avoid prey that has been killed with shot which contains lead. Finally in the spring of 2002, the first wild Californian condor was born. By 2020, the scientists expect to achieve the goal of removing the condor from the endangered species list when they will have established two stable wild populations of 150 birds each, as well as maintaining a captive population of 150. See Kaplan (2002).

It is also interesting to note that the reintroduction would not have been possible without the introduction of an exotic related species, namely the Andean condor. It is true that when the programme of reintroducing the Californian condor had stabilised by 1991, the exotics were recaptured and returned to their South American habitat. Nevertheless, this shows that the scientists and related professionals were prepared to import an exotic species in order to learn how condors behave in the wild as well as to enlist these exotics to help the captive-bred juvenile native condors to learn condor culture. (For details, see http://species.fws.gov/species_accounts/bio_cond.html [09/02/05].) This demonstrates the point, which will be discussed in greater detail a little later, that not all conservation scientists fully grasp the true significance of what Chapter 1 of this book has called the zoological conception of an animal, namely that a species in the wild and its individual members, in all aspects of their behaviour and their culture, are the product not only of the processes of natural evolution and the mechanism of natural selection, but also of the complex interrelations among themselves as well as between them and their habitat.

20. For an account of the problems facing captive-born and reared animals for release in nature, see Price (2002), ch. 19, but especially his conclusions on p. 202.

21. For the former, see, for example, http://www.animalinfo.org/species/ artiperi/elapdavi.htm [09/02/05]; and for the latter, see Hancocks (2001). As a compromise, one could suggest that the animal became extinct in the wild roughly 1500 years ago.

22. On all these changes, see Price (2002); on genetic changes see also Crandall (1964, 377) and Blomquist (1995, 178–85). Regarding phenotypical changes, Price cites one longitudinal study of wild Norway rats (*R. norvegicus*) over the first 25 generations in captivity. The study reports increase in 'body weight, percentage of mated pairs that produced offspring, number of litters born and length of the reproductive lifespan. The investigators also reported that the tendency to escape and the resistance to handling declined over generations . . .' (Price, 2002, p. 16).

23. See Baratay and Hardouin-Fugier (2002, pp. 273–4).

24. Note that in the case of the Père David deer, no one has a clue not only about its original habitat in the wild but also what the genetic variability in the original wild population would be.

25. An example when conservation scientists/managers forget the equally important precondition of conserving genetic variability concerns the Española tortoise in the Galapagos Islands. From 14 individuals in 1965, the population has increased to over 800 today. Indeed the species would have become extinct by now without the acclaimed success of this captive breeding programme for reintroduction to the wild. However, according to a genetic study conducted by the Free University of Brussels (Belgium), the population lacks genetic diversity. In 1965, the two remaining males and 12 females were transferred to the Charles Darwin Research Station on Santa Cruz Island and captive breeding began with reintroduction in the wild in 1975. However from a study of 134 of this population, the scientist has found that its genetic diversity is equivalent to roughly 11 unrelated individuals when it should have been 300. Nearly 80 of the individuals sampled turn out to have been sired by a single male from San Diego Zoo, nicknamed Super Macho. If this finding is correct, then the revived population may, nevertheless, face extinction in the long run in view of the problems and difficulties facing a population with such a small number of founding members. See Anderson (2004). For a brief summary of why genetic variability is important, see *WZCS* (1999, ch. 6.2).

26. For the term 'lifeline', see Steven Rose, especially ch. 6. It may be appropriate to cite the last paragraph from that chapter:

> Lifelines . . . are not embedded in genes: their existence implies homeodynamics. Their four dimensions [of space and time] are autopoietically

constructed through the interplay of physical forces, the intrinsic chemistry of lipids and proteins, the self-organizing and stabilizing properties of complex metabolic webs, and the specificity of genes which permit the plasticity of ontogeny. The organism is both the weaver and the pattern it weaves, the choreographer and the dance that is danced. That is the fundamental message of this chapter, and therefore in many ways of this entire book. And it provides the framework within which I turn now to consider the mechanisms of evolution.

(1997, p. 171)

27. In July 2004, the Natural Science Museum (London) announced what is dubbed its Frozen Ark Project. The aim of the museum, the Zoological Society of London and the University of Nottingham together with other likeminded institutions throughout the world is to freeze DNA and tissue samples of animals facing extinction, so that scientists would be able to continue to study them from the evolutionary point of view, as well as in the hope that, very soon, advanced cloning techniques would enable the re-creation of extinct animals, using surrogates. The Project will begin with animals from zoos, captive breeding and other research programmes. See Sample (2004) and *New Scientist* (31 July 2004, p. 5).

28. The Indian scientists claim that they have had some breakthroughs recently in overcoming the problems associated with the lack of genetic variability inherent in cloning from cells of a small number of animals.

29. See Ramesh (2004) for the details cited of the Indian cloning project of the Asiatic lion and cheetah.

30. One would not like to carp, but it does seem surprising to read in the two quotations just cited from Hancocks (2001) that its author appears to think that zoos display their animals in natural settings rather than simulated naturalistic settings, that is to say that such exhibition enclosures are constructed entirely with the help of technology, as shown earlier in Chapters 4 and 5.

31. See *WZCS* (1999, ch. 31).

32. According to Kreger et al.:

If an animal or group of animals is intended to serve an educational role, then . . . a premium is placed on the naturalistic appearance of both the exhibit and the animals it contains.

(1998, p. 62)

According to Jones and Jones in their 1985 Kansas City Master Plan, their message of structured zoo conservation-education programme is as follows:

A new approach to zoo design begins with presentation of animals in such a way that their right to exist is self-evident. The educational message accompanying this presentation should be clear and persuasive. Whole habitats should be exhibited, with rock and soil substrates and vegetation supporting communities of species typical of the environment and logically associated. Visitors should feel they are passing through a natural environment, with a feeling of intense involvement. The point should be made that animals live in habitats, and it is the destruction of those habitats that is the principal cause of wildlife extinction today.

(As cited by Mullan and Marvin, 1999, p. 60)

33. For instance, according to figures of the *International Zoo Yearbook*, in North America, over 100 million people – that is, just under 50 per cent of the population – visit zoos on an annual basis; the figures for Europe and Japan are similarly high. See *WZCS* (1999, ch. 3.2).
34. The figures are cited in Mullan and Marvin (1999, p. 133).
35. See also ibid., p. 136.
36. This is, however, not to deny that seeing a living exotic animal does have unique appeal (as we shall see in the next chapter); it is just not obvious that it has that special transformative power to educate the public which zoo advocates claim it does.
37. For instance, David Hancocks has written:

    careful application of landscape-immersion philosophy, with attention to concealed barriers and authentically replicated forms of the natural landscape and use of borrowed vistas and studied sight lines, can all combine to create a memorably evocative experience in which zoo visitors associate wild animals with appropriate wild habitats. It achieves two important goals. Zoo visitors, even if they don't read the interpretive graphics, can learn by associative intuition that certain animals and certain habitats are inextricable. And they can by similar association gain more respect for wildlife.

    (2001, pp. 147–8)

38. The resemblance is not absolute; immurated animals over the years, as we have already mentioned, would display morphological/anatomical and other biological features which are somewhat different from their counterparts in the wild. These differences, however, would not be readily detectable to the passing eye of the ordinary visitor at a distance.
39. See Martin Mere: http://mm.eyelook.co.uk/edu/edu.html [16/02/05].
40. There is another method which is used on the pelicans in St James's Park (London) precisely because the Royal Family found evidence of mutilation, caused by pinioning, stressful. This involves removing one or two strips of the extensor tendons on the leading edge of the wing. This means that the bird would not be able to thrust downwards into the wind to fly. However, this method of rendering it captive but without ostensible mutilation is not foolproof; occasionally in a strong gale, it could be lifted off. See Jones (2002, p. 139).
41. In the case of tamed birds, morphological reductionism does not work so well, as already pointed out the visitors could see readily for themselves that the birds do not fly, and therefore do not resemble in one essential way birds in the wild; they might also observe that one of their wings had been mutilated.
42. One could also get to this conclusion, that a certain dissonance is inherent in the zoo experience itself, even without the benefit of having visited zoos. The experiment one is engaged in is, after all, only a thought experiment, and that is all which is needed here.

43. See Mellen et al. (1998, p. 198).
44. See Baer (1998, p. 293).
45. See Clubb and Mason (2002, ch. 5).
46. One could perhaps say in their defence that they do not distort reality quite as blatantly as Carl Hagenbeck's attempts to create theatrical spectacles of his exhibits at Stellingen at the beginning of the nineteenth century. Hagenbeck exhibited them in the open without cages; he aimed to create the illusion that there was no separation between the human visitors and the captive animals, and indeed between the various species of animals on display. He did not hesitate to put predator next to prey; he created invisible (that is, to the human visitors) barriers which the animals could not cross between the groups of animals. Today's arrangements may spring from different motives and the effects aimed at may not be exactly identical to Hagenbeck's, but in spirit they are akin to his. That is to say that both types of arrangements do not portray animals-in-the-wild, as they manifestly claim to do, but to present immurated animals as exhibits.
47. Note that this author differs from Hancocks in the vital matter of the ontological status of zoo animals, namely, that they are not wild but artefactual in character; on the other hand, Hancocks seems to imply that they are wild, but that, unfortunately, zoos have not succeeded or made the right efforts in presenting them in their full wildness.
48. Such children would be those very same children who see their families buy milk in cartons from supermarkets and, as a result, infer that milk comes from supermarket shelves, and that domesticated animals called cows have nothing to do with the provenance of milk.
49. See Hancocks (2001, p. 249).
50. See Andersen (2003, p. 79).
51. In order to give visitors a taste of carnivores hunting their prey, the Panaewa Zoo in Hawaii has installed clay figures in the tiger enclosure to be operated by a computer but which the public can manipulate, whenever they like, in order to see the tiger 'hunt' a rabbit or squirrel (in clay) which could then be 'saved' by a computer – see Baratay and Hargouin-Fugier (2002, p. 268). Some zoos in non-Western parts of the world are apparently 'bolder' in their policy of what may be fed to their animals in public. This author has been told by a recent visitor to Harbin Zoo that visitors could choose a cow, pay for it, and get the zoo staff to feed it to the tiger for all to behold. Now, this supposedly 'barbaric' practice would at least have the decided merit of teaching the public by allowing them to see what carnivores really eat and how they eat in the wild; most assuredly, they would not leave the zoo thinking that wild tigers in the wild dine off zoo pellets! (This author has no means of checking whether by now such a practice has ceased in Harbin Zoo.)

10 Justifications deemed frivolous

1. Of course, adherents of the philosophy of animal rights would have objections to zoos, but zoo professionals, in the main, are not known to be supporters of animal rights as far as this author knows.
2. These numbers are cited by *WZCS* (1999, ch. 3.2).

3. Ironically, those who may stand a chance of success in bringing a case under the Trade Descriptions Act would be those visitors who can genuinely claim that zoos do not trade fairly only in the light of their zoo experience – that is to say when they fully grasp the dissonance between the exotic exhibits under captivity on the one hand, and wild animals and their behaviour in the wild on the other, and they have been lured by zoos into seeing animals billed as wild when in reality these are not wild at all.

4. Shepherdson writes:

> An animal in the wild basically does only a few things. It hunts, eats, sleeps, often plays and breeds. But when they don't have to hunt for food or engage in normal activities like playing or exploring, they can become bored, even morose. http://www.zooregon.org/ConservationResearch/environm. htm [16/02/05].

5. See Poole (1998, p. 84).

6. Singapore Zoo advertises this experience on the web: http://www.sightseeing-tours.net/web/_li-n73p-2.html [22/02/05]. Tickets start from £13 (Sterling) per person. The highlight of the experience is described as follows: 'Witness the orang utan descend from a naturalistic backdrop filled with vines and branches, followed by a once-in-a-lifetime opportunity for interaction and photography with these intelligent and iconic symbols of the disappearance of the tropical rainforests.' This is followed (without irony) by the line: 'With a commentary highlighting the plight of this magnificent creature, the Zoo hopes to inspire in its guests a respect and deep appreciation of nature.'

7. Possibilities include London Zoo, Singapore Zoo and the Smithsonian National Zoological Park. For the former, see http://www.londonzoo.co.uk/weddings/wedding.html [22/02/05]. Its brochure (received September 2004) says: 'Hold your event at London Zoo and you are helping to support our vital conservation work throughout the world.' Animal Houses: Bear Mountain on the Mappin Pavillion (holds 150 people); Happy Families (200); Lion Terraces (200); Reptile House (350) B.U.G.S! (350); Komodo Dragons (100). 'Unique Features: View some of the world's rarest animals after the zoo closes, in one of these six unique venues. Perfect for canapé receptions, pre-dinner drinks, parties and barbecues.' For Singapore Zoo, see: http://www.singaporebrides.com/venue/wrs/ [22/02/05]. For the Smithsonian Zoological Park, see http://nationalzoo.si.edu/ActivitiesAndEvents/Celebrations/birthday-parties.cfm [25/02/05]. This zoo as well as the San Diego Zoo, amongst others, offer sleepovers for children and adults, marketed as Snore and Roar. http://www. sandiegozoo.org/calendar/cal_sleepovers.html [22/02/05].

8. Shamu is just one of three names – the other two being Kandu and Nandu – dreamt up by the marketing boys to refer indiscriminately to any one of the three performing killer whales in the zoo's entourage, although it is true that Shamu is more often used than the other two. Their real, behind-the-stage, off-duty names are known only to the Sea World staff – see Mullan and Marvin (1999, 23).

11 Philosophy and policy

1. Their philosophical preoccupation focuses on captivity denying the animals the right to be free. Although it is true that not to be free to roam is part of the denial to be wild, nevertheless the source of their concern lies primarily in political philosophy rather than in ontology.
2. Professionally endorsed zoos, at least according to their mission statements, have renounced the policy of capturing wild animals to replenish their stock; instead, they do so through captive breeding.
3. As for the goal of *ex situ* conservation, recall that the last chapter has argued that it is an activity which sits ill with the defining characteristic of a zoo as a collection of animal exhibits open to the public. Therefore, even if such an activity were to be deemed to be desirable, it should be conducted elsewhere in a different space. However, this book has said enough on the subject for the conclusion to be drawn that *ex situ* conservation, because of the ontological risks and the heavy financial resources involved, has little or no merit and that such resources should be channelled towards *in situ* conservation instead.
4. This, however, as an earlier remark has already made clear, should not be interpreted to mean that zoos should not support *in situ* conservation financially or by way of resources which are considered scientifically as well as ontologically relevant to the projects of serving species and their habitats in the wild.
5. It is true that, at the moment, the success rate of such techniques is not very high. However, scientists working in these technologies hope to improve it and are confident that improvement would come in the light of further experimentation and research.
6. The two types of biodiversity – natural and artefactual – have different values because of the difference in their ontological status. See Lee (1999) and Lee (2004).

Appendix: Environmental enrichment or enrichment

1. Shepherdson (2001) gives two other justifications, namely, enriched environments are more interesting to zoo visitors; they help conservation by improving reproductive rates, psychological behaviours when the captive-born grow up to become adults, and survival rates upon introduction to the wild in the case of animals taking part in *ex situ* conservation programmes.
2. However, one should be wary of the claim that boredom is at the bottom of all stereotypies. There may be a genetic 'predisposition for the development of stereotypies' in some cases – see Price (2002, p. 220); furthermore, '[t]here is increasing evidence that many stereotyped behaviours reflect feeding ... problems ... that premature weaning and low weaning weight result in the development of relatively high levels of stereotyped wire-gnawing on the lids of their cages' in the case of laboratory mice (*m. musculus*).' In other words, '[s]tress early in life could predispose animals to stereotypy by affecting the persistence of behaviours exhibited at that time' (Price, 2002, pp. 218–19).
3. Physical distress may not manifest itself as stereotypic behaviour but could be readily ascertained by, say, abnormal condition of the feet in the case of

animals with soft pads when they are made to stand on concrete most of the time.

We are, in this discussion, however, concerned primarily with the amelioration of psychological distress.

4. The third, however, namely to aid conservation, only applies to a tiny portion of zoo animals. Furthermore, that small proportion, which does take part in *ex situ* conservation, would have to be protected from such enrichment schemes, as Chapters 7 and 9 have shown.

References and Select Bibliography

Andersen, L. L. (2003) 'Zoo Education: From Formal School Programmes to Exhibit Design and Interpretation', in *Zoo Challenges: Past, Present and Future* (*International Zoo Yearbook*, Vol. 38) ed. P. J. S. Olney and Fiona A. Fisken (London: The Zoological Society of London).

Anderson, James R. (2004) 'Conservation Plans Are Fatally Flawed'. *New Scientist* (24 January 2004): 13. http://www.newscientist.com/article.ns?id= mg18124312.000 [10/02/05].

Baer, Janet F. (1998) 'A Veterinary Perspective of Potential Risk Factors in Environmental Enrichment', in David Shepherdson, J. Mellen and J. Hutchins (eds), *Second Nature: Environmental Enrichments for Captive Animals* (Washington DC: Smithsonian Institution Press).

Baratay, Eric, and Elisabeth Hardouin-Fugier (2002) *Zoo: A History of Zoological Gardens in the West*, trans. Oliver Welsh (London: Reaktion Books).

Baron, Marcia, Philip Pettit and Michael Slote (1997) *Three Methods in Ethics: A Debate*. (Malden, MA, and Oxford: Blackwell).

Blomquist, L. (1995) 'Three Decades of Snow Leopard in Captivity', in *International Zoo Yearbook*, Vol. 34 (1995).

Bököyni, Sandor. (1989) 'Definitions of Animal Domestication', in J. Clutton-Brock (ed.), *The Walking Larder: Patterns of Domestication, Pastoralism, and Predation* (London: Unwin Hyman).

Botkin, David (1990) *Discordant Harmonies: A New Ecology for the Twenty-first Century*. (Oxford and New York: Oxford University Press).

Botkin, David, and Edward Keller (1995) *Environmental Science: Earth as a Living Planet*. (New York and Chichester: Wiley).

Bottema, Sytze (1989) 'Some Observations on Modern Domestication Processes', in *The Walking Larder*, J. Clutton-Brock (ed.) (London: Unwin Hyman).

Bratman, Michael E. (1999) *Intention, Plans and Practical Reason*. (Cambridge, MA.: Harvard University Press).

Budiansky, S. (1994) 'A Special Relationship: The Coevolution of Human Beings and Domesticted Animals'. *Journal of the American Veterinary Medical Society* 204 (1994): 365–8.

California Condor: http://species.fws.gov/species_accounts/bio_cond.html [09/02/05].

Carrico, Christine K. (2001) 'In Search of Pharmacology'. *Molecular Interventions* 1 (2001): 64–5. http://molinterv.aspetjournals.org/cgi/content/full/1/1/64 [10/01/05].

Casamitjana, Jordi (2003) *Enclosure Size in Captive Wild Animals: A Comparison Between UK Zoological Collections and the Wild* (October 2003). http://www. captiveanimals.org/zoos/enclosures.pdf [05/01/05].

Cheetah. http://members.aol.com/cattrust/cheetah.htm. [11/01/05].

——. http://nationalzoo.si.edu/Animals/AfricanSavanna/fact-cheetah.cfm [05/01/05].

——. http://www.hlla.com/reference/anafr-cheetahs.html [06/01/05].

Clubb, Ros, and Georgina Mason (2002) *A Review of the Welfare of Zoo Elephants in Europe* (London: RSPCA). See http://users.ox.ac.uk/~abrg/elephants.pdf [06/01/05].

——. (2003) 'Captivity Effects on Wide-ranging Carnivores', *Nature* 425 (2 October 2003): 473–4. See http://www.nature.com/cgi-taf/DynaPage.taf?file=/nature/journal/ v425/n6957/full/425473a_ fs.html&content_filetype=PDF [06/01/05].

Clutton-Brock, J. (ed.) (1989) *The Walking Larder: Patterns of Domestication, Pastoralism, and Predation.* (London: Unwin Hymaan).

——. (1999) *A Natural History of Domesticated Mammals.* (Cambridge: Cambridge University Press).

Crandall, Lee S. (1964) *The Management of Wild Animals in Captivity.* (Chicago: Chicago University Press).

Crockett, Carolyn M. (1998) 'Psychological Well-being of Captive Nonhuman Primates: Lessons from Laboratory Studies', in *Second Nature: Environmental Enrichments for Captive Animals*, D. Shepherd, J. Mellen and J. Hutchins (eds) (Washington DC: Smithsonian Institution Press).

Department of Farming and Rural Affairs (Defra), UK. http://www.defra.gov.uk/wildlife-countryside/gwd/zoosforum/handbook/2.pdf [03/02/05].

Denver Zoo. http://www.denverzoo.org/animalsplants/mammal01.htm [10/01/05].

Dickens, W., and J. Flynn (2001) 'Nature or Nurture', *New Scientist* (21 April 2001): 44.

Dipert, Randall R. (1993) *Artifacts, Artworks and Agency.* (Philadelphia: Temple University Press).

Ducos, Pierre. (1989) 'Defining Domestication: A Clarification' in *The Walking Larder*, J. Clutton-Brock (ed.) (London: Unwin Hyman).

Easterbrook, Gregg (1996) *A Moment on the Earth: The Coming Age of Environmental Optimism* (London: Penguin).

Ereshefsky, Marc (1998) 'Eliminative Pluralism', in *Philosophy of Biology*, David Hull and Michael Ruse (eds) (Oxford and New York: Oxford University Press).

Good Zoos. http://www.goodzoos.com/Animals/small.htm [last modified 16 January 2000].

Gorilla. http://www.worldwildlife.org/gorillas/ecology.cfm [06/01/05].

——. http://www.cotf.edu/ete/modules/mgorilla/mgbiology.html [11/01/05].

Hagenbeck, Carl (1910) *Beasts and Men* (London: Longmans and Green).

Hancocks, David (2001) *A Different Nature: The Paradoxical World of Zoos and Their Uncertain Future* (Berkeley, Los Angeles and London: University of California).

Hediger, Heinrich (1950) *Wild Animals in Captivity: An Outline of the Biology of Zoological Gardens*, trans. G. Sircom (London: Butterworths Scientific Publications).

——. (1968) *The Psychology and Behaviour of Animals in Zoos and Circuses*, trans. G. Sircom (New York: Dover Publications).

——. (1970) *Man and Animal in the Zoo* (London: Routledge and Kegan Paul).

Hilpinen, Rosto (1995) 'Belief Systems as Artifacts'. *Monist*, 28 (1995): 136–55.

Hull, David, and Michael Ruse (eds) (1998) *The Philosophy of Biology* (Oxford and New York: Oxford University Press).

Irwin, Aisling (2001) 'Wild at Heart'. *New Scientist* (3 March 2001): 28–31.

Isaac, E (1970) *Geography of Domestication* (Englewood Cliffs, NJ: Prentice Hall).

Jablonka, Eva, and Marion Lamb (2005) *Evolution in Four Dimensions: Genetic, Epigenetic, Behavioral and Symbolic Variation in the History of Life* (Cambridge, MA: MIT Press).

Jones, Oliver Graham (2002) *Zoo Tails* (London: Bantam Books).
Kaplan, Matt (2002) 'The Plight of the Condors'. *New Scientist* (5 October 2002): 34–6.
Kellert, Stephen R., and Joyce K. Berry (1981) *Knowledge, Affection and Basic Attitudes Toward Animals in American Society* (Washington, D.C.: US Government Printing Office).
Kleiman, Devra, Mary E. Allen, Katerina V. Thompson and Susan Lumpkin (eds) (1996) *Wild Animals in Captivity: Principles and Techniques* (Chicago: University of Chicago Press).
Kreger, Michael D., Michael Hutchins and Nina Fascione (1998) 'Context, Ethics, and Environmental Enrichment in Zoos and Aquariums', in *Second Nature: Environmental Enrichments for Captive Animals*, D. Shepherd, J. Mellen and J. Hutchins (eds) (Washington DC: Smithsonian Institution Press).
Kuczai II, Stan A., C. Thad Lacinak and Ted N. Turner (1998) 'Environmental Enrichment for Marine Mammals at Sea World', in *Second Nature: Environmental Enrichments for Captive Animals*, D. Shepherd, J. Mellen and J. Hutchins (eds) (Washington DC: Smithsonian Institution Press).
Kuroda, Suehisa (1997) 'Possible Use of Medicinal Plants by Western Lowland Gorillas (*G. g. gorilla*) and Tschego Chimpanzees (*Pan. t. troglodytes*) in the Ndoki Forest and Pygmy Chimpanzees (*P. paniscus*) in Wamba. http://www.shc.usp.ac.jp/kuroda/medicinalplants.html [11/01/05].
Laporte, Joseph (2004) *Natural Kinds and Conceptual Change* (Cambridge: Cambridge University Press).
Lee, Keekok (1969) 'Popper's Falsifiability and Darwin's Natural Selection', *Philosophy* (1969): 291–302.
——. (1989) *Social Philosophy and Ecological Scarcity* (London: Routledge).
——. (1997a) 'Biodiversity', in *The Encyclopedia of Applied Ethics*, Vol.1, ed. Ruth Chadwick (San Diego: Academic Press), pp. 285–304).
——. (1997b) 'An Animal: What is it?'. *Environmental Values* 6 (1997b): 393–410.
——. (1999) *The Natural and the Artefactual: The Implications of Deep Science and Deep Technology for Environmental Philosophy* (Lanham, MD: Lexington Books/Rowman and Littlefield).
——. (2000) 'The Taj Mahal and the Spider's Web', in *Ethics of the Built Environment*, Warwick Fox (ed.) (London: Routledge).
——. (2004) 'There is Biodiversity and Biodiversity', in *Philosophy and Biodiversity*, ed. Markku Oksanen and Juhani Pieterinen in the series *Philosophy and Biology*, series ed. Michael Ruse (Cambridge: Cambridge University Press).
——. (2005a) *Philosophy and Revolutions in Genetics: Deep Science and Deep Technology*, 2nd edn (Basingstoke: Palgrave Macmillan).
——. (2005b) 'Is Nature Autonomous?', in *Recognizing Nature's Autonomy*, Thomas Heyd (ed.) (New York: Columbia University Press).
MacFarland, D. (1981) *The Oxford Companion to Animals* (Oxford: Oxford University Press).
Markowitz, Hal (1982) *Behavioral Enrichment in the Zoo* (New York: Van Nostrand Reinhold).
Maturana, Humberto, and Francisco Varela (1980) *Autopoiesis and Cognition: The Realization of the Living* (London; Dordrecht: D. Reidel).
Mayr, Ernst. (1982) *The Growth of Biological Thought: Diversity, Evolution and Inheritance*. (Cambridge, MA, and London: The Belknap Press of Harvard University Press).

——. (1988) *Toward A New Philosophy of Biology: Observations of an Evolutionist* (Cambridge, MA, and London: The Belknap Press of Harvard University Press).

Meek, James (2001) 'Cage Life May Drive Lab Animals So Insane that Experiments are Invalid'. *The Guardian* (28 August 2001). http://www. guardian.co.uk/uk_news/story/0,,543234,00.html [03/02/2005].

Mellen, J, and M. S. Macphee (2001) 'Philosophy of Environmental Enrichment: Past, Present and Future'. *Zoo Biology* 20 (2001): 211–26.

Mellen, Jill, Marc P. Hayes, and David J. Shepherdson (1998) 'Captive Environments for Small Felids', in *Second Nature: Environmental Enrichments for Captive Animals*, David Shepherdson, J. Mellen and J. Hutchins (eds) (Washington DC: Smithsonian Institution Press).

Morgan, Kathleen N., Scott W. Line and Hal Markowitz (1998) 'Zoos, Enrichment and the Sceptical Observer', in *Second Nature: Environmental Enrichments for Captive Animals*, D. Shepherd, J. Mellen and J. Hutchins (eds) (Washington DC: Smithsonian Institution Press).

Mullan, Bob, and Garry Marvin (1999) *Zoo Culture*, 2nd edn (Urbana and Chicago: University of Illinois Press).

New Scientist (2004) 'Noah's Freezer' (31 July 2004): 5. http://www.newscientist. com/article.ns?id=mg18324580.700 [10/02/05].

Norton, Bryan et al (eds) (1995) *Ethics on the Ark: Zoos, Animal Welfare and Wildlife Conservation* (Washington and London: Smithsonian Institution Press).

Olney, P. J. S., and Fiona A. Fisken (eds) (2003) *Zoo Challenges: Past, Present and Future* (*International Zoo Yearbook*, Vol. 38) (London: The Zoological Society of London).

Père David deer. http://www.animalinfo.org/species/artiperi/elapdavi.htm [09/02/05].

Plotkin, Mark J. (2000) *Medicine Quest* (New York: Viking).

Polar Bears. http://www.polarbearsinternational.org [01/12/04].

——. http://www.nwf.org/wildlife/polarbear/[05/01/05].

——.http://www.seaworld.org/infobooks/PolarBears/home.html [05/01/05].

Poole, Trevor B. (1998) 'Meeting a Mammal's Psychological Needs', in *Second Nature: Environmental Enrichments for Captive Animals*, David Shepherdson, J. Mellen and J. Hutchins (eds) (Washington DC: Smithsonian Institution Press).

Price, E. O. (1984) 'Behavioural Aspects of Animal Domestication'. *Quarterly Review of Biology* 59 (1984): 1–32.

——. (2002) *Animal Domestication and Behavior* (New York: CABI Publishing).

Rachels, James (1991) *Created From Animals: The Moral Implications of Darwinism*. (Oxford and New York: Oxford University Press).

Ramesh, Randeep (2004) 'Cloning breeds hope for India's big cats'. *The Guardian* (18 August 2004). http://www.guardian.co.uk/international/story/0,,1285280,00. html [09/02/05].

Regan, Tom (1983) *The Case for Animal Rights* (Los Angeles: University of California Press).

Rockcliffe, Wendy, and Graham Robertson (2004) 'Emperor Penguins: Winter Survivors'. http://www.aad.gov.au/default.asp?casid=3524 [01/12/04].

Rolston, Holmes, III (1988) *Environmental Ethics: Duties to and Values in the Natural World* (Philadelphia: Temple University Press).

Rose, Steven P. R. (1997) *Lifelines: Biology, Freedom, Determinism* (London: Penguin).

Routley (Sylvan), Richard (1973) 'Is There a Need for a New, an Environmental Ethic?'. *Proceedings of the XVth World Congress of Philosophy* (1973): I, 205–10.

Ruse, Michael (1988) *Philosophy of Biology Today* (New York: State University of New York Press).

170   *Zoos*

Sample, Ian (2004) 'Frozen Ark to Save Rare Species', in *The Guardian* (27 July 2004). http://www.guardian.co.uk/uk_news/story/0,1269747,00.html [10/02/05].
——. (2005) 'What Is This Woman Doing?'. *The Guardian*, Life Section (7 April 2005).
Shepherdson, D. J. (2005) 'Environmental Enrichment'. http://www.oregonzoo.org/ ConservationResearch/environm.htm [16/02/05].
——. (2001) *A Guide to Improving Animal Husbandry Through Environmental Enrichment* (June 2001). http://zcog.org/zcog%20frames/A%20Guide% 20for% 20Improving%20Animal%20Husbandry%20Through% 20Environmental% 20Enrichment/Guia%20Enriquecimiento--Oregon%20Zoo.htm [16/02/05].
——. (2003) 'Environmental Enrichment: Past, Present and Future', in *Zoo Challenges: Past, Present and Future* (*International Zoo Yearbook*, Vol. 38), ed. Olney and Fisken (London: The Zoological Society of London).
Shepherdson, D., J. Mellen and J. Hutchins (eds) (1998) *Second Nature: Environmental Enrichments for Captive Animals* (Washington DC: Smithsonian Institution Press).
Siipi, Helena. (2005) *Naturalness, Unnaturalness, and Artifactuality in Bioethical Argumentation*. (Turku, Finland: Department of Philosophy, University of Turku).
Singer, Peter (1976) *Animal Liberation: Towards an End of Man's Inhumanity to Animals* (London: Jonathan Cape).
Smith, J. C., and Brian Hogan (1996) *The Criminal Law* (London: Butterworth).
Stevens, P. M. C., and E. McAlister (2003) 'Ethics in Zoos', in *Zoo Challenges: Past, Present and Future* (*International Zoo Yearbook*, Vol. 38) ed. P. J. S. Olney and Fiona A. Fisken (London: The Zoological Society of London).
The European Union (EU) Zoos Directive (1999). http://europa.eu.int/eurlex/pri/en/oj/ dat/1999/l_094/l_09419990409en00240026.pdf [03/02/05].
The Guardian (2005) 'Sunbed for Rhinos at Dutch Zoo' (12 February 2005): 18.
Tudge, Colin (1992) *Last Animals at the Zoo: How Mass Extinction Can Be Stopped* (Oxford: Oxford University Press).
Ward, David (2005) 'Parlez-vous baboon? Zoo Keepers Bridge Gap'. *The Guardian* (22 January 2005).
Weilenmann, P., and E. Isenbrugel (1992) 'Keeping and Breeding the Asian Elephant at Zurich Zoo', in *The Asian Elephant: Ecology, Biology, Disease, Conservation and Management*, E. G. Silas, Krishnan Nair, M. and G. Nirmalan (eds) (Kerala, India: Kerala Agricultural University).
Weisse, Robert J., and Kevin Willis (2004) 'Calculation of Longevity and Life Expectancy in Captive Elephants'. *Zoo Biology* 23, 4 (August 2004): 365–73. http://www.aza.org/Newsroom/PR_elephantlonglives/ [12/12/04].
Wilkerson, T. E. (1995) *Natural Kinds* (Aldershot: Avebury).
Wilson, E. O. (1994) *The Diversity of Life* (London: Penguin).
The World Zoo Conservation Strategy (1993) sponsored by The World Zoo Organization, The Captive Specialist Group of The World Conservation Union's Species Survival Commission.
Worstell, Carlyn (2003) *Reconciling User Needs in Animal Exhibition Design: Gorilla Exhibits as a Case Study* (ZooLex Zoo Design Organization). http://www. zoolex.org/publication/worstell/gorilla/content.html [revised 14/01/2004].
Wright, Sue (2001) 'What's it like to hand-rear a wild animal?', *Best* (10 July 2001).
Young, Robert (2003) *Environmental Enrichment for Captive Animals*. (Oxford: Blackwell).

Index

To

Donna &

Mark +

a true courage

Survivor!. +

Stay Strong +

Approach

each situation with

courage, Strenght &

Positivity, you will

Discover New Opportunities

for yourself!.

Happy Joyful Reading

Olga Smaly
et.

The *Joy* Of CANCER

Praise for *The Joy of Cancer*

The Joy of Cancer will go well beyond inspiring cancer patients; it has dealt with an amalgam of complex emotional, psychological and social issues, which most people have experienced but have not had the courage to express! It will give readers hope, courage and most of all, the feeling that change is possible, regardless to how insurmountable it may seem.

—Sonia Djevalikian
division head, Kirkland Library

The Joy of Cancer is a deeply inspiring story of empowerment and personal transformation. Even when faced with one of the greatest adversaries known to man, the deadly disease of cancer, Olga reminds us that we have the power to choose how we respond. If illness is an invitation from the body to change, *The Joy of Cancer* shows us how to transform: with faith, openness, trust, determination, perseverance, and a positive outlook. I am certain this book will provide that much-needed light in what can be a lonely and terrifying dark time for those diagnosed with any serious illness.

—Maria Amore

The Joy of Cancer recounts how one woman's quest for happiness and fulfillment was unexpectedly realized through her cancer passage. Olga takes you on an intimate journey of self-discovery and hope. Her frank and honest storytelling coupled with little pearls of wisdom on the impact of a positive attitude creates a compelling and inspiring story.

—Sylvia Rabinovitch

After being diagnosed with cancer, Olga takes charge of her life and refuses to see her glass as half-empty. Olga teaches us to live one day at a time and to appreciate all that life has to offer. This is an inspiring story of one woman's remarkable journey of self-awareness. My outlook on life is changed after reading this book. *The Joy of Cancer* will inspire you to take a closer look at your life.

—Kathy Sassano

The universe handpicks certain people to share their gifts to inspire others. Olga is one of these people. *The Joy of Cancer* illustrates how cancer, with the right attitude, does not have to be a death sentence. The courage to share her story has resulted in a book that is both moving and inspiring.

—Janice Auger

Riveting, honest, raw with emotion, this book gives us a day-by-day description of living with a cancer diagnosis. Brave, strong, and determined, Olga inspires us with her fighting spirit as she faces and conquers the challenges this brings. A motivational must-read for all of us.

—Maria Varvarikos
head librarian, Lower Canada College

I was moved to tears every chapter! Kim Mecca's heartfelt account of Olga's brave bout with cancer was a book I couldn't put down. It left me feeling inspired and hopeful. Opportunities for change and a better life show up in strange and scary packages. Olga unraveled hers to find a life of beautiful gifts! Thank you, Kim, for taking me along on Olga's uplifting journey.

—Vered Haiun

OLGA MUNARI ASSALY
& KIM MECCA

The *Joy* OF
CANCER

A Journey of Self-Discovery

iUniverse, Inc.
Bloomington

The Joy Of Cancer
A Journey of Self-Discovery

iUniverse books may be ordered through booksellers or by contacting:

iUniverse
1663 Liberty Drive
Bloomington, IN 47403
www.iuniverse.com
1-800-Authors (1-800-288-4677)

ISBN: 978-1-4759-5194-3 (sc)
ISBN: 978-1-4759-5195-0 (hc)
ISBN: 978-1-4759-5196-7 (e)

Library of Congress Control Number: 2012917888

Printed in the United States of America

iUniverse rev. date: 10/29/2012

I dedicate this book to my loving and devoted husband, Leonard. You are my best friend and an outstanding father to our children. You have always supported me and continue to, in all that I choose to endeavor. You understand me like no other person can. Our strong connection unites as one. With each day that passes, I learn to trust you more and more, and my love for you continues to grow. We stood by each other through many years of struggle. I am so lucky to have a man who loves me unconditionally and stands by me steadfastly. I thank you from the bottom of my heart.

—All my love, Schmoo

xox

To my children—Andrew, Laurie, and Jeremy—for being strong and standing by me in difficult times. Your smiles, your encouragement, and your joie de vivre touch every aspect of my life.

You are the reason I am who I am today

—Love, Mom

For Jade

all of it, every moment till the end, was a
labor of love without any labor
the wisdom you brought into my life
I could live another lifetime and not amass as much

you could tell me a million times I was your blessing
but you are mine

may I continue to be a blessing
and may our souls remain intertwined

if I am truly blessed
in my next life we will play in sandboxes
on swings
in open grassy fields near giant willow trees
again, as sisters
in spirit, soul, and blood

in the meantime
may the cracks in your heaven continue to be big
enough for me reach through and hold your hand
and long enough for our voices to echo back and forth
and deep enough to match the love held
deep in my heart and yours

that we may continue to be a testimony
to the art of friendship
the art of life
the art of love

—Sprout

Contents

Foreword by Kim Fraser

When Olga first walked into my studio, it was difficult to believe that she was battling breast cancer. She was full of energy and infinitely more nervous about being on the radio than about fighting a disease that kills almost forty-five thousand North American women a year. I soon learned that she was neither naïve nor in denial. After a brief tearing-up as she described telling her family about her illness, she very quickly made me understand that to her, cancer was a challenge that left her little time for self-pity. Instead, she intended to roll up her sleeves and face it head-on. As one of her "butterflies" for the Weekend to End Breast Cancer, I watched as Olga, who was still in treatment, discovered a newfound ability to motivate and lead others. These are qualities she continues to develop and share today, using her cancer not only to change her own life but also to help change the lives of those around her. I have interviewed many cancer survivors over the course of my broadcasting career but have never come across someone like Olga. Like the butterfly she so loves, she has been transformed by her cancer into a woman whose life has deep meaning and who faces the future with confidence and joy. Her journey is an inspiration to us all—and I, for one, feel honored that she chose to share it. I hope you will as well.

Kim Fraser
Host, *The Kim Fraser Show*
CJAD 800, Montreal, Canada

Foreword by Jennifer Campbell

In this life, we all have our fair share of hardships. But what ultimately marks the value of a person is how he or she chooses to deal with them. When faced with the greatest challenge of her life—a harrowing cancer diagnosis—**Olga Munari Assaly,** devoted wife, mother, daughter, friend, and activist, chose to do everything in her power to turn her adversity into a positive—a life-altering experience with the potential to unite and strengthen, rather than desecrate and tear down. Through an extraordinary journey fraught with peaks and valleys, Olga managed to find the joy in cancer, and with this very honest and real account of her personal life story, she is an inspiration to us all. G-d bless.

Jennifer Campbell, social columnist, *The Montreal Gazette,* and publisher/editor-in-chief, *Diary of a Social Gal*

A Note from Olga

I want to stress the importance of early detection because it really *is* the ultimate lifesaver. My journey was wonderful, and I would not take back one minute of it. Nonetheless, going for a mammogram, as uncomfortable as it is, is much less painful than the process of chemotherapy. Don't let fear of the mammogram—or worse, fear of the diagnosis—stop you from getting screened.

My wish for this book is that it will help you get through the rough times and see the light at the end of the tunnel. You can embrace the unexpected and step outside of your comfort zone. You can learn to become comfortable with uncertainty and difficult times. You can face your worst fear and make good come from it. More importantly, you do not have to be a victim of circumstance; but rather, you can choose to rise above, as I did with my cancer diagnosis and create a more empowering story for yourself.

Use your challenges as an opportunity to love and accept yourself and the people around you. Our challenges are a doorway to opportunity and love if we allow ourselves to be open to change. You may be pleasantly surprised to find that you indeed have been granted a second chance at life. Most of all, I hope that you will learn to appreciate what life has given you, and make the best of it.

I believe that there are no coincidences in life. The reason, *my* reason, for getting cancer is becoming clearer and clearer: I want to help people. I have always wanted to do so, and now more than ever! I used my cancer as an opportunity to do that on a higher level than I had ever done.

At the end of the day, I wrote the book because I wanted to help one more person. That person may be you. If that is the reason I got cancer, then my journey, and this book, will have served its purpose.

With love,
Olga

Preface

Olga's story is, to say the least, an inspiring one. It attests to how a potentially deadly diagnosis can make you feel more alive than you ever have been—free to be whoever you want to be, to live however you want to live, and to do things you hardly even dreamed you could. It's not about sickness and struggle. It's about life. It's about transformation and, believe it or not, joy. It's a story that will transform your notion of cancer and shed light on the gifts and opportunities hidden (not so deeply) in something most people consider to be devastating.

Olga's story is about the many opportunities that lie within the experiences we believe to be our greatest challenges. These challenges come bearing gifts if we are willing to be open to them. They come with the potential to change us, to change our lives, and even to reach far beyond that and help us change other people's lives. They are merely instruments serving a purpose greater than we can imagine. And what you will find in this story is a clear wisdom that demonstrates that the attitude we adopt in the face of these challenges is the key that unlocks the door to all that potential.

Kim Mecca

Acknowledgments

From Olga

Kim, from the first day we met, you were like an angel who came from the sky, and I knew that you were the right person for my book. We connected in ways that only you and I can understand. You made my dream come true. Together, we worked as a team. I am so thankful for our new friendship. You are so special in your own way; your caring, commitment, and sensitivity helped me through our journey. You never judged, you always grasped what I meant (even with my broken English and my sometimes inaccurate French translations), and you made it happen! I commend you for your dedication, hard work, and willingness to commit yourself to the book. I truly enjoyed our journey together, and I am blessed for having met you. Through our collaboration, you recreated my "joy of cancer" in a way that will help many others. I am certain that this is just the beginning for you. Congratulations on a job well done! Kim, you are awesome!

To all my wonderful friends and family, I am grateful that you helped me through my journey of self-discovery—the time when I needed you the most.

To all the members of my "Olga's Butterflies" team: Together we walked with pride and made history! We raised money for the cause in memorable ways. All of you are the reason I was so driven to survive and to continue to fight. Your constant support gave me courage, hope, strength, and the joy of living. I could not have done it without all of you; you have been my pillars of strength.

I would like to thank my very caring doctors and nurses for their commitment and dedication to the Jewish General Hospital and the well-being of all their patients. Because of you, I now have a second chance at life and so do many others. You are lifesavers!

I would like to extend a special thank-you to my Lower Canada College family. Your heartfelt support meant the world to me. The school has been my home away from home. It has been a wonderful playground for me to come out of my cocoon and acquire invaluable hands-on experience as a volunteer. My time with you has allowed me to grow and has served me well in all the fundraising endeavors that I have since taken on.

Thank you to my "partner in crime," Kathy Sassano, who has always been a phone call and an e-mail away. You gave me all the support that I needed.

Teresa Nash, thank you for being my confidante, my friend, and the one who always made me look at things clearly. Your calmness helped me to slow down a little.

To Stephanie Rossy-Beauchamp, thank you for guiding me to the right medical staff at the very beginning of my journey.

Sonia Djevalikian, you have been there since the day I was first diagnosed and you continue to stick by me. Thank you for all your support and for believing in this book. I am so grateful that you have chosen me to give my first public appearance as the author of *The Joy of Cancer.*

To our interviewees, Francine, Debbie, Stephanie, Anna, Joan, Mary, Teresa, Beth and Kim, thank you for your time and more importantly, your invaluable contributions. Thanks to Sylvia Rabinovitch, Maria Varvarikos, Janice Auger, Beth Stutman, Kathy Sassano, Jennifer Campbell, and Kim Fraser for reading the book and giving their very touching testimonials and continuous support.

Isabelle Paradis, I am so happy that our paths crossed. You immediately understood the vision that I had for the book cover and you put your ingenious creativity to work. The end result is magical. Thank you for creating such an outstanding and striking book cover. Both Kim and I absolutely love it!

To Abria Mattina, thank you for your editorial contribution.

Ronald Arceo, you recognized the power in my story from the get-go, and you allowed me to touch more lives by having me speak at Tedx the Human Revolution. That was my biggest speaker accomplishment yet, and from there, I have never looked back. Through your brave endeavor, you brought Virlane (the missing piece of the puzzle) to Olga and me.

Virlane Torbit, when we spoke on the phone for the first time, there was an instant connection. When I say that there are no coincidences in life, this is the perfect example. There I was at Tedx, and you were watching me. Little did I know that the next day you would pay your time, passion, expertise, and love forward to edit my book. Thank you!

Most of all, thank you to my mom and dad for bringing me into this world. Without you both, I would not be here today. Thank you for believing in me and all that I can accomplish.

Thanks to my mother-in-law, my sisters-in-law, and my brothers-in-law for their helpfulness and their goodwill.

To my sister, who is like a daughter to me—I love you, Anna. I made it!

To my brother, Mario, you are always on my mind and in my heart. xox

Finally, saving the best for last, Leonard, thank you for your contribution to the book. You helped us to connect the dots with your insightful ideas. Kim and I truly appreciate the wisdom you shared with us.

From Kim

Beautiful and beloved Olga, you have believed in me and my creative voice from the very beginning. You believed in this miraculous connection and how all the pieces would fit so snugly together. It is your faith that allowed us to make this dream come true. This collaboration has birthed a very special friendship that was founded in values that are true to the core of who I am: authenticity, contribution, expansion, and best of all, love. There are not enough words or heartfelt thank-yous to express my gratitude. The process of creating this book has been a blessing for the direction of my life, and I am forever grateful to you.

Olga and Leonard, you have changed my life. My time with you has been memorable from start to finish. Thank you for welcoming me into your home and private lives with such warmth, and for bravely putting your trust in me with the very heart of your personal journeys. By allowing me to carry out this project and be an intimate part of your lives, you have allowed me and many others to discover how much more is possible in our lives.

Virlane Torbit, your contribution to this project has been a guiding light for me. You were not only our editor but also a source of wisdom, strength, and clarity. When I could not tell the forest from the trees, you were a source of great comfort and peace. Indeed, you have been my own very special witness, watching this project come to its full fruition with bountiful love and encouragement. I look forward to greeting you in person one day with a warm hug and a grateful heart.

Mom and Dad, thank you for your unconditional belief in me, for celebrating every milestone and honoring my work. Mom, you have been my best friend, my confidante, my pillar of strength, a fountain of wisdom, a source of comfort, and my biggest cheerleader. At the moments when I was exhausted and discouraged by what still lay ahead of me, you helped me to stay focused and continue seeing the light at the end of the tunnel. Your love, support, and belief in me, and everything you have taught me, have been vital to the realization of this book. I thank you and love you with all of my heart.

Riccardo and Grace, thank you for bravely and warmly opening your hearts to me so early in your grieving journeys and for your loving contribution to this work. Riccardo, thank you for allowing me to honor Jade by telling some parts of her journey; it has been a vital part of my own grieving process. Your loving ways continue to touch my life and the lives of others as if it were Jade herself.

And to the rest of my family—my brother, sister, and uncle—thank you for sharing in my joy and celebrating this wonderful accomplishment with me. Fabio, my beloved cousin and kindred spirit from so far away, you are close to my heart for honoring me, celebrating me, believing in me, and showering me with your love and appreciation.

Vered and Maria, thank you for your time, support, and unconditional love. Vered, for the countless times I called on you to bounce ideas off you, for celebrating me and nudging me with love and confidence into the spotlight; and Maria, for your unconditional belief in me, your faith that never tires as you help me to remember who I am at the moments when it feels so out of reach. You have both been angels who have encouraged me to shine my light brighter and brighter.

Paula Engels, you have empowered me through your coaching. You have taught me about value, and you have helped me to build a toolbox overflowing with resources to make the experience of writing this book wonderful from moment to moment. Your patience, skill, and love are unique gifts to this world, from which I have been lucky to benefit.

Ronald Arceo, our special connection is a blessing in my life. You have often seen the potential in me before I even recognize it in myself. You are a visionary who has held a strong vision for me, for Olga and for this book. You and James Archambault have been a source of strength for me over the last two years. I carry your unwavering love and depth of presence within my heart, wherever life takes me. Thank you to both of you.

Patty, thank you for your genuine interest and for always taking a moment to ask how things were going. Such a small gesture had a huge impact on my journey. Thanks to Caroline (Pebbles) for reading me and celebrating me and sharing in my reverence for love, relationships, and writing. Anna, thanks for recognizing that I don't fit the mold and for nudging me away from it and toward my own path since the beginning of time. Teri, thank you for sharing my joy and for letting me hash things over with you the way writers need to. Thanks to Piero, for believing in me and honoring me from the first day we met and every day that followed; Laura, her husband Anthony, and my beloved kindred spirit, Deryk for celebrating all of my accomplishments, for their steadfast love, and for their belief in me; and Lynn Walden, Nathalie Pequet, and Antonella for helping me make decisions that brought me to this opportunity and for being living threads that allow me to stay connected to all that Jade had taught me. Steven, thank you for celebrating every small step for the huge leaps that they were, and mostly for seeing me, my light, and my love when I was blind to my own self.

Walter, thank you for your ongoing support so early in our friendship and for helping me crack the cancer synthesis line— something I had been working on for over a year. Thank you to Karine, for putting up with me and for empathizing with me day after day, and Julie, for your strong support, encouragement, and infallible belief in me. My dragon-boat girls have my gratitude for celebrating my guts. Suzana, thank you for celebrating this accomplishment with me, and Elmi, thank-you for making me dance during the last leg of this journey, when I needed it the most. Dominique, thank you for being one of my biggest fans from as far back as I can remember, and Daniel, thank you for helping me shape the foundation of this book and lay the building blocks of each chapter.

My gratitude and appreciation go out to *all* the people who believed in me and in the purpose of this book. Your contributions were invaluable to the realization of *The Joy of Cancer.* We could not have done it without you all. To all the old friends and new friends and acquaintances I have made along the way—you have touched my life by helping to make this book a reality. Thank you to each and every one of you for touching Olga's and my life. You are the reason for *The Joy of Cancer.*

Finally, Olga and I would like to thank Vered for being the force of love that connected us. This incredible collaboration would not have been possible without you, Vered. That small seed you planted by reaching out to initiate a friendship with me is blossoming into a beautiful garden of love and opportunity. You always wanted Olga to write her story; you pushed her and connected her and brought us together to create the perfect team to write this book. You are a big part of making Olga's dream come true. Thank you!

Chapter 1

The Butterfly Effect:
The Joy of Beginning

The silence was deafening as my mind reeled. I made the call at 6:20 a.m. after a message from a close friend indicated to call her urgently. The ringtone roared like thunder in my ears against the morning hush. When she finally answered, I spoke in my groggy morning voice. "What do you want? It's six o'clock on a Monday morning. What could possibly be so urgent?"

"Kim, someone is interested in your writing," my friend Vered replied.

"Come again?"

"I know someone who is interested in your writing skills," she repeated. "A longtime friend and client has wanted to write a book for a few years now, but she doesn't write. We got into a conversation last night, and I spoke to her about you. She wants to meet you this Saturday."

Silence.

"Vered?" I said slowly. "Are you serious? What do you mean?"

"I mean she wants to meet you, and she may want you to write her book."

More silence.

It was the Monday morning after I had turned down the job of my dreams. Writing was a lifelong hobby for me, and I worked as an IT systems consultant to support myself. I was successful and enjoyed my day job, but I still turned to writing in my free time as the most natural outlet for my creativity. Writing allowed me to feel fulfilled. I had a flash of wisdom as to why I had felt in my gut that I had to stay in my current position and decline the corporate opportunity of a lifetime. Everyone around me thought I was crazy to step away from a career opening that would take me to glamorous spots like L.A., New York, and Miami on a regular basis. I could hear it in their voices, pretending to understand and support my choice. "Good for you!" many told me, but their tone suggested the words were followed by three little dots rather than three huge exclamation marks.

As much as I agonized over making the decision, in the end I blindly trusted my gut. And now, not even seventy-two hours later, I possibly had the answer to why I'd had that nagging "don't do it" feeling from the start. My current job was demanding and involved a lot of responsibility, but taking on that new position would have been even more demanding and would have made it much more difficult to commit to a major writing project. I had been tinkering with the idea of writing a book for a few months, but after this brief exchange with Vered, I wondered if I was meant to write someone else's book instead.

I didn't know much about Olga, except that she was a

breast-cancer survivor who, according to Vered, had quite an interesting story and had been wanting to write a memoir for some time. Vered believed my writing skills would be a great match for Olga's story. I quickly found Olga herself to be an incredibly interesting character, and I came to understand later that it was the way Olga participated in her story that had made her journey so remarkable.

Initially, the coincidental timing of our paths crossing intrigued me. Not only was it a few days after the decision I'd been agonizing over for months, but more significantly, I had lost my best friend, Jade, to breast cancer four months earlier. Jade was a remarkable woman who left a wealth of wisdom to her friends, family, students, and clients. The many who were blessed to cross paths with her were never the same. After my first meeting with Olga, it was clear that she had the same impact on people, and this made me want to help her realize this project.

Jade had wanted to write the book of her journey with cancer, except that her journey ended before she had time to leave behind a written legacy. I helped her edit her first chapter less than a year before she died. I remember leaning over Jade's bed, holding her hand as she took her last breaths, thinking, *I'm almost certain this wasn't what you had in mind as the ending for your book, now, was it?* I saw working with Olga as an opportunity to pay homage to Jade, to Olga, and to all the other women out there who were handed a ticket on the big C journey and traveled it in their own way.

Vered set up a meeting for the following Saturday. Olga would come to Vered's morning Zumba class, and then the three of us would have lunch together. On Saturday morning, I walked into the dance studio and found Olga standing by the reception counter talking with Vered. She had short blonde hair and sparkling blue eyes. I remembered immediately noticing how healthy she looked and how beautiful she was. If cancer had taken any toll on her physically, there was no sign of it in her appearance. She was tall with a slim figure, and her skin glowed. Dressed in black workout pants with a matching top, she seemed like a simple, down-to-earth woman.

After the class ended, the three of us made small talk for a few minutes before zipping over to our favorite spot, Le Commensal, for vegetarian eats. It was a damp, nippy day, but the sun was shining brightly. On that cold January morning, I was warmly invited to be privy to this woman's personal journey. Over the next hour, Olga recounted the story of how she came to be diagnosed with breast cancer and what happened in the months that followed.

I couldn't judge her reaction when I told her that I had lost one of my closest friends to breast cancer just four months earlier. I quietly wondered if this made her feel uncomfortable. It was the last thing I intended. Later, Olga told me that not only had she not taken offense, she believed that it was Jade who brought us together and was helping us bring this project to fruition. According to her, it was no coincidence that I had met our mutual friend, Vered, in early December, just a little over two months after Jade's passing.

And then it happened: the moment I knew for sure that this was right. Olga was telling us about how the butterfly had been her chosen symbol throughout her journey. She talked about butterflies the same way I'd heard so many others do over the past two years. "The butterfly is about transformation and rebirth," she said as our eyes met. "They are very special to me." The spoon I was using to eat my tapioca pudding stuck to the roof of my mouth, and then my jaw dropped as I made the connection.

For the last year or two, butterflies had been following me everywhere I went, and I was not able to understand why until that day. They were in the bookstore when I turned around; they were giants on the walls when I walked into church for the first time in several years; they were papier-mâché in a little boy's hand at a yoga studio where I practiced; and finally, they were Olga's token throughout her journey. As I shared my experience, all three of us were covered in goose bumps.

It was a sign.

After the butterfly coincidence, it was decided that I would be the torchbearer for Olga's story.

Chapter 2

Something Bad Happened:
The Joy of Murphy's Law

For our second meeting, Olga warmly received me in her home. Over a savory homemade lunch, she shared the ins and outs of the beginnings of her cancer journey. We sat at her dining-room table, talking as we nibbled. Olga began, "It was early morning, and I was at the gym talking to my friend and personal trainer, Janet."

The sound of the treadmill barely covered the sobs that Tuesday morning—March 28, 2006. "I was crying," Olga recounted. "Leonard and I had an argument the day before. It was the current routine. We were either disagreeing about the construction we were doing on the house, about work, or about the kids. I felt overwhelmed and confused. My life felt out of control. We had been doing construction work on the house for almost a year, which is an eternity to be living in that kind of chaos day to day."

I silently agreed with her. I could barely handle the mess my dad made putting up a set of shelves in my bedroom. I pictured a small team of men tearing out kitchen tiles, ripping out flooring, and breaking down walls, and I could imagine how difficult it was to entertain a renovation party for so long.

"I was chair of the school store at Lower Canada College (LCC), where my children attended," Olga continued. "I had been chair for seven years. I wanted to step down, but I was paralyzed by my inability to say no and afraid of disappointing all the people who had become my friends there. I just kept it up, even though I knew it was time for me to let it go."

As I listened to Olga go on, I nodded in silent agreement, thinking of the many women who must be in similar situations—feeling overwhelmed, burdened, or ready for some kind of change, but unable to take action for fear of disappointing the people they care about. It is baffling to find that for many people, it can take the diagnosis of a life-threatening illness to develop the courage to affirm their own needs.

"The kids were, for the most part, normal kids," Olga explained to me. "Andrew, at sixteen, was immersed in his studies, planning for his graduation party and college—his next chapter. Laurie, at fifteen, was my princess, a little social butterfly, now diving headfirst into the world of boys. And Jeremy, my youngest, at nine, was Momma's baby. He was in grade four and had just been diagnosed with attention deficit disorder. I battled with myself, trying to understand what I had done to create this problem for him. I worked hard to get Jeremy the support he needed. We were going to weekly appointments with a psychologist who specialized in treating ADD, on top of my weekly sessions with my own therapist to try to get some order in my life.

"We were not spending a lot of quality family time together.

The kids all had their own bedrooms equipped with plasma TVs, granted by their tech-savvy dad. And so, between being wrapped up in their normal day-to-day childhood life and having their own personal entertainment suites, who ever saw them? I couldn't say no, and I couldn't set boundaries. I couldn't just be myself without the fear of not being liked if I said no to someone or something. I felt like I had to take everything on, and it was breaking me. I was frustrated, and the plasma TV at the gym—ironically mounted above the treadmill I was standing on—only frustrated me more. 'I'm all over the place!' I told Janet."

Olga's facial expression, as she recounted this to me, reflected the frustration she felt at the time. She was being worn thin, bearing the physical and emotional toll of everything that was going on. She was often very tired, and her patience had run dry. "I was running around like a chicken without a head, totally spent at the end of the day and not feeling fulfilled. Len was working the required long hours for the huge IT changes at work. We weren't getting much quality time together as a couple. Our marriage was strained by the demands of a very assiduous everyday life. I felt disconnected from him and alone. I wasn't getting any quality time with the kids either. I felt like I was always being set aside. My life was full, yes! But at the end of the day, I felt like I was coming up short, like I wasn't enough. I was doing everything, running here and there, taking the kids everywhere, food shopping, cooking, construction, doctors, volunteering … I felt pulled in a million directions without any real purpose. I was not happy with what I was doing and how I was living. I thought, *This is not what I want! This is not what I signed up for!*"

Olga's response to her feelings of being overwhelmed and unhappy was born out of desperation: "I wished that something bad would happen to me, really bad, like getting sick, to try to get my family close again, or to see how Leonard would react." Janet, of course, had been appalled at Olga's words. She spat on the floor, in the typical Jewish custom, to try to erase the bad omen of the words that had just been spoken. It was useless.

The following Monday, Olga was diagnosed with breast cancer.

I thought of all the things I say in my own day-to-day life. As a writer, I have a natural tendency to be extra-spirited, sometimes even dramatic, in my self-expression. I can, and do, use the phrase, "I would die if that happened!" in response to a variety of experiences, whether joyful, embarrassing, exciting, or frightening. The language we use has a huge impact on how we experience life. *What if it did happen, and I would die?* I thought to myself. Isn't that what Olga did? She voiced the need for something of impact to happen, and it did. I was hooked. I needed to know more about Olga's journey and was eager for her to continue sharing how her story unfolded.

Two years prior to Olga's diagnosis, her gynecologist had sent her for her first mammogram. It was a routine checkup. "Unless you have history of breast cancer in your family, most doctors don't send you for a routine mammogram until the age of fifty," Olga explained. "My doctor insists on sending her patients when they are forty." Olga's results came back very good, except for a small irregularity in her left breast. Right underneath the areola, the test revealed a cluster of cysts, approximately five to seven stuck together. She was told there was nothing to worry about.

And worry she didn't, not for a while anyway, especially because she felt no lump whatsoever. She felt nothing at all. About six months following the mammogram, Olga began to feel a throbbing discomfort in her left breast. One night, while trying to find a comfortable position, she discovered a tiny lump in that breast. *Interesting,* she thought to herself. *Must be the cysts growing.* Knowing there was no history of breast cancer in her family, she made nothing of it. But the throbbing continued and increased over time.

A year later, the throbbing had become difficult to ignore. It was preventing her from sleeping on her left side like she normally would. She had noticed that the lump was getting bigger. Olga thought about it now and then, but she did not focus on it. *I'm due for my next mammogram in six months,* she said to herself. *I am going to wait.* She continued to assume it was the cysts growing or acting up.

Six months later, she woke up suddenly in the middle of a chilly fall night. *Maybe I should have this checked out.* The thought came out of nowhere, and it had a sense of urgency. She felt like someone was sending her a message—maybe her grandmother, whom Olga believed had been looking out for her since passing away sixteen years earlier. The following morning, Olga contacted her gynecologist.

The office secretary had always told her to send them a fax with the details of what she needed. This would ensure that she got through to them and that she would get a quicker response. The fax read, "Discovered a lump and need an appointment." Less than two minutes after she sent the fax, the phone rang.

"Olga!" The voice of Ellen, the office secretary, blasted into the phone. "Come down to the clinic this afternoon. What time can you be here?" There was concern in her voice.

"I can't come this afternoon, I have to pick up the kids at school," Olga replied.

Olga explained to me that Ellen had lost her sister to breast cancer the year prior and, as a result, was all too cognizant of the possible dangers of a tiny lump. The secretary's tone became firm, like that of a mother who wasn't taking no for an answer.

"Olga! You know your health comes first, and you are the only person who can really take care of yourself. Make other arrangements; we will be expecting you today." *Click.*

At the appointment Olga dutifully made time for, Dr. Madder scribbled a note in her file, flipped her chart closed, and stepped over to the sink. "Everything seems fine, Olga, there is nothing to worry about, it's just the cysts that you are feeling," she said as Olga buttoned up her white blouse and stepped off the examining table to gather her purse and jacket from the chair placed across from the doctor's desk. "But since you are almost due for your next mammogram, let's have it checked out anyway," Dr. Madder added as she casually handed Olga the requisition.

Although a part of her was dissatisfied with this response, Olga proceeded to put the doctor's recommendation in her purse and leave it there ... for months.

As Olga told me this, I got shivers. I remembered Jade telling me that she had a questionable exam a year or two prior to her cancer diagnosis, but she had not made anything of it and put off further screening. Both Olga and Jade had the same response: "There is no history of breast cancer in my family. I was sure it was fine."

The requisition went from sitting in Olga's purse to sitting on her kitchen table. She always intended to book an appointment. But it wasn't until four months later, February of the following year, that Olga paid any attention to that requisition again.

Leonard had planned a long weekend for just the two of

them as a Valentine's Day treat. After they got back, Olga decided to call for an appointment. She was to do an ultrasound and a mammogram. She also booked an appointment with her general practitioner for the following day to get a second opinion. A feeling of concern was beginning to emerge, and she wanted to see someone immediately.

Later that afternoon, she hesitantly mentioned her concern to her sister-in-law, who had called to check in and catch up. Olga had not made mention of the pain in her breast and the mammogram requisition to any friends or family yet.

"What!" Jackie yelled on the other end of the line. Having lost an aunt with whom she was very close to breast cancer a few years prior, she was immediately alarmed. "And they're only seeing you in March? No, Olga, you have to go sooner. I will call you back." Jackie had worked as a technician in the radiology department at the Jewish General Hospital (JGH). Luckily, she still had friends over at the JGH Breast Cancer Center (sometimes called the CRID, for Centre de référence et d'investigation désigné). Their facility is one of Quebec's designated screening and early diagnostic centers for breast cancer. Jackie called back ten minutes later with an appointment for Thursday of that week.

The following day, Olga saw her general practitioner as planned. She left the doctor's office feeling dissatisfied. "Everything seems fine," her general practitioner had told her. "It's the cysts you are feeling." This was not the answer Olga wanted to hear. These supposedly benign cysts were causing quite a stir. She immediately contacted her plastic surgeon, one of the best in the city, and arranged for a visit the following day. She wanted the lump removed, period. Her intuition told her that something was not right, and she could no longer ignore it, despite her doctor's reassurance that everything was fine.

The following day, though, Olga found herself reliving a similar scenario in her plastic surgeon's office. "I'm sure there's nothing to worry about," Dr. Sheldon said as he examined her breast. *There are those words again*, Olga thought to herself with frustration.

"I will gladly remove the lump for you," Dr. Sheldon said, "but first, I need you to have all the necessary testing taken care of." Luckily, her testing was scheduled for the following day. She was finally going to get to the bottom of this.

Olga underwent testing as scheduled at the JGH Breast Cancer Center. "Have a seat." Dr. Kowalski signaled to Olga as she was brought into the room. After the mammogram and ultrasound, she had been asked to stay in her gown and sit in a waiting room while the doctor reviewed her test results. Thirty minutes later, when she was finally called upon, she was greeted by two radiologists. Dr. Kowalski and Dr. Bergmann just looked at her.

"I think they were trying to figure out who was going to speak first," she said to me. "And as I sat there waiting for them to call me in, I knew I had cancer."

"We need to do a biopsy immediately," Dr. Kowalski said.

"No!" Olga said, alarmed. "I'm leaving tomorrow for a family vacation."

"That's not an issue," both doctors reassured her. "We can still do the biopsy, and we'll have the results ready for when you get back."

This alarmed Olga even more. She shook her head and assertively reaffirmed that she did not want them to perform the biopsy. Years later, as we sat having lunch at her massive dining-room table one Saturday afternoon, Olga explained to me that she was terrified of having a biopsy done. It was the biggest fear she faced during her journey, she admitted. "I was frozen with fear at the thought of that needle."

"Olga, we really need to have this looked at as soon as possible," Dr. Kowalski said sternly.

"What if it gets infected?" she asked the doctors. "I am not having a biopsy done before I leave." They looked at her in disbelief. "What are the chances of me developing an infection?" she asked.

"Very minor," Dr. Kowalski assured her.

"It's too much. I'm not doing it," she told them. "I'm going on vacation. I will deal with it when I get back!" Olga felt confused by their pressing demeanor, since three competent doctors had reassured her that she had nothing to worry about.

She sat there waiting for the doctors to respond to her request to postpone the biopsy. "Why did you wait so long?" Dr. Kowalski said with a concerned tone.

Oh shit! This is serious! she thought to herself. The concern on their faces was giving rise to feelings of angst within her.

"Maybe because I was afraid?" The words unexpectedly spilled out of her mouth. Olga was surprised; until then, she had not felt afraid. But as those words erupted from her mouth, she realized this new conflicting reaction from Dr. Kowalski and Dr. Bergmann had her feeling insecure. Her mind was reeling with this sense of urgency that the doctors standing before her were emitting. She quickly gathered her thoughts. Although she'd had a nagging sense that something was wrong with her breast, she also had felt calm about it until today. Now, she simply did not want to be given the third degree about something she had repeatedly been told was a nonissue. "Maybe because all the other doctors who felt the lump said that I had nothing to worry about."

As she told me later, "Sitting across from them in the office that day, I knew. I didn't need a biopsy to confirm anything. I knew I had cancer. I just wanted them to leave me alone. I wanted to go away, and I would deal with it when I got back."

I was amazed. It takes a lot of courage and self-assurance to look not one, but two intimidating doctors in the face and tell them how you want things to be done and not the other way around. Olga assumed control of her own journey before she had even been officially diagnosed.

Finally, Dr. Bergmann looked at Dr. Kowalski. "At this point, she's waited this long, an extra two weeks won't make a difference."

It was obvious that Dr. Kowalski did not agree, but she shrugged her shoulders and said to Olga, "As you wish. But Olga, we are scheduling you for first thing Monday morning after you get back."

Olga had just bought herself two weeks to mentally prepare for the biopsy.

Chapter 3

Everything Happens for a Reason:
The Joy of Friendship

The snow that blanketed the city that Monday morning left Olga feeling small and frail as she made her way through the white streets to the JGH for her scheduled biopsy. "For the first time in my life, I was so nervous, I felt like I was going to vomit," Olga told me. The plan was for Jackie to drop her daughters off at school and be at the hospital in time to be there with Olga for moral support.

The waiting room was cold and empty. Olga was the only patient there. She sat waiting, feeling completely terrified. Carole, the nurse who was assigned to Olga the day of her diagnostic testing, kept coming in and out of the waiting room to ask if she was okay. Carole wore a black skirt and pink wool knit sweater under her white overcoat and smiled warmly.

"I'm really nervous," Olga told the seemingly trustworthy nurse. "I'm a fainter. Can you give me something to calm me down?"

As Carole took Olga's blood pressure to make sure it was high enough for her to be able to take a sedative, she gently asked, "Is your husband, a family member, or a friend here with you?"

Olga explained that her sister-in-law was on the way, and they continued to chat. Carole's caring demeanor made it easy for Olga to immediately feel at ease. Carole showed genuine interest and empathy as they talked, and Olga felt a connection emerging between them.

Olga sat there wondering where Jackie was; it wasn't like her sister-in-law to be tardy. She began to worry. "Olga Munari," Dr. Kowalski called. Olga took a deep breath, closed her eyes, and dug deep into herself, looking for some courage. Just as she opened them and looked up, she found Carole's eyes looking back at her. "I am here. I will come in with you," Carole reassured her with a gentle pat on the shoulder. Olga exhaled. She was relieved. Support had found a way.

Carole held Olga's hand throughout the entire procedure. She was gentle and reassuring with Olga. "I couldn't imagine a more genuine and caring person, or anyone more adept at what she does for a living," Olga told me. "Carole was an angel sent by God for me that day." As Olga recounted these events, there was so much emotion in her face. Fear, relief, and thankfulness poured out of her.

"I found Jackie waiting for me outside the biopsy room, talking with her friend who happened to be the head of the radiology technicians at the hospital." As Olga approached, they both stopped talking abruptly. Olga felt like something was wrong; she felt that they were talking about her. "To me, it was yet another clue hinting that I really did have breast cancer." And again, for some reason, she felt no fear. As Jackie apologized for missing the appointment (she had car trouble because of the cold), Olga thought that Jackie just was not meant to be there and was silently grateful because she had made a new friend. A special bond had formed between Carole and Olga from sharing the biopsy experience; it felt as though the two had been friends for years, even though they had only met a few times.

"That was the first great thing that happened to me, making a very good friend through sickness. It was the first of many blessings," Olga affirmed to me with a smile on her face.

As they were preparing to leave the clinic, Carole called Olga into her office. "It will take a week or two before the results of the pathology come in. Once the results are in, I will call you to set an appointment to look over the results," she told Olga in her typical soft-spoken manner as they sat facing each other.

"Please! I will not drive here knowing that there is a possibility that I have breast cancer. You call me when you have the results and give them to me over the phone, because to tell you the truth, I believe that I do have breast cancer, given the way everyone is behaving around me."

Carole insisted, "Olga, I have never given any results over the phone. I prefer to talk to my patients in person."

Olga was equally insistent. "There is always a first time in life. Don't you worry about telling me over the phone. Just tell me the truth when you call, and I will be fine."

Again, here she was setting her own rules of play, owning her journey and setting the tone for it. I was baffled, and then I laughed, amazed by Olga's nerve.

"Carole gave me a big hug and told me that everything was going to be fine," Olga recalled to me. "She gave me her phone number and told me to call if I had questions or if I wanted to check up on the results."

Later, Olga and Jackie were sitting at Second Cup, sipping hot coffee as they debriefed the morning's events. "How are you feeling?" Jackie asked Olga gently. Olga was nibbling on a cheese Danish as Jackie watched her, waiting for a response.

"I feel good," Olga said. "Honestly. But you know what, Jackie? I have a feeling that I do have cancer."

Jackie noticed no trace of fear or anxiety in her sister-in-law's voice or face. Her eyes welled with tears in silent agreement, but she said nothing.

"I just wanted to get my results and start dealing with whatever was coming my way," Olga explained as she and I talked.

The Monday following the biopsy, Olga was coming out of her weekly session with her psychologist and was heading to school to pick up the kids. As she got into her car, she considered driving by the hospital to inquire about her results. She still had an hour before school let out. She was antsy to get the results because she wanted to move forward. "Until they confirmed it, I could do nothing but wait. But once the cancer was officially diagnosed, I could move forward with treatment. I would have a goal, and I would know exactly what direction I was headed in." Every day since the biopsy, she had been mentally rehearsing what would come next after the diagnosis was confirmed. She was preparing herself for the journey.

Olga had called a few times during the week, and each time Carole gently and reassuringly told her, "No, my love, they aren't in yet, but as soon as they come in I will call you. And if you need to talk in the meantime, no worries, you can call me."

Just as Olga reminded herself that Carole would call when the pathology report came in, her cell phone rang. The caller identification indicated that it was an unknown number. Olga immediately knew it was the hospital calling with her results.

"Hello?"

"Olga." It was Carole on the phone. "How are you? Are you in the car? Are you parked?"

Olga had pulled over. "Yes, I'm parked, and I have cancer."

"I'm uncomfortable, Olga. I've never told anyone over the phone."

"It's fine, I know I have cancer, just tell me."

"Yes, it's cancer, Olga. But I already booked all the appointments. I got you the best surgeon. Your appointment with him, Dr. Woodman, is booked for Thursday of this week, and you will be meeting the oncologist shortly after. Are you okay, Olga?"

"Yes." There was a pause. "Yes, I'm okay."

Olga told me later that "as a child, I had very few fears. Few things would scare me. Spiders. I was afraid of spiders. I was terrified of them. I still am, actually. My son Andrew handles the spiders in our house," Olga said smiling. "And breast cancer. For some reason still unknown to me, from a very young age, I had a fear of breast cancer. I remember saying to myself one day when I was just a teenager, if I ever get cancer, it will be breast cancer."

I got shivers. How eerie is that? But then again, it made sense. The thing we fear the most always has a way of finding us.

"Immediately, my biggest concern," Olga told me, "was how I was going to tell my children. How was I going to tell my family? How do you tell your parents that their daughter has cancer? Usually you would expect it to be the other way around. How does a child tell her mother that she has cancer?" Although she had rehearsed what would follow next on her journey to recovery, Olga had, perhaps deliberately, omitted the scene where she would tell her family. "How was I going to tell my teenage daughter? How could I say those words to her? 'Mommy has cancer.' I could not fathom seeing her face when I pronounced the words."

She continued, "As I drove to pick up the kids that afternoon, I ran through the scenario in my mind. Tears streamed down my face, I was anguished by the thought of telling the people I loved. I knew they would be hurt, and I did not want to be the cause of their pain."

As Olga talked, I thought of Jade telling her husband and sons. I imagined her beautiful boys, the rug pulled out from under them. My heart ached a little bit as I remembered how sad the boys had been after Jade was diagnosed. My heart went out to Olga's family, backward in time to the day she had to tell them.

"I asked Andrew to drive," Olga told me. "I needed to think through my strategy. Unfortunately, I wasn't given much time. My cell phone rang immediately; it was Leonard. The first time he called, I did not pick up. Laurie intuitively knew something was wrong."

"Are you okay, Mom?" Laurie asked.

"Yeah. Mommy's okay," Olga quietly answered her.

"How come you're not answering Dad's call?"

Olga told me, "The second time he called I had to answer because Laurie was onto something, and I did not want her to think Leonard and I were fighting or that I was ignoring him."

"So?" Leonard asked with a curious tone, "how are you? Any news from your doctor?"

"Uh, yeah," Olga answered after a short pause.

"Uh-oh." He knew instantly. "I can tell by the tone of your voice, it's not good news."

"Well …" Olga stuttered, "the kids are in the car … I can't really talk right now." Olga tried to cut him off before he could speak any further.

"You have cancer?"

"Yeah."

He started to cry and hung up the phone.

Olga's heart sank. *Great!* she thought to herself.

"Mom," Laurie asked again as she observed her mother carefully, "are you *sure* you're okay?"

"Yeah, Mommy's fine," Olga answered her, fighting back her own tears. Andrew just kept driving, looking dead ahead.

Olga told me she was sure Andrew knew something was wrong, but he did not flinch—he never took his attention off the road, not even for a second. And little Jeremy, only nine years old, sat in the backseat just observing everyone.

Heavyhearted, Olga called Leonard back. "Len, are you okay?"

"I don't know. How are we going to do this? How am I going to deal with this? I can't believe it! You have cancer?" There was a short pause, and he ended the conversation abruptly. "I'll see you at home." *Click.*

Great! Olga thought to herself. *This is just great.*

When they got home at four fifteen, the boys both headed up to their rooms to do homework and watch some after-school television. Laurie headed to the neighbors' to do some homework with her friend. Alone, Olga sat in the den—

typically Leonard's room. She had given Andrew and Jeremy specific instructions to say she was not home if their grandma called. Olga was not ready to tell her mother yet. She was waiting for Leonard to come home, and she could not talk to anyone until she spoke to him. So she sat there, planning out how she was going to tell everyone the news. Finally, she picked up the phone to call her sister.

"Anna."

"Yes?"

"I have something to tell you, but you have to promise not to tell Mommy. I have cancer. Breast cancer."

"You have what? Oh my God, Olga!" Crying and frantic, Anna went on, "Are you going to die? Are you going to die? Promise me! Please tell me you're not going to die!"

Great! More hysteria, Olga said to herself as her sister cried on the other end of the line. "Anna, stop! I'm fine! I'm not going to die! Just don't tell Mom, I'm not ready to tell her yet. She knows I had a biopsy done, but I told her it was just precautionary so she wouldn't get too worked up. So if she asks you anything, please don't tell her. Can you do that for me?"

Anna finally calmed down, and Olga hung up the phone. She continued to sit there, waiting for Leonard to come home. *You would think he would rush home*, she thought to herself as the phone rang. It was Jackie.

"Did you get the results?" Jackie asked with a serious tone.

"Yes," Olga answered. "Yes I did, and I have cancer." Jackie started to cry on the other end of the line. "You kind of knew, right?" she asked Jackie, already knowing the answer.

"Yes," Jackie answered. "Tom knew. Dr. Kowalski told him, 'It looks like it might be cancer' after the ultrasound and mammogram." Tom, being a radiologist as well, had seen the test results, but could not say for certain until the pathology report came in.

Leonard walked in at a quarter to eight, devastated. Olga was sitting in his chair. He took one look at her and, through his tears, mustered up the strength to speak. "I can't believe this. That's it? After twenty years, this is how it ends?" There was so much anguish in his piercing blue eyes. He could barely look at her as the tears streamed down his face.

"Leonard, I'm going to be fine, don't worry about it."

Leonard moved to the kitchen, pacing. "That's it! Twenty years down the drain. Our life is over."

"Whoa! Whoa! Wait a minute." Olga cut him off. "I'm not dying here. I have cancer. Life is not over because I have cancer. Come on, Leonard. Look at me. I am going to have surgery and treatment, and I will be cured. And I will come back even stronger," Olga told him. "And then you'd better watch it!"

Now as well as then, Olga's attitude—calm, steady, certain—amazed me. It inspired me. It made me want to be more like her. I thought of all the challenges I had faced. In my most trying moments, I had not always seen that there were other choices I could have made in the face of those challenges. She was so certain in the face of such uncertainty. I was blown away by her self-assuredness.

"Leonard, I will be fine," she continued to reassure him. "Don't worry about it. What I'm worried about right now is telling the kids. Telling my mom, your mom, and the rest of the family—this is my biggest concern right now. I'm not worried about me going through the cancer. I just don't want people to pity me, or worse, feel alienated from me because I have cancer."

Leonard just stood across the long island in the middle of their kitchen looking at Olga without speaking. Tears continued to stream down his face, and Olga walked over and gently moved in to hug him. He took her in his arms in silence. After a few moments, Olga looked up at her husband. "I'm ready to tell the kids," she said. He nodded back at her in agreement.

Olga went to phone Karen, their neighbor, asking her to send Laurie home.

"Olga, are you okay?" Karen asked.

"Yes, yes, I'm okay. Why?"

"Laurie came in concerned. She feels that there is something wrong."

"Actually, yes, Karen, there is something wrong. I have breast cancer, but please don't tell Laurie. We are going to tell the kids now." Karen started crying on the other end of the line.

Olga explained to me that the emotionally charged responses from the people around her indicated that she would have to step up and be the leader for her close friends and family. This only strengthened her resolve to move through cancer with graceful empowerment. Olga calmly instructed Karen, "If she sees you crying, she'll know there is something wrong. So could you please not say anything to her? Just tell her I want to speak to her and send her over."

"Okay, Olga," Karen said, gathering herself. "I'll send her right over."

Olga recalled that "Laurie walked in with a petrified look on her face. She looked at me like I was dying. My heart still aches today when I remember the look on her face." As she spoke, Olga's face revealed how painful it was to see her daughter so terrified.

The family gathered in Laurie's room, where Leonard was pacing madly. Jeremy sat on a small couch in the corner of the room. Both Andrew and Laurie were sitting on the bed. Laurie was curled up with her arms and legs crossed. She had a gray sweatshirt on with the hood pulled over her head. "All I could see were her big blue eyes," Olga recounts. "She just sat there staring at me, waiting for the bomb to drop."

Looking at her family, Olga said, "Mommy has something to tell you. I have to have surgery. They found a lump on my breast, and it has to be removed."

"Is it cancer?" Laurie asked.

Not wanting to hide anything from the children, she told them the truth immediately. "Yes, yes, it is cancer, but I am going to be fine; don't worry about it. They are going to do surgery to remove the lump, and then Mommy will go through chemotherapy and radiation."

As Olga told me, "I wanted the kids to know about cancer and to know that it wasn't necessarily a death sentence. My desire was for them to understand that it does not have to instill terror. I intended to communicate to them, through my attitude and actions, that it is possible to be positive in the face of challenges; that you could have bad news and make the best out of it. In the face of the strong emotional reactions of the people closest to me, I realized I wanted to be an example of the fact that we always have a choice as to how we respond to the challenges life brings us. I felt it was a challenge that myself and those around me would greatly benefit from, especially my children. This was the perfect opportunity to teach them experientially."

That night, as Olga explained things to her family, Laurie started tearing up. Andrew looked at his little sister and said,

"Laurie, Mommy's going to be fine. They'll just remove the lump and that's it."

"That's exactly it, Laurie," Olga joined in after him with the same calm and certain tone as Andrew.

"Andrew!" Laurie raised her voice in anger. "Do you know what cancer is?"

Everyone was silent.

"Yeah, it's a lump. They just have to remove it," he answered.

"No! Cancers are bad cells eating all the good cells in your body. She could die. Mommy could die!"

Olga's heart sank. *Oh my God!* she thought to herself. *This kid's in for trouble.*

"Laurie, listen to me. I promise you, Mommy is not dying." Olga wasn't sure what else she could say or do to convince her daughter, but fortunately Leonard jumped in to put things into perspective.

"Mommy is going to need your help," he said. "There will be days when she might be tired. There may be days when she won't feel like cooking. She may have days where she will just want to stay in bed." Not knowing how Olga would react to the treatment, he wanted to prepare the kids to be supportive through her recovery.

"I am going to be fine, guys!" Olga insisted. "Now, do you have any questions?"

"No," they all chimed, needing time to digest the news. Olga took each of them in her arms and hugged her children tightly. The boys went back to their rooms, and Leonard went downstairs. Olga stayed behind with Laurie. She gave her another hug.

"Mom, can I be alone?" Laurie asked.

Olga walked out, and Laurie closed the door behind her.

As she walked away, Olga immediately heard Laurie sobbing uncontrollably behind the door. Olga knocked on the door and went back in. "Laurie, are you going to be okay?"

"Mom, you're going to die! You're going to die! I'm so afraid!"

Tormented, Olga looked at her daughter. "Laurie, listen to me, I'm not dying. Laurie, I promise you! Mommy will always be here for you." She took Laurie in her arms and held on to her tightly.

"As difficult as that day was," Olga said to me as her eyes welled up, "that was the day I created a very sacred bond with my children. I had held them and hugged them but never with such depth. It brought us much closer than we ever were."

The following day, a very brave Olga called her mother to give her the news. In line with many other reactions, her mother burst into tears and hung up on her. *Great!* Olga said to herself once again. *This is just great!* She picked up the phone to call her mother back. "Mom, listen to me, calm down! I am going to be fine."

"Olga, do you want me to come over? I will be there right away."

"No, Mom, I'm fine. Everything is going to be fine," she assured her mother. "You can come tomorrow if you like. I will be here."

Having told all of her immediate family was a huge relief, and Olga slept soundly that night compared to the night before, when she did not close an eye. She had tossed and turned, agonizing over how she was going to break the news to her parents. But now, she was ready to face her mother and father and let them know what she needed from them from this point on.

Olga sat her parents down, pulling out her take-charge

attitude as they both sat across from her in devastation. "Listen, I am going to be fine. What I need right now is not people crying in front of me. I need support; I need sources of strength to draw from on top of my own. I don't need you guys to cry every time you are going to visit," Olga told them seriously. "I am warning you now, I have a long journey ahead of me, and I want this journey to be positive. I don't want any negativity in my life. If you're going to walk through this door and cry and be negative, I will send you back home and close the door behind you."

Listening to this story, I was in awe. I kept thinking about how many families must have been devastated and torn in the face of such a diagnosis. Olga could have easily felt like a victim of disease, of circumstance, of life itself. She could have easily been devastated and afraid and passed that energy on to her husband, her kids, and her parents—but she chose to respond with confidence and self-assuredness.

"You know what, Kim?" Olga said to me. "My mother never again cried in devastation through the entire journey. When I would come home from a treatment, she would instead shed tears of joy because she was happy that it was one treatment less for me. She would compliment me on how great I looked with my wig; she would encourage me. But there were no tears of pity.

"I had to train everybody around me to be positive. And one by one, I made sure they would adopt a positive attitude in the light of the recent news." Once everyone was aware, Olga said she felt like she was already cured. "It was so challenging," she explained. "I felt like I had a huge job to do, and once it was done, the weight of the world was lifted from my shoulders. I could breathe again.

"I realized immediately that I would have to be the

stronger one in the face of this," she went on. "I would be the one to support the others around me. Their fear was greater than mine. I had no fear. And once everyone knew, I felt like I could direct my energy toward making the best out of this journey."

Olga continued, "Not long after I was diagnosed, I had an appointment scheduled at the JGH to review what the next steps were. The doctor explained to me that before they could start the treatment protocol, I had to first undergo surgery to have the lump removed. Carole and I were alone in the changing room for a few moments after I was done. As I finished getting dressed, I looked at her and asked her what my chances were.

"She looked me straight in the eye and put her hand on my shoulder. 'My gosh, this high!' she said, gesturing, her hand way above her head. 'I don't know you very well yet, but I can tell that you're a fighter and you will be fine.'" Again, Carole took Olga into her arms and hugged her with warm and genuine affection.

"From then on," Olga said, "I set fear aside and looked straight ahead. *How can I take this and make good from it?* I asked myself."

Olga knew she wasn't going to die. The question was, how was she going to live?

Chapter 4

The Journey of a Thousand Miles Begins with a Single Step:
The Joy of Daring

"This, Olga, I'm doing this for you!" Janet smiled as she pointed to the pamphlet for the Weekend to End Breast Cancer benefiting the Jewish General Hospital. "I signed up!" It was a few days after Olga's diagnosis, and her friend and trainer, Janet, was visiting to share the news and offer support.

"You did what?" Olga was confused and touched at the same time. "What do you mean, Janet?" Olga asked from across the dining-room table, buying herself time to respond.

Branded with this new label—the *cancer* label—she had to reassess what role she wanted to play on this new team onto which she had been unwillingly enlisted. Learning to receive, to be supported, to be cared for, to ask for what she needed, to *claim* it, would indeed be a part of this journey, as she would discover along the way. But she would also have to stay true to herself. She wondered how could she do that and still gracefully accept all the resources available to her. She knew "patient," "victim," "martyr," or any of those other similarly disempowering labels didn't suit her. But she had yet to determine what role she was going to play in all this.

Just four days after her diagnosis, she found an opportunity that matched both her need for a challenge and her love of giving back to others. Olga made a choice. "Well, if you're walking for me, then I'm walking with you," she said, smiling as she snatched the pamphlet from Janet's hands.

Later that week, Olga sat at the computer and completed her online registration. Each participant was required to raise $2,000. Olga drafted a list of possible sponsors. She composed a short e-mail explaining her recent diagnosis and her choice to participate in the Weekend to End Breast Cancer. Off it went, as she hoped for the best.

It was now the beginning of April. Judging by the length of her list, she figured she would have the money raised just in time for the walk at the end of August. She got up and made her way through the dining room and was heading toward the staircase. Just as she started walking up the stairs, her computer sounded the e-mail alert. She made her way back, assuming an e-mail had bounced back. In her inbox sat a message from the Weekend to End Breast Cancer. It read, "Congratulations, you have reached your goal!"

Reached my goal? But I just signed up! she thought. The e-mail said that one couple had donated the $2,000 in full. "Leonard!" she called out. "Come here, come look at this, it must be a mistake. I have to call them to straighten this out."

She immediately picked up the phone to call their friends. "Beverly, I think there's a mistake, I just got an e-mail from the

Weekend to End Breast Cancer indicating that you donated $2,000?"

"Olga, you had Stewart and me in tears! We were so moved by your story. We never give money, but we couldn't think of a better cause to support."

Olga was amazed. *How could this be?* she thought. She had just signed up—had just one response.

Within a week, a team had been formed, and nineteen women were christened Olga's Butterflies. Janet and Olga had spent a day brainstorming a suitable name for the newly formed team. As they sat around throwing out a variety of names, including the likes of "Olga's Boobies," Olga had an idea. "What about Olga's Butterflies?" she suggested. "I like it," she said, answering her own question. "Me, walking, surrounded by all my butterflies as they support me on my journey."

Olga felt a deep connection to butterflies, as they evoked childhood memories of time spent with her grandmother. White and yellow butterflies had always made her grandma's yard their home. When she spent summer afternoons there, she would find herself surrounded by their majestic beauty. They anchored within her the sense of inspiration she felt during those moments. They were a token perfectly suited to her journey.

Olga was a natural fundraiser. She approached potential sponsors with ease and grace and had a knack for making people feel inclined to contribute. She generated funds from a variety of people. The bottom line to her success, to use the words of Olga's friend, was, "I'm giving because of the way you ask." Of course, the fact that Olga was currently affected by the cause created an excellent impetus to motivate donors.

With this new project at hand, Olga was even more excited

to get the surgery over and done with. "Leonard and I were both nervous the morning of my lumpectomy," she confessed to me. "It was very quiet on the drive to the hospital early that morning; we stuck to small talk. I think Leonard was more nervous than I was.

"As they wheeled me away on a gurney, Leonard leaned over and kissed me. 'I love you,' he said. 'I'll see you when you wake up.'

"The surgery schedule was running late, but because they had put a marker in my breast to identify where the lump was, I was sure to have my turn. Instead of going in the operating room at ten a.m., I ended up having my turn at two in the afternoon. Carole from the Breast Clinic kept popping in to check on me.

"'Are you okay, my love?' she gently asked me as she held my hand. She also relayed updates over to Leonard. 'I spoke to Leonard,' she told me. 'I explained to him that there is some delay and that you are fine.'

"And I was," Olga affirmed. "The minute I was lying down on the gurney, I felt calm and peaceful. The next thing I knew, my turn was up, and I was counting backward from ten.

"I woke up to Leonard sitting by my side, crying. 'The doctor said everything went well. They removed the lump and ten lymph nodes,' he said as he held my hand. 'But they may need to go back in if the margins are not clear.'

"I felt so much better after the surgery was done," Olga told me. "For me, the cancer was gone."

One bright and sunny morning just a few days after the operation, Olga sat in her dining room sipping a cup of coffee. Now that the surgery was behind her, she intended to kick her fundraising efforts into high gear. She was quickly taking the lead among registered participants. Olga had individually raised over $5,000 in just the few weeks following her registration. Naturally, she was causing a stir over at the Weekend to End Breast Cancer's headquarters. The amount of money she raised was well above average among both current and past participants. Olga had now set a new goal for herself: she was aiming to raise an audacious $20,000. As she sat going through her stack of business cards and e-mail contacts, preparing for another fundraising blitz, the phone rang.

"Hello, I'm looking for Olga Munari."

"Yes, this is she."

"My name is Fiona. I'm calling from the Weekend to End Breast Cancer."

"Yes."

"We're very curious: you registered as a participant in the Weekend to End Breast Cancer on April fifth?" Fiona asked. "It's only been a few weeks, and you've already raised over $5,000. What are you doing to raise all this money?"

"I don't know," Olga answered. "I just identified a list of people I believed would donate and sent them an e-mail." There was a short pause. "Maybe because I have cancer?"

"You have cancer?" the woman on the end of the line said.

"Yes, I have cancer, and I had surgery just a few days ago."

There was a short pause.

"Listen," Fiona said, even more impressed. "We're looking for someone to give a speech at the Queen Elizabeth Hotel this coming Sunday, to talk about raising money and motivating other walkers and so on. We would like you to be that person."

"You want me to speak where?" Olga said in shock. "In

front of how many people? I can't talk in public. I've always wanted to do something like that, but I can't do it." There was a sense of hesitation in Olga's final words.

Olga told me that she had always dreamed of being in the public eye, driven by a purposeful mission. It was part of the reason she was so committed to actively fundraising for causes and charities of her choice. She had always wanted to inspire people and reach them somehow, but she had not had the courage to step up and do so until then. She felt pulled by this new opportunity, but fear was holding her back from accepting Fiona's proposal.

"Olga, please," Fiona insisted. "People need to hear your story."

Olga still hesitated. "I can't go up there. What would I say?"

"Just tell them your story. It's very powerful. You'll be great."

Silence.

"No," Olga replied again with a quiver in her voice. "I can't do it." And so they wrapped up their conversation.

"Well, thank you for your time," Fiona said, "and keep up the great work. If you have any questions, please get in touch with me."

I could relate to the way Olga felt that day on the phone with Fiona. There were times in my life when I was given the opportunity to step up and shine in ways that were outside of my comfort zone. Often, this stepping up was an opportunity to be a leader by way of example, to move and inspire others, simply by sharing the story of my own experiences and endeavors. Sometimes, to my own disappointment, I had declined. I was very capable of leading on a one-on-one basis, often inspiring others to action and change by way of my own example, but I felt unable to do this in a bigger way. Even if I felt a strong desire or had always wanted to—to use the same words Olga said to Fiona—I had been too paralyzed by my own fear of stepping outside of my comfort zone. Or, rather, as I understand now, I didn't have enough incentive to push through my own fear. Sometimes, incentive that we consider big enough comes in shapes and forms we least expect. Cancer was the leverage Olga needed to take a leap and move out of her comfort zone.

Two hours later, a very persistent Fiona phoned Olga back, asking her to be on CJAD with the Kim Fraser radio show. "Yes!" Olga replied enthusiastically before she even realized what she was saying. She was shocked at how quickly the words slipped out of her mouth. She told me later, "That I could do."

Olga confessed to me, "I remember thinking to myself as I uttered the words, 'What the hell am I doing?'" Kim had contacted the Weekend to End Breast Cancer asking for someone she could have on the air to share a personal story and promote the event, as well as to encourage listeners to participate in the event.

Now that she had Olga back on the phone, Fiona was not about to waste the opportunity to possibly get her up onstage at the Queen Elizabeth. "Olga, we still would like to have you be our guest speaker at the Queen Elizabeth Hotel next Sunday. Would you please reconsider?"

There was a moment of silence.

"Okay, I'll do it!" It was a moment of impulse in which Olga's courage overrode her fear. In a flash, she felt she had nothing to lose, given her circumstances. If she was going to go, she would go all the way. Olga hung up the phone feeling slightly confused. *What did I just do?* she asked herself. *This is a little nuts.*

She called Leonard. "Len, I'm going to be on the radio, on Kim Fraser's show, and I'm also going to be the guest speaker at the Queen Elizabeth Hotel next Sunday night for the Weekend to End Breast Cancer."

"You're doing what?" he asked, confused.

"They called me to ask me what I was doing to raise all this money, and they asked me to give a speech, so I said yes. Len, I don't know what I'm doing," she giggled nervously. "I need you to help me write a speech."

"Okay, we'll work on it together when I get home this evening," he said, sounding confused and proud at the same time.

Nervous and excited, Olga went from chair to chair after they hung up, trying to understand what she had just done and how it happened. "Finally," she told me, "I justified that if the words came out of my mouth, it meant that I was up for the challenge, and I got straight to work developing the draft."

I loved Olga's response. Rather than question herself and move into fear and doubt, she chose to attach an empowering meaning to the choice she had made, thereby encouraging herself to move forward with confidence.

With Leonard's help, Olga put a speech together that would briefly retell the events leading her to that point. She recounted being at home the day before the speech, rehearsing in the dining room and tearing up over and over again. Her mother-in-law sat across the table, watching her. "How am I ever going to do this in front of all those people without breaking down?" Olga asked her. "This is ridiculous."

Her mother-in-law leaned over and gave her a knowing look. "God will give you the strength."

Well said, Olga thought to herself. *I will be fine.*

Her nineteen butterflies were there at the Queen Elizabeth Hotel to cheer her on, as were her close friends and family. Seated in the front row, Olga listened attentively to the two speakers who went before her. She was to go third. Olga turned her head to look back at the rows behind her as the master of ceremonies began to introduce her. She was taken aback by the large mass of people. Rows and rows of people were there waiting for her to speak. She felt a rush of emotion and decided that the only way she would get through this was to not look at her family members' faces. As she made her way up to the podium, feeling the heat of the spotlights on the ceiling above her, she remembered the old trick Kim Fraser had suggested about imagining all the listeners sitting in front of her in their undergarments. With sweaty palms, she faced the three hundred spectators before her, took three deep breaths, and began speaking.

She told them about her diagnosis, made less than a month ago. She told them about her first surgery, and how she was now waiting to see the doctor to determine when she would be starting her chemotherapy. As Olga spoke, she noticed how many people in the massive room were moved to tears. She told them about the $2,000 donation on the first day and how quickly her fundraising efforts were moving. As she stood there that day, she had already raised $12,750. She announced her new goal to raise $20,000 by the time the weekend of the walk came around. "My journey is just beginning," she told them. "I believe that everything happens for a reason, and that even in the bad things you can always find something good." With a reminder to keep a positive attitude, she ended her speech by expressing gratitude for the bountiful support she was receiving from her loved ones, and underlined the importance of community and the importance of reaching out to friends, family, and coworkers.

Olga received a very animated standing ovation. People were beaming, crying, and cheering. Olga stood there mesmerized. She could not believe that all these people were there to hear her. Not only had they actually listened to her story—that in itself amazed her—they were incredibly moved.

"I had always dreamed of being up onstage—of doing something that influenced people. It's a dream I came back to again and again in my life. It felt like this could be the beginning of something new for me. Like the shaping of a new life."

As that day came to a close, she thought to herself, "If this is what cancer is bringing me so early on, I can't imagine what the rest of the journey has in store for me."

Chapter 5

Everybody Loves a Winner:
The Joy of Taking Initiative

.

"I have good news and I have bad news," Dr. Woodman said. "Which do you want first?"

Olga flinched; she wasn't prepared for bad news. "Whichever you want, Dr. Woodman. It really doesn't matter," she said, trying to hide her disappointment. She and Leonard were sitting in her surgeon's office to discuss the results of the surgery and to review the next steps. The surgery now behind her, Olga wanted to press forward with treatment as quickly as possible.

Leonard stepped up. "Well, why don't you start with the good news."

"The good news is, the surgery went really well, there is no sign of infection, we removed ten lymph nodes, and two of them were affected. The cut was clean, and the scarring is minimal as you can tell. I'm very pleased with the results, and we were able to save most of your breast. The bad news is," he paused as he took a breath, "the margins were not clear. We have to go back in for another round of surgery to make sure we get it all out."

Olga's heart sank. Leonard must have felt it, because he turned to look at her. Reading each other's minds, they exchanged a quick glance that screamed, *What the hell? How could this be?*

Olga sat there only half-listening as Dr. Woodman explained how common a second surgery was and that sometimes even a third one was required. She was more concerned about the feeling of her chest closing in as the lump in her throat swelled until her eyes watered. She took a deep breath, trying to calm herself. She managed to stay composed even though she felt as though she were walking in quicksand. Then she realized that the time it took for her second surgery to take place and heal meant her chemotherapy would be pushed back a month. She felt a surge of anger rise from within her as she remembered telling Dr. Woodman on the operating table before the anesthesia kicked in, "If you have to take out the entire breast, go for it! Do what you need to do, Dr. Woodman. I'll deal with it afterward."

His response as he looked down at her had been, "We're not in the business of removing breasts, we're in the business of saving lives." She wondered now if she'd still need a second surgery if he'd heeded her request.

The drive home was quiet. Neither knew what to say or feel; they were still digesting the news. Olga was in disbelief. The only thing she could do was be quiet and let the news sink in slowly. Her mind vacillated from numb to a series of racing thoughts as she moved between disappointment and frustration. She had felt so good about getting treatment started. She wanted to keep moving forward; this felt to her like a giant leap backward in her plans.

Back at home, as she sat with the news and noticed the effect it had on her feelings, she quickly realized that she needed to keep herself in a good place and that sitting idle while waiting would not serve her. She decided to move into action; she was not going to sit around and sulk. She threw herself into her fundraising efforts with more vigor than anyone could have imagined. The days moving forward after her first surgery became a merry-go-round of brainstorming and broadcasting e-mails to recruit more sponsors. She would sit in her dining room with her laptop and go through her Rolodex and e-mail contacts, making lists of people and businesses she could approach. She rummaged through drawers looking for every single business card she'd ever been given. She sent out e-mails to whoever she felt would have the faintest inclination to make a donation. She asked donors to forward her request to anyone they thought might have an interest in her cause. As a result, she often would get donations from strangers. "I had sent an e-mail to the woman who ran our local tanning salon. She in turn had forwarded the e-mail to her friends, and one of them made a hundred-dollar donation," Olga recounted as an example. She reached out to every nook and cranny of her network, from her closest friends to acquaintances to the board of directors at her children's school to the owners of her favorite restaurants.

People were incredibly generous, both with their money and their support. It did not matter to her the amount of the donation; it was the intention that touched her. "I profoundly believe that it's not about the amount that you give but the intention with which you give," she said to me.

"For example," she said, "Ginette, the manager of one of our stores, drove down from Cornwall to visit me one day after my diagnosis. Before volunteering at the supply store at LCC, I had been the buyer and supervisor for seven Dollarama stores. Ginette managed the first store I had been put in charge of when our family business was sold to the large chain; she was special to me, and I had a good relationship with her. She came not only to visit but also to hand me an envelope with two-hundred-dollars' worth of donations. Employees from the store and their families had wanted to pitch in, so Ginette had coordinated the effort of gathering the donations and was there to deliver what they amassed. Most of the donations were coming from cashiers and stockers; they were small amounts given with an enormous amount of compassion. Every single employee had left a note. Many of them wrote, 'This is all I can give, but I wanted to support you.' I was incredibly touched. Every dollar donated counted, especially the ones given with so much love."

With each fundraising milestone she reached, Olga defined a new one for herself. Always substantially increasing her goal, she routinely kept her followers and supporters updated. She sent them regular e-mail reports on how she was doing and what the next steps of her journey were. In those initiatives, she found a sense of purpose that invigorated and inspired her. It kept her focused and driven. People were incredibly present and supportive. They loved her enthusiasm and exchanged with her often and willingly. Not only was she receiving more and more collaboration, she also got e-mail after e-mail from people stating how inspiring she was. Checking e-mail responses from her supporters was like a part-time job during that period, and one that Olga took to heart. Every interaction was important to her. It wasn't just the nineteen women who had been inspired to join the Olga's Butterflies team—it seemed everyone who came upon Olga's story was infected by her attitude. The support and encouragement continued to pour in as her journey progressed, and so did the testimonials of the impact she was having on others. Her attitude was causing a wave of inspiration and enthusiasm within her community. It was more than Olga could ever have imagined.

The more I spoke with Olga, the more evident it became that we both feel a strong pull to support people who have a positive mental attitude in the face of challenges. We want to cheer them on; we want to see them win. I realized it is for this reason that athletes are so enticing. Running a marathon is nothing short of a huge challenge, and the person who takes that journey has much courage and determination. They do not sign up and say, "I can't do it! It's too hard! I will not make it. It's going to kill me." They train, day in day out. They prepare relentlessly, they keep their mental game on and stay focused. And we, as followers, watch them and cheer them on because it's natural to want to see the people around us win. Hearing Olga's story made me understand clearly why some people are easier to support than others. It's their own drive to succeed that ignites us. Like any success-driven athlete, Olga had enlisted, with her strong attitude, a crowd of people to cheer her on every step of the way.

The way I saw it, being diagnosed with cancer is like being handed a runner's package for a marathon you didn't sign up for. You find yourself thrown off guard and you need to act quickly and decisively. Questions like "What attitude am I going to adopt?" and "How am I going to show up day in day out on this challenging journey?" need to be addressed quickly in order to secure a successful endeavor.

For Olga, it meant treating this like she treated most of the challenges in her life: cancer was a project, and she was the manager.

"I believe everything happens for a reason," Olga would affirm repeatedly. "I would tell myself, *There is a reason why this happened to me.* And the question I posed was: *How can I take this thing and make something good out of it?*" She was constantly exploring cancer from this perspective. And explored through this lens, cancer became Olga's very personal journey of self-discovery and evolution.

Facing challenges is an undeniable part of life. Challenges of all sorts are a platform for growth. A huge challenge, like cancer, is a colossal platform for growth. The size of the challenge is proportionate to the amount of growth and expansion you can draw from it. Challenges are experiences that can be capitalized upon, platforms from which to bounce off in order to propel ourselves forward. And if you've ever watched a child experience growing pains, you know that even though they may be painful, they always leave the child a little taller, bigger, smarter, and better. I don't think Olga or any cancer patient knew this from the start, but it seemed to me that the ones who put one foot in front of the other with a strong focused attitude have a better journey. By "better," I mean the quality of their experience varies vastly from that of someone with a bleak mental attitude.

I thought of a friend I had met just a few months prior to meeting Olga. He had survived testicular cancer in his late twenties. Curious as to how the experience had impacted his life, I asked him, as he looked back now from age thirty-six, how did he think surviving cancer had changed him? I was taken aback by his response. "It didn't," he said simply. "It was a very dark and lonely time in my life. I didn't learn much from it." Then I thought of Simon, an amazing man I had worked side by side with for three years. He also had been treated for testicular cancer at twenty-eight, the same age as my new friend. Simon's account of his experience took a very different tone. "It was scary and lonely, no doubt," Simon told me. He didn't have family at hand, didn't have the grace of a pampered life—he was a gypsy, working just to get by so he could continue to do what he really loved, which was play jazz. He told me how there were days during treatment where he was so sick that he could barely hold an instrument, but even that inspired hope within him. "I'm lucky to be alive," he told me. "My view on life has changed. I don't care about what normal people care about. My priorities have changed—a glimpse of death does that to you." He would check in with work on a regular basis, and even though his treatment left him exhausted and sick, he always kept his chin up, making jokes when the team called him and always asserting how he would get through it and come back even stronger. "What made it even scarier was that I had just been hired. My health benefits had not kicked in yet. I got lucky that our manager pushed for me to get some sort of compensation to get me through it financially." Looking back now, after connecting the dots with Olga's story, I noticed the different attitude these men had and contrasted it with the quality of their experience. Simon's story profoundly affected me. He and I are the same age; that meant I was not immune. I wondered, *Could I be that brave? That strong?*

I thought back to my own difficult times, the times when people were present and the times when they weren't. How was my attitude? What was my mind-set? What was I focused on? Was I participating in my own journey? Was I simply sitting there waiting to be saved? Or did I put one foot in front of the other and move forward? Did I believe in myself, and thereby attract the support of others who mirrored that back to me? I thought of times when support just showed up, like it did for

Olga, versus times when I felt like it wasn't there. What was different about those experiences? Was it as simple as my own attitude in the face of the challenge? I made a mental note to pay attention to this in myself and others as I moved forward.

A few weeks after her second surgery, Olga once again found herself sitting in a doctor's office. This time, armed with more than just willpower and determination, she had a strong sense of purpose and commitment to her fundraising cause. Whatever the doctor told her, she had bigger plans ahead for herself. In the end, the second surgery proved to be more than necessary, as they found another lymph node that was affected. Having it taken out may have saved her from further spreading and possibly even relapse.

Dr. Carroll, her oncologist, wanted her to start chemotherapy that Thursday. On Friday, her oldest son, Andrew, had his high-school graduation. "That's impossible," Olga told the doctor. "I'm throwing a party for my son's graduation on Friday, I'm hosting thirty people, and I am not willing to be indisposed from the effects of chemotherapy. We do not know how I will react. We have to postpone it."

When Olga recounted this to me, my eyes grew big and I laughed—this time not surprised by such a daring response.

"It was Andrew's graduation," Olga explained to me as she noticed my reaction. "There was no way I was going to forfeit my son's one-time-only graduation for chemotherapy."

Apparently Dr. Carroll's eyes grew big as well. He attempted to reason with her. "Olga, we need to get started as soon as possible."

"I know, Dr. Carroll, but I waited this long for both surgeries, it can wait another week or two."

Since Olga wasn't budging, Dr. Carroll had no choice. "The next available space we have is June twenty-ninth."

Her appearance on the Kim Fraser show was scheduled for the day before her first treatment. The following week would be full of action, she thought to herself, secretly feeling grateful that they had found an appointment slot for her in the very near future.

Here it was again, another example of how she empowered herself by appropriating the journey to what felt right for her and not others.

"I was not going to change my life because I had cancer," Olga repeatedly asserted to me. "I wanted to keep doing what I was doing: hosting, entertaining, socializing. I even continued exercising and helping out with the school store when I felt up to it. I listened to my body as my doctor recommended, but I continued my daily activities. People would often tell me things like 'Don't go out too much. Be careful where and who you spend time with—you don't want to be exposed to unnecessary germs or stresses.' I thought this was nuts! Being around people is the equivalent of food to me. I needed and wanted to keep participating in life. And as chemotherapy progressed, I stayed true to this intention. In fact, I lived with more passion and inspiration through my chemo than I had ever done in my life."

As Olga explained to me, "The common belief when diagnosed with cancer is that doctors are like gods, and if you do everything they tell you to do at all times, this will save your life. You have no say in things, no power. I did not want to feel like I was at the mercy of the doctors. This was *my* journey, and it was important for me to feel like I had some control over it."

In the case of mankind versus the often scary and intimidating medical system, I felt like Olga had just taken leaps and bounds on behalf of all of us.

Chapter 6

From Fear to Trust:
The Joy of Chemotherapy

"So you're getting chemo tomorrow?" Kim Fraser asked Olga off the air during a commercial break. "Aren't you nervous?"

"Well, right now, I'm here. I'm not thinking about tomorrow. When I've stopped to think about it, I've felt some apprehension, because I have no idea what to expect. I will be stepping into the unknown. In this moment, I'm too excited about being here, on the radio. This is so cool!" Olga answered.

She actually had been nervous about being on the radio and having so many people listening to her. To alleviate the nervousness, she had come prepared with notes to make sure she could answer any question that was posed. When Kim saw Olga sitting across from her in the studio, with supersized earphones on, anxiously sifting through papers, blue eyes shining with nervous excitement, she felt compelled to say something. She reached for Olga's hand. "It's okay, relax, no one can see you, it's going to be great, don't worry! Put those papers away."

Olga smiled at her, remembering to breathe.

The interview went well—so well, in fact, that Kim asked Olga if she could stay the full hour instead of the intended half hour. When they were done, the radio station's receptionist came into the studio and handed Kim a piece of paper. "Can you have Olga call these people back?" she asked.

Two women had called to speak with Olga while she was on the air. Because of time constraints, they were not able to answer all of the callers' questions. "Sure," Olga said as she looked at Kim and the receptionist. "I can do that."

Lots of calls came in from friends and family who had been listening. Everyone was, again, amazed and inspired. Leonard was the first person she spoke to. "Wow," he said, "you did a great job, you had everyone in tears! I'm very proud of you!" Olga felt inspired and excited.

As she drove home, she went over the experience in her mind. "I wondered if maybe this was the reason cancer had come into my life. Am I meant to inspire others more actively?" It was an open-ended question, and Olga remained open in the face of any response that would provide some guidance.

Back at home, Andrew was happy to see her. "Mom, I heard you on the radio today. You were great!" he said. "I heard you mention my name and talk about our family." He flashed her a proud smile.

Olga smiled back, feeling so grateful that Andrew had heard her on the air. These were small experiences that would contribute to weaving the fabric of his values. Above and beyond the wave of inspiration she was creating in her community, she was laying a solid foundation for her children. She was teaching them about thriving in spite of adversity, about finding opportunity in challenge, about making a positive contribution to causes outside of themselves.

The day after her appearance on the Kim Fraser show, Olga woke up to a knot in her stomach. She hadn't slept very well, waking up every hour. "On one hand, I was looking forward to chemotherapy; I was anxious to see what it would be like, to get the ball rolling. But I was also nervous because I did not know what to expect."

Now the apprehension she had felt about chemotherapy had blossomed into full-blown fear. "I'm a control freak," Olga explained to me. "I had no control over what they were going to do to me, or how my body would react." I imagined that from her perspective, chemotherapy was a control freak's worst nightmare.

"I finally just reasoned with myself," she said to me. "I thought to myself, *Let's just go there with an open mind and see what happens.*" And off she went, with Leonard as her chaperone. The drive was quiet. Olga was busy working on keeping herself in a good mind-space.

Olga's oncologist accompanied her to the chemotherapy unit to introduce her to the woman who would be her assigned nurse. There were a number of people waiting in the hall outside the suite; without counting, Olga assumed there was somewhere between fifteen and twenty people. As the nurse opened the door and walked Olga down the center of the room to the very last cubicle, Olga saw over a dozen people who were having chemotherapy administered to them. Some were sitting in chairs, others were lying in bed, some were having breakfast, and others were playing cards—but what they all had in common was an IV. As she observed that some looked well and others looked rather ill, she reminded herself not to jump to conclusions about how she would respond to the treatment.

The first thing she asked the nurses was how they were going to administer the treatment. "Well," one of the nurses explained, "the drugs need to be administered intravenously, and since the treatment is ongoing in most cases, we install a butterfly under the skin that will allow us to hook up the IV without having to poke you every time."

Olga had heard about the implications of this type of procedure, and she decidedly was not going to let them cut her to install a butterfly. She knew that she would have to have the cut cleaned, and that there was a possibility that it would get infected. She just did not want to deal with the hassle of it all. She felt like she had been cut enough. "No, no, no!" Olga said. "You're not going to cut me. You're going to put in a new IV every time. We'll use my foot if we have to, but I do not want to be cut again."

The nurses on duty just looked at her, taken aback by her request. "Well," one of them responded, "if we run out of usable veins, we will have no choice, we will have to put in a butterfly."

"You just keep poking me," Olga replied. They had not yet been exposed to the force to be reckoned with that was Olga Munari. No doubt, she would find a way to avoid that procedure.

"Chemotherapy was the starting point in my lesson on learning to trust," Olga explained to me. "As the nurse set up the IV, and the magic potion that had been concocted for me began to flow into my veins, I had to let go and trust. There was nothing I could do—the treatment had to be administered, and I had to trust the people who chose it for me, trust the person who was administering it to me, trust myself to be okay. I had to trust that everything was going to be okay. It was that nurse, that day, who opened me up to trusting. My life was in her hands, and so I let go and trusted her to do her job. I decided that, whatever the outcome, I was going to be all right.

"I went home that day unsure of what to expect. Jeremy and my dad were impatiently pacing the driveway. My parents and my mother-in-law were there to welcome me. I was quiet and kept a hesitant demeanor. I kept waiting for it to hit me. I was hungry and asked for chicken soup with liver and fried onions, my favorite comfort food. I figured the timeless remedy was best for anything that might be coming. Or I rationalized that liver was healthy for me, and it became my standard meal after every treatment.

"I had been given Gravol to counteract any possible nauseating side effects of the drugs. When a headache came on, my mother-in-law called the nurses over at the treatment center to ask what to do. After they confirmed it was normal, I took two Tylenol and excused myself for a nap.

"The sickness never came. I waited and waited. The headache subsided, and the side effects were minimal. Over the days that followed, I felt more tired than normal; the third day after my treatment was often the worst. My legs felt very heavy, like they weighed a thousand pounds. I would go food shopping and have to go straight back home to rest because I felt drained. As a result of the tiredness, everything I did took longer than it normally would. I learned to listen to my body and respect its limits. But I never got sick. In that way, I was truly blessed."

Later that day, she returned the calls from her interview the day before. She was able to reach one of the women, Romi, a teacher in her mid-fifties who was also battling cancer. "My story is a little bit different," she explained to Olga. "After several surgeries, my cancer has metastasized near my lungs. They can only give me chemo in small doses. But they've been able to stabilize it, and the tumors are shrinking."

As Olga told me about their conversation, I was shocked that this woman had shared the complications of her illness with a recently diagnosed cancer patient. "Olga," I asked her, "didn't she terrify you?"

"No," Olga replied, "actually, she didn't. Romi had asked me to meet for lunch the following week, and as I sat there listening while she told me her story in depth, I asked myself what I was doing there sitting with this woman who was telling me what some would consider a horror story. I felt a lot of compassion for her, and as she talked, I noticed that she seemed like a happy woman and was managing well. So I deduced that, even if I were ever to get to that point, there were still options and hope. Mostly, it made me very grateful for my situation. I felt lucky that my cancer had been caught early enough that I would not have to go through the difficulties she was living with. And I felt even luckier that although my cancer was aggressive, it was treatable. Over the course of my recovery, I heard many cancer stories, and for some reason I never felt afraid. I knew I was going to make it through just fine. I just felt more and more grateful for my situation."

Again, I was amazed at how Olga did not buy into the fear of the "C-word" and its implications. I remember people's reactions when I told them about Jade's diagnosis. There was so much terror attached to the mention of cancer. Before then, I had never been so intimately exposed to cancer, and I was surprised by people's responses. Through all my time spent with Olga, she never seemed to be terrified by cancer itself. Needles, biopsies, and chemotherapy, yes, but when it came to cancer itself, she was grounded, always keeping the attitude of, *Okay, we know what it is, let's move forward and deal with it already.* I remembered her saying to me once, "Acceptance is primordial; denial will not help you move forward." Olga had accepted the reality of her diagnosis and decided she was going to be okay. She asked herself empowering questions that would lead to empowering deductions. Just out of her first chemotherapy session, she was able to sit with Romi, who had been through so much, and see this woman's story as a blessing in regard to her own situation rather than moving into fear.

She went on, "As chemotherapy progressed and my fear subsided, I realized that what I was most afraid of was the unknown. On that first day, not knowing what to expect kept me in a fearful place. I learned that the antidote to fear is knowledge. Now, when I am afraid of something, I realize that it is usually because I don't have enough information."

I found her approach to be such a powerful reframing of fear. It incites curiosity. I, myself, have always been known for my inquisitive nature, always wanting to know more. I believe that curiosity keeps judgment at bay because when we are curious, we generate more openness and less judgment. I had not yet, though, determined that curiosity was also an attitude that could be used as a tool to alleviate fear. But Olga's perspective made so much sense; it integrated well with my own belief. It was easy for me to expand my perspective to include curiosity as one of fear's antidotes. I thought of all the things in my life that I felt fearful about and how I could shift into curiosity to help myself overcome fear.

Although she did not get sick from the chemotherapy, Olga did begin to lose her hair after the second treatment, as the nurse had predicted when they chatted during the first session. But Olga was prepared. Her wigs had been chosen back in April. Weeks before she even started chemotherapy, she had picked them up from their styling stay over at her hairdresser's. She had made a fun day out of wig shopping, taking some good female support—her friend and decorator, Janice—along with her. Afterward, they had bonded and created some good memories, debriefing over lunch.

"Losing my hair was a tough blow to my ego," Olga admitted to me. "Much like any woman, I was so attached to it and wondered who I would be without my hair. Although it was tough for me to accept the eventual total loss of hair, I felt better equipped to handle it once I was armed with a wig. What was really difficult to handle," she explained, "was Jeremy's reaction. Jeremy was so attached to my hair. Every night he would sit on my bed, or I'd sit on his, and he would stroke my hair. One night, after my second chemotherapy treatment, he was sitting on the bed with me and a clump of hair stayed behind in his hand after he ran his fingers through it.

"'Mommy,' he said with a frightened look on his face, 'you're losing your hair!'

"It did not only affect me, it also affected my child," Olga recalled. "It hurt me to see the pained look on Jeremy's face.

"There were clumps in my hand every time I touched my hair. I finally just wished it would all fall out. I called my hairdresser and asked him to make a special house visit, to cut my hair as short as he possibly could." But the hair loss progressed, and even though it was very short, as short as Paolo could cut it without using a clipper, Olga had started to lose whole patches, and it was very uneven.

Janice was over making a decorator house-call one day. "She wanted to show us some samples of fabric for chair covers," Olga explained. "We ended up chatting about how my chemo was progressing."

"Would you show me your hair?" Janice asked.

Olga removed her wig. As Janice ran her hand through the little hair Olga had left, patches of it stayed behind in her hand. "My gosh, Olga, your hair is really going," she said. "You know, you have the wig now, why don't you just shave it all off? I could do it for you if you have a clipper."

Olga just looked at her. "Really, you would do that? It doesn't bother you?"

"No, not if it doesn't bother you," she answered. "I shaved my sister's hair when she had breast cancer six months ago. I'm a veteran now," she added with a comforting wink.

Clipper in hand, they stood in the boys' bathroom as Janice shaved Olga's head bald.

Warm tears slowly streamed down Olga's face as Janice ran the clipper over her head. She felt a mix of emotions. She was worried about Jeremy's, and the rest of her family's, reaction. She was touched that Janice would do this for her. And she was coming to terms with her reality.

When they went back downstairs, no one said a word. They barely even looked at her. Everyone, including Olga, needed to adjust. Jeremy walked in the kitchen, looked at his mom in shock, and walked back out. She thought of their special bedtime ritual and how upset he had been at the hair she had lost in his hand. Olga's heart ached a little to know that her cancer was turning his world upside down.

"You know," Leonard said to her, "I think you look great! I actually find it makes you look really sexy." He smiled at her. She laughed and smiled back. Finally, she put her wig back on and cracked a couple of jokes to lighten the mood. The day went on as normal, without further discussion of what she and Janice had done. It wasn't until that night, before bed, that Olga met her new self. Alone in her bathroom, she removed her wig and stared at her reflection in the mirror. Her eyes welled up, and tears streamed down her face. *Shit, now I really have cancer. It's real.* She just stood there absorbing her feelings.

The next morning, she stood facing the mirror, once again in shock. *I really look like a sick person now,* she thought to herself. She stood there for a few minutes and finally just reasoned with herself the way she always did when she needed inspiration. *This is your reality, you have to accept it. You've got the wig, just keep going, and it will grow back. This is just part of my journey.*

From then on, she wore her wig constantly.

"My looks are important to me, and I wanted to look good. I also wanted how I looked to be a reflection of how I felt. I wore it to support me in my intention to journey from a place of feeling good. Looking good helped me feel good. Moreover, I was in a good place, I was dealing well, and I did not want any pitying looks to make me feel sick or handicapped." She was committed to guarding her disposition and was not allowing her circumstances to define her as a victim of cancer. Beyond what people would think, what was important to her was how she felt. Olga was taking responsibility for how she wanted to feel. She recognized her own sensitivity to the attitudes of others and took the steps necessary to safeguard her stance. She took the notion of not caring about the opinions of others to a whole new level, putting more precedence on how she felt and how she wanted to continue to feel along her journey.

As chemotherapy continued, so did Olga's fundraising efforts. One morning, while going through her routine check-in with her collaborators, she had a flash of inspiration. She would place an ad in her local newspaper, the *Chronicle*, to raise awareness for the walk and reach out to her community for donations. She picked up the phone to make the call.

"Hi, I'm participating in the Weekend to End Breast Cancer, and I would like to place an ad to raise funds."

"Well …" The woman on the line hesitated. "We've already had several ads placed."

"Well, my story is a little bit different, because I have cancer," Olga persisted, determined to get her ad spot.

"You have cancer?" the woman on the other end of the line responded with a dumbfounded tone, "and you're doing the walk?"

"Yes, I have breast cancer, and I'm participating in the Weekend to End Breast Cancer, and I would really like to put an ad in your newspaper to help raise more funds. I'm not asking for a free ad. I will pay for it."

"No, no," the woman responded, "you have cancer, you shouldn't be paying for an ad in the paper. Listen, can I have someone call you back?" she asked with a tone of urgent excitement.

"Yes, sure," Olga responded, a little confused. All she wanted was to place an ad.

Later that day, she got a call back. "Hi, I'm calling from the *Chronicle*. We would like to write a piece about you. Can we interview you now?"

Olga's jaw dropped. She was going to be in the newspaper!

The following week, a man was at her door, camera in hand, to take pictures of her for the article. She opened the door and warmly welcomed him in, smiling. He had a confused look on his face. "You're Olga?" he asked. "You don't really look like a sick person. I can't take pictures of you like this, you look too healthy."

"Well, umm, okay," Olga responded with hesitation. She felt uncomfortable about removing her wig in front of this stranger, and even more so in front of all the construction workers they had coming and going in the house. Their home renovations were still in full swing, and workers moved freely within their home. Olga often chatted with them and served them refreshments, and none of them had ever seen her without her wig. Although, as a rule, she never went out in public without her wig, she knew that the image of her bald head would have a powerful impact on the readers. She mustered up some courage and said, "I can remove my wig if you think it will help."

"Okay, show me what you got," the photographer said with a confused shrug.

That's the thing about Olga's story. It breaks down our preconceived notions of cancer. I was so happy that the upside of cancer had been exposed by the article and photograph in the newspaper. Of course, that doesn't mean the raw and real downside doesn't exist. Having held my best friend's hand in her last breaths after a three-year struggle with breast cancer, I knew it existed. I also knew that even in the calamity, there was an upside. Jade's journey had been our journey. *Ours*, meaning all of us around her. How well she traveled her journey affected all of us. It had a tremendous impact on my life, but her children, her husband, and the people closest to her were forever changed. I'd seen Riccardo, Jade's husband, evolve and expand into a man with a heart deep and wide, with more wisdom and compassion. The experience, which could have made him hard and bitter, made him wise and gentle, yet still unmistakably strong. Despite the indubitable and palpable loss, there were also gains. The amount of wisdom and love that Jade's journey created for those of us closest to her would never fill the void of her departure, but it was just as concrete as the loss. There was a lot of wholesomeness that came from her journey, and it was because of how Jade had traveled. It was because, like Olga, she had chosen to look for the gifts in cancer, and she had intentionally shared them with those around her. Both of them had always been people who touched lives, generous with their warmth and love, and they both chose to let cancer open them up to greatness. Cancer created a beneficence in their lives that they allowed to overflow onto others. Their physical end result may have been very different, but the impact on the lives of those around them was still tangible and memorable. Despite the relentless efforts to find a cure, it still eludes us. The big C has a very marked and growing presence in today's society. The depth of the struggle and sadness that comes with cancer is irrefutable, but the possibilities for growth, love, and joy are just as obvious if we choose to see them. And Olga standing there, looking too healthy, being photographed as a cancer patient walking to end cancer, was a tangible example of the positive opportunities manifested as a result of a cancer diagnosis. Olga was the walking demonstration of the upside of cancer.

"You're a beautiful woman with no hair," the photographer said to her.

"Thank you," she responded with her warm smile.

In the end, he took her picture with no wig but a beautiful smile. The picture itself created a wave of confusion when Olga showed it to me. She seemed so elegant and graceful. The baldness blatantly emphasizes the cancer, which immediately brought up feelings of sadness and compassion, but the bright smile breaking through that preconceived notion creates a short circuit in the brain. I imagine that's what Olga's chosen title for this memoir, *The Joy of Cancer*, will do to many readers.

A few days later, Olga and her son Jeremy had spent the afternoon over at the neighbors, and as the sun came down, Karen, their neighbor, asked them to stay for dinner. By the time they left, it was dark out, so Karen turned on the porch light when letting them out. Jeremy stepped out onto something. They all looked down, and then stood there, jaws dropped. Olga's bright smile and bald head was staring back at all of them. She was on the front page.

"Oh shit!" she let out, before she could stop herself.

They all went back inside. Olga took the paper in with her.

"My understanding was that they would run a small ad," she told me. "The woman said headline, not front page. So I never expected to see myself on the front cover of our local newspaper. I was shocked."

Jeremy was shocked too, and he panicked. He was still adjusting to his mom's bare head. She had been working so hard to get him to be comfortable with it—being playful and having fun, letting everyone try her wig on and taking pictures to create happy memories of them playing. But this was beyond his comfort zone. "Mom, I'm going to go to every house and pick up all the papers, right now!" he exclaimed with loyal determination.

"Jeremy, it's okay. It's okay," she repeated in a soothing tone. "Mommy's okay. I never go out without my wig. Let them see me like this, Jeremy, it's a good thing." She wanted to soothe him, but she also did not want to shelter him. Olga knew it was good for the upside of cancer to get some exposure. Despite being a little shocked, she still felt good about her decision to accept the interview; she knew that eventually, as Jeremy grew up, he would benefit from her attitude. She felt it was better to support him than to shelter him. Still, she had to admit, seeing herself on the front page had been a shock.

Chapter 7

Discovering the Spotlight:
The Joy of Saying Yes

"I knew it was you, but I didn't know you had cancer!" Olga's friend Dale called the morning after the paper had been released. She had seen and read the article on the front page. The two friends had not spoken in months, and so Dale was not yet aware of Olga's diagnosis. But now the word was out. "Those big blue eyes of yours were staring back at me while I sipped my coffee this morning."

The first e-mail Olga received that morning was from Caroline, a good friend of the family. She wrote, "I just stared at those blue eyes thinking to myself, 'I know this person,' and then it clicked—it was you! We've never seen you without your wig, so it took a minute for me to realize it was you. Even with no hair, you look beautiful, and you don't look sick. It's pretty gutsy of you to put yourself out there with not even a single hair on your head. Wow!"

That week Fiona, from the Weekend to End Breast Cancer,

called. "Olga, we would like you to be the spokesperson for our opening ceremony."

Olga was dumbfounded. "You want me to be what? For when?"

"Spokesperson," she replied. "There will be over ten thousand walkers, and we need someone to say a few words at the opening ceremony to kick off the two days. We want you to be that person. You fit the profile. You're bilingual, you're actively fundraising, and you're going through cancer. Moreover, your enthusiasm and determination are inspiring to others. You are the perfect person to take this on. Are you interested?"

"Wow, that's a big challenge. Really big. Can you give me more details? How will those segments unfold? Do I have to write my own speech?" she asked Fiona, her head buzzing with questions.

"Actually, we provide the words for the speech, and there is a special poem to be read. All you have to do is get up there and do what you do best: shine! What do you say?"

"I don't know, Fiona, this is huge. I have to think about this. Why don't you call me back tomorrow, and I'll give you an answer."

All day she pondered Fiona's proposition. She felt a nervous kind of excitement at the thought of being up there, at Olympic Stadium, in front of thousands and thousands of people. But then she would feel like it was too big of a challenge for her and tell herself she was crazy to take this on. Then she realized her excitement was stronger than her fear, and a feeling of certainty began to emerge from within her.

Leonard was away with the kids in Florida, on the family vacation they had planned a year ago. Even though she could not travel, she was firm that they should go without her. She wanted the children to live their lives normally. She did not want them to be held back by what she was experiencing. Jeremy, however, had insisted on staying behind. At the tender age of nine, he was particularly protective of his mother. He wouldn't budge; if she was staying, so was he.

"Jeremy was young, but still old enough to understand that there was big stuff going on with me. I think his protective behavior was also a way of staying close to me so he could feel comforted," Olga explained.

The family kept in touch daily over the phone and through iChat. Every night, Olga would tell Leonard the ins and outs of the day. Here she was, back at home with events happening in domino effect. One night, she was telling him about having her picture taken for the newspaper; the next, she was showing him the *Chronicle*, with her blatant bald head on the front page; and tonight, she was telling him about being asked to be the spokesperson for the Weekend to End Breast Cancer.

"I go away for a week and you're a superstar?" Leonard exclaimed. There was a tone of pride in his voice.

"I'm gonna do it, Len," she said with excitement in her voice and a twinkle in her eye. "When she calls me back tomorrow, I'm going to say yes." Leonard, as always, was very supportive.

When Fiona called the next day, Olga asked for more details on the unfolding of the weekend. As she listened to Fiona describe how the weekend would play out for the spokesperson, she felt butterflies in her stomach. Finally, Fiona asked again, "So, what do you say? Are you in?"

"Yes," Olga replied. "Yes, I'll do it. It's an honor!"

She hung up the phone, her head spinning with excitement. When she called Leonard back to say that it was confirmed, she was the official spokesperson for the 2006 Weekend to

End Breast Cancer, he responded with proud enthusiasm, "Amazing, amazing, amazing! That's you!"

Shortly afterward, Janice, Olga's decorator, called to check in, and Olga told her what she had just agreed to.

"You're unbelievable, you know!" Janice exclaimed with excitement. "You're the only person I know who gets diagnosed with cancer and becomes an overnight local star. The Queen Elizabeth, Kim Fraser's radio show, now this. What's next? You're going to be on television?"

Lo and behold, Olga got a call the following day from Beverly Kravitz, the chair of the Weekend to End to Breast Cancer. They needed someone for a television interview that was being conducted by CBC News. The Ville Marie Medical and Woman's Health Center had just received a new MRI machine that would allow for early detection by radiologists through clearer image resolution. The project cost $3 million and was made possible thanks to funds raised for breast-cancer research. Beverly wanted Olga to be their representative breast-cancer patient for the interview.

"Tomorrow?" Olga said. "Wow, that's really short notice."

"Olga," Beverly pleaded, "we really need you. You're going through chemo right now, you're a symbol of the importance of early detection."

"Okay, okay, I'll do it," she answered. After all, she had always dreamed of being on television.

When they hung up, she dialed Janice's number. "Jan!" she exclaimed into the telephone. "Guess what? I'm going to be on TV."

"What?" Janice replied with excitement in her voice.

"You called it," Olga exclaimed. "The Weekend to End

Breast cancer called again, this time asking me to be part of a television interview for some new MRI technology."

Janice laughed in amazement. "You kill me, Olga. You're the real deal! How do you do it?"

"I don't know, Jan, they called me. I didn't do anything, they came to me."

The following day, she presented herself for the interview in a pink blazer and her breast-cancer-ribbon brooch. It was a short interview, and surprisingly she didn't feel very nervous. Instead, she was beginning to embrace her love of being in the spotlight. If anything, these short appearances were a great way to get used to being in the public eye and prepare herself for more. These were things she had always dreamed of, but they had always felt outside her reach. Now, here they were, in a time that most would consider incredibly challenging, effortlessly falling into her lap. All she had done was open up and follow her heart.

Seeking an answer to the question "How do we bring what we want into our experience?" is like searching for the Holy Grail. What I came to understand through Olga's story is that the first step to manifesting anything is to be open. You can want something with all your heart, but you must be unbarred to it if it is to find its way into your experience. And, on the opposite end of the spectrum, you can want nothing in particular and just have a general intention, but be very open and bring a lot of stuff into your experience, even things you had wanted for a very long time but didn't think they would be a by-product of certain experiences. This is what Olga's journey through cancer did for her. Olga had so often dreamed of being in the limelight. It wasn't until she was diagnosed that these opportunities presented themselves to her. Cancer allowed Olga to disclose herself to opportunity.

Through this openness, Olga made many childhood dreams come true. Cancer gave her a sense of purpose and the courage and leverage she needed to expand and say yes to all the opportunities that she was manifesting. She could have turned away from these opportunities, but Olga had established her intention from the get-go: "How can I take this thing and turn it into something good?" She didn't have a plan, but she was open. When opportunity knocked, she responded with an enthusiastic yes, even though she hadn't planned for it and didn't know exactly how it would play out. Even when fear reared its ugly head—like the first time Fiona asked her to speak at the Queen Elizabeth—she chose to act in spite of it and continued to do so through her journey.

Being on the front page of the newspaper was great for Olga's fundraising campaign. She amassed $2,000 as a result of that alone. Olga had already raised over $35,000. The most amazing part of it was that she only started in April. The walk was just around the corner, and she finally just said to herself, "Screw it, I'm going for fifty!" She had nothing to lose. If she didn't meet her goal, she still would have raised thousands of dollars. Truth be told, she felt a rush of exhilaration from her new goal. "I knew I had set the bar pretty high and that it would be quite a challenge. But I also knew that I'm a go-getter, and when I set a goal, I *have* to accomplish it and will do whatever it takes!" It was so typical of Olga to take something on and excel at it; she had done this her whole life. Whatever she took on, she dedicated herself completely to it and ensured a successful outcome. But this cause combined with her diagnosis created an even deeper level of dedication.

I observed that where most people would focus on cancer as the big challenge, Olga was giving herself her own, more empowering challenges over and above eradicating her cancer.

In one way, she made her own cancer smaller by focusing on things that were bigger to her. In that way, she did not give the disease more power over her than it could have. In another way, she also made her cancer way bigger by using it as a means of expanding her experience of life. Many people, when diagnosed, will wage a war against the massiveness of the disease and create even more resistance against it, thereby depleting themselves. Wars have definitely been known to deplete resources and create casualties. In her own war against cancer, Olga disarmed her cancer by making it her ally in creating a better life for herself. She used its massiveness to give her energy instead of taking away from her own energy.

The more I learned about how she handled it, the more intrigued I was with this concept. I thought back to my high-school and college years and the science courses we were forced to take—which, of course, I believed would never be useful. We learned how all matter is energy and energy cannot be destroyed, only transformed. Probably the only concept I remembered thoroughly from my high-school science class was photosynthesis. I had always been in awe of how plants used carbon dioxide (toxic to human beings) to release waste matter in the form of oxygen (vital to our survival). When I looked at it through that lens, the positive or negative consequence of something depended on what we do with it or how we use it. From this perspective, then, cancer becomes just another form of energy, like carbon dioxide for the plant; if used effectively, it could produce a positive result, like oxygen, propelling growth rather than hindering it. Olga may not have known it, but she had created her own very simple and miraculous form of photosynthesis. She took something toxic and used it to produce her own form of oxygen, which she lavishly shared with others. I coined the term "cancer-synthesis" to describe what she did.

I have learned that, in the realm of happiness, there are two quests that bring the most fulfillment to human beings. The first is to grow and the second is to give. By growing, I mean to expand, to take ourselves beyond the limits of our comfort zone—whether physically, financially, emotionally, or spiritually, in the quality of our relationships or in our work.

There are many ways that people can be thrown outside of their comfort zones. For some, it could be losing a job or a marriage, for others it could be the loss of a limb or being diagnosed with a heart condition and needing to make radical changes in diet and lifestyle. Writing this book took me outside of my comfort zone in many ways. Cancer definitely throws people outside of their comfort zone. On the other side of the safety fence lie many opportunities that often cannot be seen from within its confines. At any time, before being diagnosed with cancer, Olga *could* have joined the Weekend to End Breast Cancer, and she *could* have put her magical fundraising skills and fierce determination to use to amass thousands of dollars, and she *could* have become a fundraising star and discovered something that gave her a sense of purpose and had a significant impact on the lives of others. But that opportunity was not in her field of awareness without cancer as the catalyst. She also could have let Janet walk for her and not joined the team. But she chose to embrace the occasion and all the other occasions that followed. She found that she enjoyed life outside of her comfort zone, even though it took some growing and stretching on her part. She could have simply chosen to focus on getting better and sidestepped all the doors that opened to her, wanting to go back to her life as it was. But there she was, with cancer—in uncharted territory—and she used it to get herself moving forward because she needed something bigger than cancer as leverage to get better. For her, this leverage came through welcoming the opportunity to make an impact on the lives of others while making her dreams come true. A by-product of this was the huge personal growth she experienced. When they say cancer changes your life, it is not the cancer that does it, but the fact that cancer throws us into the unknown, and from *there*, life has the potential to change. When Olga said, "There is opportunity in cancer, if you choose to pursue it," she meant that people can either go back to how they lived and do nothing with the experience, or use it to expand their life beyond what it was. The person having the experience had to be ready to embrace uncertainty.

The thing about trying to go back to the life you had is that life doesn't move backward, it moves forward in time. And once you have had an experience, you've added on to the person you were and you've become more. If you've had cancer, you've gone through chemotherapy and radiation and lost your hair. You've been faced with your mortality, officially. How could you go back to the person you were? You can't undo that life experience; naturally, it will result in some change within you. But the kind of change is really up to the person who is living the experience.

I thought of my friend who had testicular cancer and said it did not change his life. I wondered what opportunities he may have missed. I wondered where his focus was.

And then I thought of Ric, Jade's husband. Losing his wife and being alone to raise his young boys was far outside of the life he'd etched out for himself. I remember him saying to me one day, "I never thought I would find myself at the age of forty-six, alone, having to raise the boys without Jade and having to start over again." In my eyes, the experience made him a stronger, more loving and compassionate man.

Over the course of her own battle, Jade often told me how her diagnosis had forced her to redefine how she related to the people around her. Jade had often referred to herself as an island, rarely reaching out to others. Cancer became her catalyst to open herself to some of the many people around her who wanted to create a deeper relationship with her. She made new friends, and many of her existing friendships deepened. Most significant was the change in her relationship with her husband. She found the courage she needed to reach out to him and express herself in ways that she had not been able to before.

In that way, she created her own form of cancer-synthesis— leaving her husband with a legacy of love and wisdom that he, in turn, used to make himself an even better man. Ric could have let the experience be a reason to close his heart and become bitter. But instead, he used it to broaden the depth of his love and openness.

The other path that I've found brings human beings the most fulfillment is contribution. Giving fills us at a deep level because it makes us feel good, and we make others feel good in the process as well. It allows us to not only support others but nourish them, and when we give we too reap the benefits.

Olga subconsciously knew this early on in life. Volunteering and supporting others has been a part of her life for a long time. "I always wanted to help in some way. I remember being in grade five and rushing to eat my lunch so that I could help the teacher with her assignment corrections. Teachers shape lives, and I felt like I was doing something valuable by helping the teacher.

"Later on in my teenage years, I would volunteer my time on Sunday mornings to coach junior figure skaters. I had been figure skating for a long time, and I loved being able to help the little girls. No one had asked me to contribute my time; it was something I was drawn to on my own. Only later on did I make official work out of it. I went on to become a figure-skating teacher before I was married. And after Leonard and I were married and had the children, I joined the board of the store at LCC. At the time the store sold only clothes— uniform apparel for the students. When I first joined the committee, I felt out of place. I was not sure what I had to contribute. For a long time, I felt like a wallflower. I was low-key, trying to find my place. While observing, I realized there was more we could do with the store. Adding on the school supplies was my idea. And then I created the supply packages that the parents could buy at the start of the year. They would no longer need to go out hunting to find all the items the kids would need. We designed complete packages, and the parents could pick them up at school. It was all volunteer work. Running the store became a full-time occupation for me; I gave forty hours a week of my time. It was my baby. I remember being in Florida on a family vacation one summer and being on conference calls with the committee to prepare for the launch of the school year. I was dedicated because we provided a service that was useful for the parents and the kids. I enjoyed my time there very much."

I was amazed at how much of herself Olga gave to others in the name of contribution. She simply took this to a whole new level in her journey with the big C.

"Cancer made me more of who I am," Olga said to me. "It took what was good in me and made it bigger. I already had the desire and potential within me to accomplish all this, but cancer was the catalyst that helped me bring it out."

She kept searching for ways to raise more funds. The friendly manager at their local grocery store let her put up a table every few weeks where she would sell Olga's Butterflies items to raise more funds. She enjoyed it very much, connecting with people, many of whom she already knew and others she came to know. It kept her busy and determined. One day as she sat, during a lull, an idea flashed through her head. She would host a breast-cancer awareness fundraiser in her home. She would ask her beloved Carole to give a workshop on self-examination and early detection.

That evening, Olga picked up the phone to start organizing the event. "Carole," Olga said as she heard someone answer the phone, "it's Olga." She was smiling. It was always such a pleasure to talk to Carole; they had such a special connection.

"Yes, my beautiful butterfly," Carole answered in her usual warm and loving way. "How are you doing? How is the treatment progressing?"

Olga filled her in on the details. "It's going very well, Carole. I feel lucky because I've managed to be spared from the most difficult side effects. I don't have any nausea or vomiting. I do feel more tired than normal, but I am listening to my body, especially when I exercise. I only do as much as I feel I can handle. I lost all my hair, but I've adapted well with the wig. I wear it all the time. It helps me feel better about myself. And of course, as you know from the e-mail updates I've been sending out regularly, I'm deeply involved in my fundraising efforts and loving it."

"Good, my little butterfly, keep those spirits up, you're doing outstanding. You are such an inspiration to others."

"Speaking of inspiration," Olga went on, "I had an inspired thought and was wondering if you might be interested in helping me out. Do you conduct workshops teaching women how to perform their own breast self-examinations?"

"Yes, I do, of course. Why? What did you have in mind?"

"Well, I want to have a fundraising evening in my home. I will invite a bunch of women and have fundraising items for sale. But I also want to give them something back. I thought it would be nice to teach them how to do their own self-examinations. Would you do this for me?"

"Of course, love, it would be my absolute pleasure. Anything for you. And," she added, "I think it's an excellent idea."

"Great, I'm so happy to hear that."

Olga had come to be notably well-regarded at the JGH, and the staff members were always happy to support her in any way they could. Olga not only brought comforting goodies to share with the staff and patients at every treatment, she also infused those around her with her positive attitude and uplifting demeanor. She was supportive and encouraging with all the other patients. I realized that her energy was as beneficial to the staff as it was to those receiving chemotherapy. Where nurses and other medical staff can easily feel drained by the nature of the work, Olga was breathing life into their morbid environment. Carole's lavish appreciation of her confirmed this. She would often remind Olga of what a ray of sunshine she was and how appreciative everyone was of her presence within their community and of her earnest fundraising efforts. Olga had once said to Dr. Madder, her oncologist, "You guys are so busy, Dr. Madder."

To which he responded, "It's because of people like you

who raise massive funds for us and make people aware of our services that we have the privilege to care for so many people." It was no wonder the staff was so ready and willing to collaborate with her.

A few weeks later, on a Tuesday evening, Olga gathered fifty women in her home. Carole, as promised, guided them through the process of doing their own self-examinations. She had a lot of props for women to practice on as she took them through the diligent details of what to look for. Women were encouraged to go for screening and given all the necessary resources to book appointments. The evening was a huge success, and Olga raised another $2,000.

But more importantly, as a result of that evening, two women were diagnosed with breast cancer. Their lives may have been saved because they caught it early. In fact, those women were still alive and kicking at the time this book was being written.

Later, one of the women who had been diagnosed thanked Olga for hosting an event that encouraged screening. She said to Olga, "You saved my life. Were it not for your evening, I would not have gone to get checked. It just was not something I was aware of. You did something amazing for us, for me. Thank you."

Chapter 8

Sixty Kilometers:
The Joy of Achievement

It was the Monday morning of her last chemotherapy treatment before the walk, also her second-to-last treatment. She was almost at the crossing of one of her milestones. Next would come several rounds of Taxol and Herceptin, with treatments every three weeks for six months, and lastly a grand finale of thirty days of radiation.

The walk was the following weekend, and so far all of her treatments had been successful. By *successful*, she meant that she had walked into the hospital with the intention of receiving chemo and walked out four hours later having done just that. She also meant that she was still thriving. Despite the loss of her hair and diminished energy levels, she was still participating in life as she knew it. She was still listening to her body, resting and not pushing herself as hard as she normally would. That was the most important part. With cancer, she finally found the courage to say no when she felt her plate was becoming too full. Being in treatment created some distance from the usual hustle and bustle of her daily life, and with that distance came new insights. She realized that in her need to please others, she had often brought herself to exhaustion, unable to turn away commitments that she felt were either too much for her or were not adding any value to her life. She had always been terrified of saying no for fear of being disapproved of, or of disappointing others. Now, with cancer, she had the leverage she needed to take on just what was right for her, and with a little bit of courage, say no to the rest. Over time, she found that this actually made a huge impact in her life and the lives of others close to her. What good was she if she committed to something where her heart was not fully participating, or where she was depleting herself? Better to let someone else take that on. She, in turn, was happier, and she discovered that a happier Olga made her loved ones happier as well. Slowly, she saw that people would still love her even if she was not always readily available to help or to connect. In the long run, this had a huge impact on her well-being, as she broke the pattern of running herself into the ground to please others in order to escape the fear of not being loved.

To Olga, *successful* also meant that the nurses had always been able to install the IV, so she had escaped the dreaded butterfly trap. Admittedly, avoiding it required some work on her part. She would pump her veins by squeezing a stress ball on the drive to the hospital. She had Maria, her chosen nurse, apply a hot wet towel and slap her arms repeatedly to get her veins to come to the surface. They were running out of places to poke, though, and Olga was hoping they would make it through the next rounds. The last time they had to poke her on the top part of her hand, which was pretty painful and cumbersome but still way, way better than having a butterfly installed.

When Olga entered the chemotherapy suite, she stiffened as she noticed that Maria was not around. She scanned every nook and cranny of the room and tried not to panic. "Apart from the fact that I loved Maria—we connected immediately and I enjoyed my time with her—I feared that another nurse might not be able to set up the IV, and it being my second-to-last treatment, I was eager to get it over with."

"Hi, Olga, my name is Angie," a nurse said, smiling and extending her hand. "Maria is not here today, so I will be administering your treatment." Olga's heart sank.

Angie began the usual routine checks before the treatment was administered, including a series of questions and the checking of the vitals. When she got to Olga's blood pressure, the nurse frowned. "Hmmm," she spoke with hesitation, "it's a little low. I'm afraid I won't be able to administer the treatment."

Olga's ear perked up. *No treatment, is she nuts?* she thought to herself. She said to the nurse, "My blood pressure is naturally always a little low. Check my chart, you will see for yourself."

Maria would know this, she thought to herself.

"I'm sorry, Olga, I can't give you the treatment with your blood pressure like this. You will have to come back next week."

Next week? Olga, determined to avoid any delays in the completion of her treatment, thought to herself, *That's not going to happen.*

"What do I need to do?" Olga asked. "Is there anything I can take or do that will allow me to get my blood pressure up so that I can get the chemo?"

The newbie nurse just looked at her. "Why don't you just come back next Monday?"

Olga just stared at her. The nurse finally got it. "Okay, hold on, let me check with the doctor."

A few moments later, Angie came back. "Some exercise should bring your blood pressure up. We can get you to walk briskly to get your blood pumping."

"Great, where do I go?" Olga asked, standing up, ready to charge. They had her do several rounds of speed walking through the palliative care unit, and half an hour later Olga found herself sitting across from Angie, watching her set up the IV. Angie had managed to find a vein, and Olga was silently so grateful. Her walking had been worth her while. She made a mental note to ask Maria to indicate in her file that she had naturally low blood pressure and that treatments should be administered anyway. Four hours later, Olga went happily on her way, with one more notch on her cancer-survivor belt.

Now, she just had to mentally prepare herself for the walk.

The week went well; she had no new adverse side effects, although she did feel a little tired. The fatigue was beginning to accumulate from compound chemotherapy sessions. Her sister Anna, who was Olga's biggest fan, was incredibly proud of Olga and also concerned about her walking, knowing that her energy levels were lower than they normally were. Anna felt that even though Olga was thriving compared to many other people who were going through cancer, she still *did* have cancer and needed to conserve her resources to heal. Anna was worried about Olga straining her immune system, which was more essential right now than ever. But Olga assured her, "Anna, it's going to be fine. I promise I will stop or slow down when I need to."

But there was none of that. The two days of the walk were a blur of excitement, enthusiasm, and adrenaline. Olga did not see the time go by, and her spirit was infused with extra positive energy from the thrill of what she had accomplished. In the past few weeks, Olga—having reached above and beyond her required goal of $2,000 and her self-determined goal of $50,000—had been helping her team members reach their goals. A few members still had not made it, and so Olga would graciously ask her sponsors to donate in someone else's name so that all of Olga's Butterflies could reach their goals and walk together. In the end, Olga's Butterflies as a team raised $113,000, and $54,000 of it was raised by Olga herself. Olga was the second top earner for that year's breast-cancer walk. "Actually," she told me, beaming with pride, "I was second after the Jewish General Hospital. I was at home the evening before the walk checking my stats, and I was disappointed because I was second for that week. Among hundreds of teams and fundraisers, three of us battled from week to week for first place. My competitive nature was showing itself, and I was so close to being first—I was less than one hundred dollars short of the number-one spot for that week. Andrew was at home with me. He disappeared for a few minutes and came back and asked me to check to my stats again.

"I just did, Andrew. I came in second, remember?"

"Check them again, Mom, you never know," he said with a wink.

Olga recalled, "I logged back into my account and found that I was now officially the top earner. Andrew had donated the one hundred dollars to make me first. For the weekend of the walk, to be the first-place fundraiser, the fact that it was my son's doing was worth a million dollars for me. My eyes welled with tears as I took him into my arms for a big hug.

My heart was filled with love; I was officially ready for the weekend ahead."

On the first morning of the walk, waiting for the opening ceremony to begin, Olga found herself in the center of Olympic Stadium, standing alone behind the curtain on the big stage facing the thousands of people she would be walking with in a few moments. Her heart was pounding with excitement, and she heard one of the coordinators say her name to call her onstage.

The opening speech would take the participants through a ceremony that would welcome many survivors into a circle at the front of the stage. The survivors were wearing pink, the token color for the breast-cancer cause. The rest of the participants were wearing blue T-shirts. Olga had been instructed to wear pink (people who had undergone surgery and were in treatment were considered survivors as well). Everyone was to hold hands, and words were offered to remember those who'd passed on from breast cancer. Olga spoke slowly and let the words sink in. When she raised her head to look at the participants, she met faces filled with so much emotion; many were tear-stricken and deeply moved. Olga watched the survivors gather onstage with pride, and her heart was full of gratitude for being a part of this huge cause. She thought to herself, *One day, I will stand in that circle and someone will be reading to me as I am to them.*

Olga's opening speech signaled the start of the walk, a sixty-kilometer journey over a span of two days. When she got offstage, her mother, sister, and mother-in-law greeted her with much emotion. They had all been very moved. They exchanged hugs as Olga and her butterflies prepared to set off. There was nervousness and excitement in the air. Her loved ones couldn't help but be concerned for her well-being, but

Olga would not entertain any negative thoughts. Her mind was made up: she would complete the walk successfully and all would go well.

And it did, although rain fell through most of the weekend. Olga found the rain to be a nuisance to the blow-dried and styled wig she wore throughout the weekend, but it helped the women manage the August heat. Olga felt so blessed to be surrounded by her butterflies, the beautiful group of women who courageously joined her on this journey. Her daughter, Laurie, who was still only fifteen at the time, walked with them on the first day. One of Laurie's friends had also been given permission to walk, and so the two girls had joined the team. Mother and daughter walked side by side at times. For the most part, Olga walked ahead of the team, and it was impossible to keep up with her. You would never know that she was battling cancer. Her sister was the only person who could almost keep up with her through the entire weekend. "I had to slow down for Anna to keep up with me, but over the span of the two days, Anna was never very far from me," Olga said smiling. Anna felt incredibly blessed to be sharing this extraordinary experience with her big sister, her hero. She also stayed very close because she knew that Olga was incredibly determined to complete this walk, and she wanted to make sure that her sister was well. Anna was not going to let anything happen to her sister; she kept a watchful eye without letting her concern show. She did not want her own concern to affect Olga's determination and enthusiasm. There were only a few times when Olga said, "I'm feeling a little tired."

To which Anna would reply, "Well, me too, why don't we take a break?" Anna used the opportunity to get Olga to slow down and give herself a rest.

Two intense days later, Olga's Butterflies crossed the finish line. Olga came in running, greeted by Leonard and her family. She extended her arms and shouted out loud, "I did the walk! I did the walk! I can't believe it! I must be crazy!" She looked over at her husband, who was filming her finish. "Leonard, I did the walk! Oh my God!" She was overcome with joy at what she had accomplished.

They had walked the last few hours in a steady rain, and Olga was drenched. While the walkers continued to come in, Olga stood in the bathroom hurrying to dry her wig under a hand dryer. She cleaned herself up and took a few moments to ground herself. Her weekend journey was not over just yet. That morning, she had been asked to give the closing speech. As they called her up onstage, Olga felt the charged energy of the group at an even higher level. Everyone was emotional. There was so much energy in the air, but being onstage seeing the thousands and thousands of women standing in front of her, Olga's sense of pride expanded like nothing she had ever experienced before. As she spoke, the air was filled with vivid emotion. Women were crying and holding each other. There was such a deep sense of triumph and solidarity, it was incredibly moving. Olga was entranced by the whole experience. When she was done speaking, she just stood there taking in the moment. The crowd was clapping like mad, women of all ages celebrating their accomplishment, honoring those who walked the cancer path with their smiles, their tears. Love poured out into the atmosphere.

"Olga, you're done ..." Brian, the master of ceremonies for the opening and closing formalities, whispered into her ear as he put his hands on her shoulders to bring her out of her daze.

"I couldn't believe it," Olga recalled. "The weekend was

done, and I had never before felt so much like I found the thing that I was meant to do. I couldn't shake the feeling that I was meant to do this. Like I always say, everything happens for a reason, and the reason for me getting cancer seemed to be getting clearer and clearer." And then Olga said, "That day, as we left the site, I signed up for the following year's walk. I was going to do it again."

Chapter 9

A Place to Call Home:
The Joy of Forging Ties

After the Weekend to End Breast Cancer, Olga's treatment progressed, and her life as a cancer patient continued. Going to the hospital had become her routine. It became like headquarters, the place where everything began and ended. The staff at the hospital grew to be like a family to her. "I felt safe there," she explained to me. "I came to develop complete trust in them. I knew that no matter what happened to me, I would be taken care of when I was in their hands."

When I stopped to think about it, I realized that the reason Olga was able to build such a level of trust with the hospital staff was because she trusted herself. I thought of the poignant times she had insisted upon having things her way instead of what the doctors or nurses had proposed. The fact that she trusted herself from the very beginning of this journey made it possible for her to open herself to trust others with respect to her treatment. This made her trust completely but not blindly—because she knew at any time, she had her own internal bearings to rely upon as well.

"The people at the hospital knew how to talk to me without me feeling like I was being pitied. They didn't walk on eggshells around me. They were not afraid. There, cancer and everything cancer-related was normal. It wasn't something to be afraid of or tiptoe around. At home, I felt like everyone was walking on eggshells. Sometimes I even felt ignored. But this was partly my own doing," Olga admitted. "I had wanted to shelter my family so much from the more challenging parts of my experience. I would come home from my treatments and act as if nothing had happened. I would not talk about it in detail or tell them about the challenges that sometimes arose. Like the time my blood pressure was too low and I had to get moving to get it high enough for them to be able to administer the treatment. My mother would have been distraught if I had told her about that. So I told them I was fine, and I *was* fine, but I had this whole other life happening outside of the home that they were not aware of. Sometimes I felt frustrated and would think that they should just understand what I was going through. But because I wanted to shelter them, there were a lot of small details relative to my experience or how I felt that they could not be aware of."

At the hospital, she had found a community of people who could relate to what she was going through. With the staff and the other patients, there was a feeling of solidarity that was absent elsewhere in her life. Everyone outside of the medical community was incredibly loving, comforting, and supportive, but she could not help but feel there was an underlying tone of pity or discomfort in their demeanor. Not knowing how to approach her, they were often awkward with her. Going through cancer is a very personal journey, and people who are not going through it may have a difficult time relating. At the hospital, she found people who were at ease with what she was going through. By seeking out relationships with those people, Olga got the support she needed to move through the treatment part of her journey. "It's so important for people who are diagnosed to not hide their cancer but rather to share with others," Olga explained to me. "The more comfortable they are with their situation, the more others will feel at ease, and the less they will feel isolated."

With Jade, there were so many times I felt uncomfortable about her cancer. I felt very sad for her and awkward at the same time. I found it difficult to express to her my own sadness about her battle when what she was going through was monumental. Moreover, I found it difficult to tune in to her needs, because I was not in her shoes and could not relate to what she was living. However, my desire to be there for her, even if only in small ways, was so important that it gave me the leverage I needed to push through the awkwardness and the sadness. I would speak to her honestly, saying, "I feel uncomfortable. I don't know how to help, but I want to do whatever I can. Tell me! Tell me how I can help. Tell me what you need." I forced myself to have the uncomfortable conversations. It did not always feel perfect, because I was not going through what she was going through, but talking about what could have been the pink elephant in her life helped to keep the lines of communication open between us, even if I was six hundred kilometers away in Montreal living my life, while she was in Toronto living with and battling cancer.

Olga had a close friend, Helen, who felt this same awkwardness, and it was a while before she acted on it. Olga got a call from Helen two months after she had been diagnosed. "Helen was crying on the phone. She kept saying how sorry she was that she had not called, explaining that she had felt awkward and afraid. It wasn't until she saw the front page of the newspaper and her husband had urged her to call that she picked up the phone. It was a rude awakening. She realized that she needed to take immediate action or she could one day regret letting discomfort and fear be the ruler of her choices," Olga explained to me.

One of Olga's best friends, Tina, had not been very present until the cancer was long over and done with. "I received a few e-mails and some donations, but no calls or visits. I was confused by her behavior. We were a huge part of each other's lives. It was so strange that she would not pick up the phone or return my calls," Olga told me. "It wasn't until two years after I had completed my treatments that I finally saw her and realized that we had grown apart. I came to understand that some people are simply not able to handle seeing sickness in real life. I also think some people understand that the possibility of death with cancer is very real. As a result, they may feel that they will suffer even more pain if they get closer to someone who then dies.

"All of these factors—pity, awkwardness, the visual aspect of the disease, and the potential face of death—have the potential to create a sense of isolation for the cancer patient," Olga said. "That is why it is so important to reach out and build ties with people in the cancer community and outside of it. The ties forged within the community will provide the support and nurturing needed to move through the months of treatment and ease the isolation felt at home with loved ones who have difficulty relating. The relationships outside the cancer community, either at home with family or with supportive friends, can provide the sense of a normal life."

She continued, "I needed to feel like I was still participating in life; it was imperative to my ability to continue moving forward. I was able to accomplish this by engaging in life outside of my treatments: spending time with friends and family, entertaining, engaging in social activities, and fundraising. Cancer patients have to take the initiative in order to help other people around them feel at ease, and they also have to take the initiative to have their own needs met." Olga continued, "I decided to shelter my family, but I also decided to cultivate other relationships in my life where I could have certain needs fulfilled."

I think this is an attitude to be cultivated everywhere in our lives, not just in the face of cancer. The ability to decide how we want to relate to life, to change, to the people around us—and to take steps to have our needs met—is valuable. It is also how we can create opportunities from even the most difficult situations. The mere fact that Olga eventually came to the awareness and acknowledgment that she had played her part in creating a situation that fostered resentment toward her most beloved meant that she could also, with conscious effort on her part, create a different outcome going forward.

I asked Olga about the flipside, how she believed those who do not have cancer could better relate to those who do. She explained that there's no set formula, but that the easiest thing to do is just to reach out and be present. People should let themselves be guided by their desire to help and support as opposed to being jaded by their discomfort or pity—or worse, by their fear of becoming close to someone who may not survive. Olga affirms that lending a helping hand to others is a life-enriching experience to which there is no equivalent. The reward is the experience in and of itself, and we all have much to learn from extending ourselves outward to others in need. Even after being recovered, Olga still continues to learn and grow by supporting and connecting with those who are still fighting cancer. Her empathy deepens over time, and her life is enhanced by each sick person who she has the gift of interacting with.

The sense of purpose she derived from her ongoing and expanding extracurricular activities, particularly fundraising and reaching out to people within the cancer community, helped her to build a vision for her future. This vision became like a magnet pulling her out of cancer and into health. I remember some conversations Jade and I had on this topic. She talked about creating a vision for her life *after* cancer. Since she had a rare and extremely painful form of inflammatory breast cancer, this was not an easy task; often, the best she could do was simply be present in a single moment. She often lived through the pain moment to moment to get through the day. Still, we occasionally discussed how she wanted to live as she looked ahead, beyond cancer, and as such, I came to understand the power of designing a future that compels us forward. This is exactly what Olga did when she signed up for the walk. It was the future, and yet still close and believable enough for her to achieve it. She continued to build on the experiences that came her way by asking herself how she could use the cancer in a positive way, by seeking out the reasons why it may have come into her life, and by engaging in a life beyond the constraints others may have observed.

Unfortunately, not everyone with cancer has the opportunity to manifest that compelling future. Jade was one of those people; so was Olga's friend and employee, Ginette. Olga got a call form Ginette's daughter explaining that her mother had been diagnosed with lung cancer and that she was not doing well. It was the weekend, and Olga was scheduled for a chemotherapy treatment on Monday. She planned to make her way to Ottawa a few days after that. On Thursday morning, Olga hopped into her car, programmed the hospital address in her GPS, and made the trek to Ottawa from Montreal. She walked into the hospital room to find Ginette, small and frail, lying in her hospital bed. Ginette was not expecting a visit, and it took her a bit of time to recognize Olga. When she spoke, her voice was faint and weak. "Olga, can you tell me what you and I did to get this dreadful cancer?"

Olga's heart flooded with sympathy. She did not answer but reached out her hand to hold Ginette's.

"I'm done, Olga," Ginette went on. "They are just giving me meds to comfort me and ease the pain." Ginette let her hand rest in the comfort of Olga's and slowly fell back into her medicated slumber. The contrast was drastic. Olga felt incredibly lucky to be in her own shoes, going through her own journey.

As Olga sat there watching Ginette sleep, it became apparent to her that she would never see her friend again. Olga cried the whole way home, partly because she knew Ginette was going to die very soon, and partly because she knew for certain that she herself would survive and had been granted a second shot at life. For this, she was immensely grateful. Ginette died a few weeks later, and Olga and Leonard were present at the funeral service. Once again, Olga was confronted with the blessing she had been granted of a continued life. With Ginette's passing, Olga also faced the experience of the sadness and void created by the loss of a treasured friend.

"Nature abhors a vacuum," Jade would often say to me when I faced any kind of loss. I would look at her confused, unable to fully integrate those words. It wasn't until after she died that I understood the profound wisdom in this message that speaks to how nature moves to fill any kind of void. I first came to a clearer understanding of this four months after Jade died, when Vered came into my life. A new friendship with substance was blossoming. Jade had been an encompassing source of wisdom for me, and her presence in my life served as a springboard for my growth. With Vered, I could pitch and catch, as we held many of the same beliefs and challenged each other to grow. Everything I had learned from Jade seemed to have prepared me for this new friendship. With other friends, I often felt like we were on different planes. But Vered and I spoke the same language, and although she and Jade were very different people, they both lived their lives routinely applying the same fundamental philosophy of love and integrity. I felt like I had the opportunity to grow this new friendship into something as beautiful as what Jade and I had shared. The void I deemed to be un-fillable had somehow been filled. Time after time, this has proven to be true in my life since I have integrated this wisdom. I now know for certain that no void goes unfilled for very long. I often think of the forest and rainforests where no patch of earth is uncovered. Grass, plants, flowers, and trees bloom continuously and quickly after one is removed; this is simply the way nature works.

And so it was true for Olga as well. She may have lost some friends, but true to form, nature filled in the gaps with new ones in very creative ways. "One day I got a lengthy e-mail from a woman named Joan, saying that she had received an e-mail that I had sent to my mailing list." Olga continued to regularly send out group e-mails to friends, family, and sponsors to keep them updated on how her fundraising was developing for the walk and how her treatment was progressing, and, most importantly, to thank them for their continued support.

The love and support had continued to pour in as the months passed. There was constant comfort and care from her friends every step of the way. Olga told me about Joan's e-mail: "In her lengthy message to me, the woman affirmed that my e-mail had touched her profoundly, and that she felt compelled to meet me. We began exchanging e-mails as if we were instant messaging, and turns out I had the wrong e-mail and had been e-mailing Joan's husband and not the woman I had wanted to include in my contact list. I had recently added a few names to my mailing list, including (or so I thought) one of the women who had been diagnosed as a result of the breast-cancer awareness evening I had hosted at home with Carole. After a few e-mails from me, Joan's husband assumed that maybe I had been trying to reach his wife, and he forwarded the messages on to her." Olga continued to explain how she had stumbled upon this new friendship. "As luck would have it, their family lived down the road from us, and Joan's husband's business was a supplier for my husband's family's business, and our kids went to school together. That is how the mistake had started; in an attempt to reach the woman who had been diagnosed at the event I hosted, I asked the school secretary for her contact information because I remembered that our kids attended the same school. She had given me the wrong person's e-mail. She had given me Joan's husband's e-mail instead, which is how Joan eventually got involved in the mix-up." More importantly, the magic of the coincidence was that Joan's life had been intimately tied to cancer. Many of her family members had battled the disease. Although Joan may not have had cancer, the life of a cancer patient was not foreign to her. In fact, hospitals and chemotherapy and all the baggage that comes with cancer was something very familiar. She'd lost several family members to cancer, including her mother.

Joan became a source of great encouragement and appreciation for Olga. Their friendship evolved and grew. She very quickly joined the Olga's Butterflies team, walking with them for the Weekend to End Breast Cancer. Joan loved Olga's attitude and fully supported the contrasting light that Olga's demeanor shed on the gloom and doom of cancer. She was well aware of the sadness and tragedy that came from cancer, but she also knew of the joy that came from it. Her belief was that cancer gave people an opportunity to appreciate each other and come to know each other more fully, because in cancer, people are reminded of the impermanence of life. In her eyes, Olga living cancer as a joyful experience was a great example to others who had been touched by the sadness of cancer. "You have it already, and it can be devastating, that's a given," Joan said. "But there are two paths you can take: you can let it take away your joy of living, or you can use it to enhance it."

Joan's perspective got me thinking about cancer and loss. Again I learned there are two paths that can be taken. I thought of Grace, my sister-in-law, and her own view on losing her mother to cancer. As Grace explained to me, "The last two months of my mother's life were a great gift to me. I poured my heart out to her, and she poured her heart out to me, during the last weeks of her life. We had conversations that we would never have had before her diagnosis. Our entire family became much closer. We opened to each other in ways that were unthought-of prior to my mother's cancer. We loved each other even more fully as a result of knowing she was dying. I developed a depth of connection with my mother that I would not have been blessed with had she died of a sudden heart attack or car accident." Grace went on, "We all know we are going to die but have no control over when or how. With cancer, even though it may be a death sentence, we are given time that we otherwise may not have had, to live life more fully." I loved Grace's perspective. It redefined cancer in the face of death, not as a monster but as a bearer of blessings, giving people an opportunity to redefine how they live, how they relate, and how they want to be remembered. Grace added, "I used to think of ⁁e things that I wanted to do when the kids were grown o⁁   ⁁tired. Now I think of how I can live to the fullest   ⁁d that cancer has the potential to change ⁁   but the lives of witnesses as well.

One day, Olga's mc⁁   ⁁rs. "Honey, your father ⁁   ⁁olon cancer," she said.

Olga was quiet ⁁   mother went on, "How co⁁   ⁁ possible? How? Why is th⁁   ⁁ies of lung cancer, then y⁁   ⁁? I just don't understand w⁁

As Olg⁁   ⁁ought back to a conversati⁁   ⁁t diagnosed. Olga remembered he⁁   ⁁hy isn't it me? Why wasn't I diagnosed? ⁁

To which Olga had res⁁   ⁁ don't know why I got it, Mom, but I know that everyt⁁ng happens for a reason. One day I will unearth the reasons why I got it. But *I know why it's not you.* Because you wouldn't have the strength to get through it. And one day you will know your own reasons why it was me and not you." The pieces of the puzzle were now falling into place for Olga. She had seen her mother grow and become a stronger person from witnessing Olga's journey.

When her mother told her the news about her father's cancer, a few moments passed before Olga finally spoke. When she did, her voice was calm and certain. "This is why you didn't get it, Mom. You saw me go through my cancer, and now you can help Dad go through his cancer and give him the support that he needs."

Olga's mother was quiet on the other end of line, as if acknowledging the wisdom in Olga's words. She was in fact better equipped to help her husband through his own cancer therapy. The family's experience with Olga's cancer helped them all deal with *his* cancer. He had strong support from them during the months that led him to recovery. At the end of his own journey, he affirmed to his daughter, "My sunshine, watching you go through cancer with such strength, confidence, and grace has helped me be stronger in the face of my own diagnosis. You have touched so many lives, especially mine. I am so proud of you."

Chapter 10

A Hand to Hold:
The Joy of Helping Others

As Olga's treatments progressed, life continued to be punctuated by magical episodes and opportunities for expansion—and fundraising continued to be the platform from which Olga's expansion sprung. Becoming a leader in a cause bigger than herself gave her the momentum she needed to keep moving forward with poised determination. It furthered her theme of not being a victim. She didn't wait until the journey was over to become an advocate for breast-cancer awareness. Olga had already made up her mind that she was cured, and her actions in choosing to walk and becoming an active fundraiser for breast cancer reflected this. This type of participation in her own journey ensured her continued positive attitude. She played the role of the survivor. It was the leverage she needed to stay determined. She had chosen to spend more time in environments where she felt purposeful and inspired, and this only enhanced her confidence.

I thought of endurance athletes, specifically runners, and

how they eat a lot of carbohydrates before long races to store energy that their body will use en route. The process is called *carbo-loading*. The body stores the carbohydrates as energy and dips into that reserve along the way. I remember Lance Armstrong writing in his book about how he and other cyclists would wolf down carbs throughout the Tour de France to keep refueling those energy reserves and preventing dips in their energy. Olga's attitude was like the carbohydrates an endurance athlete needs to journey through a course; it fed her determination, and the stored fuel prevented her from faltering as she journeyed through treatment.

As she moved smoothly through her journey, Olga felt happy and fulfilled, even as she endured radiation and more treatments. The delectation that she derived through her cancer experience far outweighed the challenges. She milked every drop of joy, using all the goodness as leverage to keep an inspired focus.

As I listened to her tell me her story, I noticed that when Olga spoke, she often put a lot more emphasis on the joyful aspects of her experience than the challenging ones. She would acknowledge the difficulties and express that they had been no picnic to transcend, but I noticed a difference in how she expressed them. She punctuated the joyful moments with a lot more details and vivid emotion than she did the arduous ones. She stacked a lot of the beautiful details of her experience and gave them a lot of weight. In comparison, I noticed that she recounted the difficulties with less emotion and less detail, keeping it more general. When she talked about the unfavorable aspects of her experience, like the panic about the biopsy, the fear of the unknown she had felt at the beginning of her chemotherapy, the anguish she had felt about telling her family, and the discomfort she felt at losing her hair, she kept a certain distance from them. She never piled up the negative emotions and experiences the way she had before she had been diagnosed. She explained that she would acknowledge how she felt and reason herself into moving forward. This is a big part of how she extracted the joy of her experience. It was an easy thing to do moment to moment that would also have an impact on her overall experience of cancer. I thought it was a powerful approach that, if practiced on a day-to-day basis, could greatly improve the quality of people's lives.

What brought Olga the most joy through her experience was helping others. From the very beginning of her journey, she had been called upon to mentor other cancer patients. Her hairdresser had contacted her shortly after she had started chemotherapy. Another one of his clients had also just been diagnosed with breast cancer and was having a really difficult time assimilating the news. He wanted to help this woman in some way.

"Olga," he said when he called one day, "I wonder if you would spend some time with a client of mine who has just been diagnosed with breast cancer as well. She's struggling, and I would love for her to see her situation through your eyes."

"Sure, Paulo, give her my number, and I'll do my best to help her out."

Less than a week later, Olga was stepping out of the grocery store on a rainy afternoon when her cell phone rang. "Hi, Olga Munari?"

"Yes, this is me."

"It's Debbie, Debbie Sheriff. Paulo said it was okay to call."

"Hi, Debbie, how are you?" Olga sheltered herself from the

pouring rain under an awning so that she could give Debbie her full attention. She knew that really listening was the key to her being able to help Debbie.

They chatted for forty-five minutes. Debbie shared the story of her diagnosis. Olga encouraged Debbie to seek out friends and family. But Debbie, unlike Olga, wanted to keep her cancer private; she resisted sharing the news with many of the people close to her, but she did manage to open herself to Olga. The two of them had lunch the following week, and so began Olga's first of many experiences supporting another woman through the breast-cancer journey.

"Debbie wasn't the only person I had met who was secretive about her cancer. This is very common," Olga explained to me. "Often people are ashamed to be sick and want to avoid everything that comes with that label. People look at you differently and treat you differently. It's easy to feel different from others in a negative way. And often people are afraid to tell their loved ones because they don't want to hurt them or, worse, feel like they have to take care of *them* now, on top of managing their own sickness. The thing is," Olga continued, "keeping it a secret is like living a lie, and it adds extra unnecessary burdens to the patient. And it is difficult to hide. Your entire life changes—your schedule, your body, your face. You go through a panoply of emotions on a regular basis. Hiding it only adds to those emotions. I encourage being very open about it so that the door is open for good support to be present. People can't get the support they need if they don't let other people in on what is going on in their life. You have to have a community to turn to." Still, some people, like Debbie, very bravely preferred to go through the experience as incognito as possible. Very few people in Debbie's life were aware of what she was going through. I imagine that, as difficult as it was for Olga to shelter her family as much as she had, it must have been a monumental challenge for Debbie to travel through her journey with so few people knowing what she was living through. She did, however, let Olga be there for her, especially in the early stages. Debbie turned to Olga with many questions, and when Olga offered to accompany her to her first treatment, Debbie gratefully accepted.

"I thought it would be comforting for her to have someone present who had already been through it," Olga said. "So I offered. Debbie was, understandably, very nervous and frightened, as we all are the first day. She still had to develop that level of trust with the staff at the hospital. I knew what she was going through, and I was so happy to be there with her, holding her hand as they inserted the IV. I kept her distracted. The nurse had a difficult time setting her up, and I especially did not want her to see the blood that was streaming down her arm and onto my hand. She would have been mortified, and it would have been even more difficult to get her set up. All in all, it went well. Debbie was very grateful to have me there, and gradually she relaxed a little as the treatment went on. Debbie's son was also wonderful in acknowledging my support to his mother. It felt good."

Olga quickly found that she enjoyed being there for others; it made her feel empowered with respect to her own treatment, which was not yet complete. As the weeks and months of treatment progressed, she continued to be called to mentor other women who were diagnosed with breast cancer. Carole and other nurses from the hospital had made a practice of referring patients who were having a difficult time to Olga. She was called upon to guide other cancer patients way beyond the end of her treatment—five years after her cancer experience, at the time this book was being written, the referrals continued to pour in. Friends would refer friends of friends, and Olga was getting calls or e-mails from people asking if they could extend her contact information to women going through breast cancer. Olga, as always, was ever open to helping others in need. She found that each experience was enriching in its own way and allowed her to discover, more and more, the gifts within her that gave her a sense of purpose. She did, however, have to be mindful of the people she chose to help. She discovered that everyone dealt with cancer differently. "I found that the people who cooped themselves up and hid their cancer had the most negative attitudes and seemed to struggle the most. I encouraged them to be open about their cancer to lessen the isolation and attract more support."

She learned that some people are more difficult to help or may not readily embrace a different attitude. One woman Olga had been asked to mentor was particularly difficult. Olga persevered from week to week, but she felt a bit drained from her time with this woman, as opposed to inspired and rejuvenated from the time she spent mentoring others. "Sara had heavy energy and a tendency toward depression. She, too, like so many women, was living her cancer in secrecy," Olga said. "Although she was particularly difficult, she wasn't the only person I saw who tended toward depression, and I kept noticing how it related to hiding the truth of what they were living." Again, it all related to openness. "If you close yourself off from the world, hiding in shame, secrecy, or fear, you are not letting in the goodness that life has to offer. This goodness is vital to healing. It's like trying to breathe with your nose and mouth closed. Even if you want to survive, the oxygen can't get in unless you find some way to let it in."

Olga continued, "I remember Sara not wanting to go to

her daughter's first day of school because she was sick. I urged her to reconsider, knowing that witnessing her daughter's first and pivotal step into the real world would bring her much joy and would alleviate some of the sadness she had been focusing on."

I agreed that closing ourselves off to the world when faced with challenges is often an automatic response as an attempt to protect ourselves from more pain, but this attitude also prevents the abundance of good stuff like love and support from pouring in. In choosing not to go, Sara would be choosing to refuse to participate in life. This action was contradictory to the desire to survive. In order to live, we must participate in life the way Olga did, especially after she had been diagnosed. Removing ourselves from life only narrows our focus, which furthers our isolation and creates more dis-ease in relation to our challenges. We must continue to interact with life if we wish to continue living it.

A question I often use to broaden my focus when facing challenges in my own life is, "What else is going on besides this challenge right now?" This allows me to see that there is other (and great) stuff going on in my life, which in turn fosters a more inspired viewpoint that helps me to see the particular challenge from a different point of view, with perspectives, solutions, and opportunities I would not have found if I had kept my focus zeroed in on the problem.

Olga wanted Sara to notice what else was going on in her life besides her cancer. "She had a beautiful daughter who was providing her with the opportunity to experience joy and pride by watching her step into her own new beginning, which would create magical memories for Sara to look back on. And Sara could not predict what would come from that day at school. Besides feeling uplifted and inspired by witnessing a special day in her daughter's life, she might also meet someone who could bring another opportunity into her life. Maybe she would make the acquaintance of another mom going through cancer and possibly have that acquaintance blossom into a new friendship. But Sara could know none of this if she kept herself hidden at home, closed off from the world," Olga explained to me. In the end, Sara took her daughter to school, and she later thanked Olga for the nudges and support, as she had felt very inspired and uplifted as a result of it. Watching her daughter step into a new world of opportunity had given her not only a sense of pride but a feeling of being connected to life. It helped her begin to see that she could still continue her life even as she moved through breast cancer.

I thought of the story of W. Mitchell, who was severely burned and lost most of his fingers when a truck turned in front of the motorcycle he was riding, resulting in a crash. Four years later, he became paralyzed from the waist down after he crashed on takeoff in a small aircraft he was piloting. I saw Mitchell speak in Montreal, and this one line from his keynote speech, titled "It's Not What Happens to You, It's What You Do About It," has stayed with me over the last decade: "Before my accident, there were ten thousand things I could do, and now there are only one thousand. I'd rather focus on the thousand left." This is, in part, what Olga was teaching others with the way she lived her journey. When she said, "Okay, I have it, now what can I do with it?" she was focusing on what she *could* do, not how she would be limited. What W. Mitchell did with his experience was to become a motivational speaker and author, spreading his message around the world. Olga, through her experience with cancer, would begin to do the same.

With the help of Olga, Sara progressed and began,

little by little, to exercise her ability to shift her focus as she journeyed through cancer. And Olga got a great lesson from her experience with Sara. She learned that every single person has different lessons to move through and progresses at different speeds. Everyone has limiting habits, and there is always room for improvement. She learned that people have to be ready to change and that you cannot force them, you can only inspire them through your own example. Working with Sara and many others, spending so much time watching people go through treatment at the hospital, she learned respect and compassion for the challenges that we know people are facing and the ones we don't know about.

On top of the mentoring, the year spanning August 2006 to August 2007 was filled with fundraising activities and more public speaking. Olga was called upon several times to speak about her experience at various engagements. She was not actively seeking out these opportunities; people were reaching out to her. A friend phoned one day to ask Olga to speak at an all-girls high school for a breast-cancer awareness day. When Olga asked the girls how many of them knew someone who was going through cancer or had been through it, almost all the hands went up. Olga spoke to them about the importance of fundraising: "It's because of people who give that I am able to be here today speaking to you girls. The donations given for cancer help provide the resources necessary for diagnostics and treatment and hopefully, one day, a cure." She also told them about how she was diagnosed and how her journey was progressing. She felt proud to be able to set an example of strength and determination for these young girls.

Her own kids followed in her footsteps, participating in the Terry Fox Run at school. Terry Fox was a Canadian humanitarian, athlete, and cancer-research activist. In 1980, after one leg had been amputated, he embarked on a cross-Canada run to raise money and awareness for cancer research. Although the spread of his cancer eventually forced him to end his quest after 143 days and 5,373 kilometers (3,339 miles), and ultimately cost him his life, his efforts resulted in a lasting, worldwide legacy. The annual Terry Fox Run, first held in 1981, has grown to involve millions of participants in over sixty countries and is now the world's largest one-day fundraiser for cancer research. Over $500 million has been raised in his name.

Olga was asked to speak at a kick-off for the run in her Montreal borough two years in a row. That event was particularly special for Olga, since Jeremy and Laurie were in the audience. "My little Jeremy sat in the front row, and he was beaming with pride at me," she told me. "I felt so good knowing my children were following in my footsteps, and that I could be a part of their experience in a meaningful way."

Not only was Olga teaching about the value of making a contribution in other people's lives, she was giving them the ability to create joy, no matter the circumstances. I could not think of a more powerful message for children, because when one has this ability, so much can be accomplished. Joy is a valuable resource in the face of any challenge, helping to alleviate fear and isolation, nurturing purpose, making a difference in people's live, and enhancing relationships. Experiencing joy piles value onto all of these things, and Olga's children had an in-house teacher laying the foundation for this ability in their own lives as they grew up. The opportunity to plant that seed in the lives of her own children would, in and of itself, have made the cancer experience worthwhile.

Besides bringing more joy into her life, propelling her forward and helping her to inspire others, Olga also experienced some exciting leisure-time surprises through her fundraising endeavors. As a Montreal Canadiens hockey fan, Olga would jump at an opportunity to meet any of her idols. She had been toying with the idea of hosting a fundraising event with a big-name superstar. Her friend Mary had a connection to Saku Koivu, then the Canadiens' team captain. Mary arranged for the two of them to have a meeting with him. Olga found herself, for the first time in her adult life, unable to utter a word. Mary, who was normally the quiet one, found herself easily chatting away with him while Olga just sat there dumbfounded by the fact that she was meeting with Saku Koivu.

"So," he asked, "what did you ladies have in mind?"

Olga snapped out of it. "I thought of having a cocktail dinner at 40 Westt Steakhouse with you present. We could split the proceeds between your foundation and the Cedar's Cancer Institute associated with the Royal Vic." Mary was being treated at the Royal Victoria Hospital, and so they would donate their portion of the funds raised to the hospital's cancer foundation. Saku loved the idea, and the evening was a huge success, with $40,000 raised in the name of finding a cure for cancer.

Olga was not going to stop at that. She wanted one-on-one time with her hockey idol. When she heard through Mary that an evening with Saku Koivu in your own home was being auctioned off, she jumped at the opportunity.

"Mom," Laurie joked. "You're going to lose the house on Saku Koivu? You already hosted an evening with him, what more do you want?" But Olga was determined. He would be hers.

When she walked into the auction room, she quickly scanned the room to determine who were her top competitors.

She was told that there was one certain gentleman who was set on walking away with the prize. Olga decided to introduce herself and voice her intentions. "Excuse me, sir. Are you Mr. Peters?"

"Yes, I am," he answered.

"I'm Olga Munari," she said. "I hear you are interested in the evening with Saku Koivu. We all know you can outbid anyone in this room," she exclaimed, "but I'm asking you a favor. You see, I'm a cancer survivor, and I would really love to have an evening with him."

"I'll tell you what, Olga," Mr. Peters said with a twinkle in his eye, "I'll let it go." And then he whispered, "But I'm a cancer survivor as well."

Olga's jaw dropped. She said nothing.

"You're a good kid, I can tell," he said to her smiling. "Enjoy your evening with him. I have your back. Go for it."

In the end, she ended up receiving Montreal Canadiens superstar Michael Cammalleri in her home for dinner because Koivu was indisposed. Olga was thrilled just the same. She never imagined cancer could be the door to such magical opportunities. And the enchanted experiences seemed to just be getting bigger and better.

As Olga moved forward with fierce commitment to her cause, what happened was unexpected. The positive feedback from the example she was setting poured in. People were moved and inspired. It was more than Olga could have ever dreamed of. As more and more people affirmed how infected they were by her attitude and her initiatives, Olga began to see herself differently. She never for the life of her imagined that she had the ability to move people so profoundly. As driven and determined as she had always been, she also, underneath it all, had low self-esteem. Through her enthusiasm and dedication, she was motivating others, and in turn, this created an opportunity for her to discover her own value. As the unfaltering support continued over the course of her treatment, she began to see how cherished she was by others. She never could have imagined to what extent her peers cared for her. She had spent so much of her life giving, just like she was doing now, but the context of her endeavors had not previously created an opportunity for her to see, in this light, how impactful she was in the lives of others. Olga's contribution to the cancer community—through fundraising, speaking engagements, mentoring, and a genuine desire to help others—was acclaimed within her community. Slowly but surely, a sense of worthiness grew inside her. She began to feel truly loved for the first time in her life. She saw the value of her own contribution in the lives of others. She developed a whole new respect for herself. She noticed the good things she did more than the bad things, and as a result, she began to cut herself some slack. She practiced being more accepting and forgiving of herself. As her feelings of self-worth expanded, she saw others in a different light as well. She understood now that you can't give to others what you have not given to yourself first. As she became more accepting of herself, she extended the same virtue to others. Where she had been critical and judgmental, now she was compassionate. She understood that she might not know the struggles that others were facing. She recognized the value and contribution of each individual she came across.

"Cancer can change you," Olga said to me with a more serious than usual tone one day. "But you have to be ready."

Olga didn't know how ready she was until she was faced with the end of her own treatment.

Chapter 11

A Sign:
The Joy of Butterflies

"The last day of her treatment was the day Olga fell off the edge of the earth," Leonard said to me. "You would think it would be the happiest day of her life—she had been given a clean bill of health and a second shot at life. Yet that was the day our world began to crumble."

Oddly enough, her last treatment was the only one she went to unchaperoned. It had been a strange feeling, not having anyone there. It also left room for a panoply of mind-chatter and unprecedented feelings to emerge. She was filled with emotion and yet unable to express any of it with words. She wanted to keep a stiff upper lip, but she just sat there and cried through the entire thing. Maria, her nurse, feeling Olga's vulnerability, quietly pulled the curtain to give her some privacy and sat beside Olga.

She looked into Olga's eyes. "You know, it's very normal," Maria said gently. "This hospital has been your home for the last year and a half." She was quiet for a few minutes, watching tears stream down Olga's face. "There is a book, *Picking Up the Pieces*. It addresses the survivorship phase of cancer very well. I think you would greatly benefit from reading it, Olga. And I give a course here at the hospital. We call it 'Back on Track.' It addresses some of the issues with reintegrating into everyday life after cancer."

Olga just looked at Maria. Her eyes welled with tears, and her heart felt like it was broken, yet she had no explanation for it.

She wasn't much of a reader, but she made a casual mental note of the book title anyway. "Thank you, Maria," Olga said, and Maria reached out her arms to hug her patient. They were wrapping up, and Olga hugged everyone and managed to keep herself together for twenty minutes' worth of good-byes. She then put her sunglasses on and made her way through the hospital corridors and down into the parking lot, tears streaming down her face. She had just completed her last treatment. She was done.

"Unexpected questions bubbled up from inside me," Olga explained. "I felt a sense of uncertainty about how things would unfold from there. I wondered what would happen to me and what was next for me." She knew she would have to come out of the cancer retreat, but she wondered if she was prepared for what would follow. She wasn't sure if she was ready to break out of her cancer cocoon. Cancer had become her world and the hospital her home. There were days when she practically had to sneak her way into and through the hospital corridors to avoid all the people who stopped to chat with her. Often she would get pulled aside and asked for more Olga's Butterflies supplies. She had made so many friends among the staff and the other patients at the hospital—they felt like a surrogate family to her. Not having all these wonderful people in her life would create a huge void. There was a sense of emptiness within her.

She got to her car, reached for the door, got in, and sat in the driver's seat without putting the key in the ignition. It was a bright sunny day. There wasn't a cloud in the sky. She just sat there, in stillness, letting the beaming sunlight soothe her. She thought of the course Maria had mentioned, "Back on Track," provided by the Hope & Cope organization in collaboration with the Jewish General Hospital (http://hopeandcope.ca). She told herself, "I don't need it. It's probably all stuff I already know. This post-partum cancer stuff is just nonsense. I'm going to beat this thing on my own." Her eyes moved upward toward the windshield. There, at eye level, was a big, beautiful Monarch butterfly. Its graceful wings were illuminated by the sun, and the outline of the black-and-gold print gleamed. It was serene, beautiful, and powerful. She exhaled. In that moment, Olga knew she would be fine, that she was ready to move forward. And just as those words trailed through her mind, the butterfly flew away. Olga put the key in the ignition knowing she was ready to move forward and create a new life. She didn't know how she would manage; she reasoned that she would just have to take it step by step.

How fitting that she had chosen the butterfly to be her token symbol along her journey. Butterflies had been special to her ever since she was a little girl playing in her grandma's yard on summer afternoons. For Olga, they symbolized transformation and rebirth. Her journey had been transformational, and in so many ways, she felt she had been reborn. She'd lost all her hair, hadn't she? By *all of it*, she very poignantly indicated to me, she meant *all* of it, hinting at much more than just the hair on her head. The week of her last treatment, her last eyelash had fallen off. She was bald from head to toe. That's not all she lost. The chemotherapy had killed so many cells, good and bad within her, that her entire body would be recreating itself day by day. The rebirth hadn't just been physical. Her definition of what was possible for her, of what she was capable of and how far outside of her comfort zone she could step, had been stretched and redefined again and again. She learned that, given the opportunity for a second shot at life, she could step much further outside of that perimeter. Her experience of how loved she was had completely been redefined by the unfailing support of her friends and family. And the rebirth was still not done. She would learn that it was in the survivorship phase that she would come to meet the rest of her new self. This journey was an ongoing process of transformation, in the same way that the caterpillar experienced many forms of transformation on its way to butterfly maturity.

Olga's life became a whirlwind of activity during her big C journey, so many amazing feats accomplished, so much happening on the outside. And quietly, on the inside, underneath all that amazing surface activity, a transformation was taking place. Life had spun layers and layers of magical activity to surround her and keep her safe along her journey. Inside, Olga's spirit, cocooned like a caterpillar during the course of treatment, took in all the nourishment the world was giving her. From deep within, she began to change. The last stage, the stage of breaking out of the cocoon, would be her most challenging. The butterfly must struggle its way out of the cocoon in order to push the fluid out of its body and into its wings so they can take shape and allow it to fly. Olga, like an emerging butterfly, would struggle to break out of her cancer cocoon. It was the stage that would make or break her.

As the weeks and months following her last treatment unfolded, Olga discovered that the whirlwind of activity and excitement from the last year and a half had died down, and everything that came with it had slowed down. She had been given a clean bill of health. As a result, the constant support she had been receiving tapered off. There were no more calls, no more care packages, no more e-mails. Olga, however, had not moved on; she was still making her way out of the cocoon. And all there was, everywhere she turned, was void. "I remember my first birthday just a few months after I had been done with my treatment," Olga said to me. "I didn't get any calls. I felt as if I no longer existed." Weeks like this passed, and Olga fell deeper and deeper into the void.

The impact of the emotional downfall was heightened by Tamoxifen, the post-treatment medication she needed to take for five years following her initial treatment. Because of the aggressive nature of her cancer, Tamoxifen was prescribed to protect against relapse. One of the side effects of the drug is depression. So basically between Tamoxifen, the cancer post-partum (as it is often called), and the common ignorance around it, the odds were stacked against her. Cancer patients quickly become aware of this post-cancer phase after they are done with treatment. The people around them, however, may take longer to realize what is happening and to come understand it. She had lost her foster family and was back at home with her real family, who thought that all was well because she was off the hook. They, and the rest of her friends and support group, were happy that she was healthy and could not understand or relate to the vast emptiness she felt in her life. Even if she tried to help herself, which she did, the side effects of the medication would take her down.

She tried to busy herself with entertaining and work. On

top of the light volunteering she was doing at school, she also volunteered at the breast clinic, doing filing and paperwork a few times a week. She even went so far as to get data-entry clerk work in an office for a couple of weeks, thinking maybe what she needed was a change in career. She quickly realized this was not something she could fix that easily. All these activities kept her distracted from the void she was feeling, but it was like rowing a boat against the current of a rushing river. She could ease the push of the current by keeping busy. But the minute she stopped rowing, the current would take her back down. She felt as if she'd taken leaps and bounds backward after all the expansion that had taken place in her life over the past year and a half. And no matter how she tried to busy herself with the ins and outs of the day-to-day hustle and bustle, with fundraising, with the kids and entertaining, the bad feelings kept reemerging. The guests would leave, and Olga would feel the void and sadness again. She would find herself crying while cleaning up or chopping vegetables for dinner. She would often be crying when she woke up in the morning with no tangible reason why. And she was angry. She zeroed her anger in on Leonard above everything and everyone else.

She became reacquainted with her original problem: she didn't feel connected to or loved by her husband. And she directed all of the sadness from the void and separation anxiety from her cancer treatment toward their relationship.

"All of it was bad for me," she explained. "I couldn't see any good in it. I rationalized that I had been given a second chance at life, and that I could do anything I wanted. Eventually, after months of finding no reprieve from my state of being, I wrote him a 'Dear John' letter, explaining that I was leaving him. I figured I wasn't going to waste this second chance. I would go see what else was out there for me."

Leonard had been trying to keep the ship steady and afloat. He reasoned that given some time and support, Olga would come to her senses. He didn't want her making any rash decisions in the state she was in. He alleviated the crisis with a clear response: "If you leave, don't expect to be coming back."

She felt an unprecedented seriousness in his tone. This made Olga think, *Maybe I can't be reckless and do whatever the hell I want.*

Slowly and patiently, Leonard worked with Olga, trying to help her see through her fog. He wanted her to see that she had chosen this life with him as well. That she also had invested twenty years and more to create the relationship, life, and family they had built together.

Over dinner the night after the "Dear John" letter, Olga lashed out in an escalating cycle of anger and sadness. She and Leonard went to bed that night feeling as if there was an ocean between them. Leonard spent the night sleepless, crying and distraught about their situation. He was clearly devastated. She had never, in their twenty years of marriage, seen him in such a state of upset. She lay awake, taking a step back from her own feelings and thoughts about Leonard. *Shit, he must really love me,* she thought to herself. *Why else would he be so disturbed?*

"I never thought Leonard truly loved me, even though he gave me every sign that he did. After I wrote that letter and saw his reaction, I saw clearly for the first time. I saw him so sad. I had never seen him that way; he was devastated. I never thought that someone could love me that much. It made me stop and think that maybe I should not have done what I did."

The next morning, he found her sitting on the ottoman by the bay window that overlooked their beautiful pool, still in her bathrobe, crying, as she had often been doing, unable to explain what she was feeling. That morning he kneeled by her side and looked into her eyes. "What do you want me to do when you're like this? What can I do? How can I help?"

She looked at him, unarmed, for the first time in months, maybe years. Somehow he had reached into her heart; it was only through being that vulnerable that she had been able to let him in. Here she was, at her worst, a complete mess, lashing out continuously and threatening to leave him, and here *he* was still by her side, holding on to her and what they had built together. It was now, with everyone else gone, and at her most vulnerable, that she needed him the most. She found the courage to express herself. "I don't know, Leonard," she said, her eyes filling with tears. "Maybe a hug would be nice?"

He took her in his arms. She relaxed and took in his affection. Somehow she had found a way to open to him. He had gotten his foot in the door, and he was going to keep it there. Leonard was their lifeboat, strong and steady, moving them forward one row at a time against this dark and heavy current she was paddling against.

"That day was probably the beginning of the long and winding road of us finding our way back to each other," Olga affirmed. "I found the courage to open myself to him, and he was paying attention to my needs. He learned to give me more affection and to express his love for me more often." They began to make progress. Sometimes it was one step forward and two steps back. Olga struggled with her state of being for a long time, and Leonard became the linchpin that held everything steady during the slow process of Olga breaking out of her cocoon. He was not going to let the process break her, or break them.

Chapter 12

Finding Your Way Back:
The Joy of Cancer

Maria, Olga's assigned nurse over the course of her treatment, had been calling every now and then to see how Olga was doing. During the summer of 2007, following the end of Olga's treatment, the two women met and chatted over a cup of coffee one morning. It was just a few weeks after the "Dear John" letter incident. Maria brought up the "Back on Track" course again. "I'm hosting another session in a few weeks," she said to Olga. She saw what Olga could not see through the fog, and that was that the cancer post-partum was a huge part of the dynamic Olga had been playing out in her life. "I have about six to eight people coming," she persisted. "There is a woman from the West Island coming as well. She's a little bit like you; I think you will get along with her. I have space for a few more people. Can I put your name down?"

"Sure, Maria. I will come," Olga answered politely as she took a sip of her coffee. But she still wasn't convinced. She still felt like she didn't need it, like she could fight this on her own.

I got through everything else on my own, didn't I? she thought to herself.

Olga shortly discovered that two of her friends who had survived breast cancer had attended the course and said it had been helpful. So a few weeks later, a very reluctant Olga showed up for Maria's session.

Walking through halls, it felt so good to be greeted by so many familiar faces. She laughed as she remembered the days when she almost had to hide to make it to her treatment on time. It was nice to be greeted so warmly again. It also made her feel sad, because she still felt like she was living with this huge void in her life.

At "Back on Track," she learned that she was not the only one going through this. In fact, there were many others just like her. She got reassurance that others felt the same grief and void she had been feeling, and that there was in fact nothing wrong with her. Until she attended the course, she felt ashamed of how she had been feeling. She wanted to hide from others and get through it on her own.

In other ways, it helped her to see that there were people who were struggling a lot more than she was. And then there were the people who relapsed, some with many reoccurrences. She was reacquainted with the feeling of being blessed. The course served as an opportunity to open her up to resources. From there, Olga was more open to the tools available to her for healing. Feeling more open and responsive, she picked up her copy of *Picking Up the Pieces*. The book, by Sherri Magee, PhD, a cancer researcher, and Kathy Scalzo, MSOD, an expert on transition management, became a source of great comfort for Olga. She made her way through it twice during her journey back to wholeness.

Olga realized that this was one battle she could not fight

on her own. As time moved her forward, and she had more and more moments of clarity of thought and feeling, she realized that Leonard was her greatest ally. He was there when it mattered the most, when things were the hardest, and when everyone was gone. Leonard had stuck by her. She learned to trust him and open up to him. For years, she had been unable to express herself to him. She would shut down when she was upset with him or when they would have an argument. "I would literally curl myself up into a little ball, unable to say a word to him. He would ask me what was wrong and try to have a discussion with me about whatever conflicts would arise, and I would just freeze or cry or sometimes just ignore him and whatever issue was at hand for days. He had never given me a reason not to trust him. He had always been a good husband. I had struggled with trust issues my entire life," she explained to me.

Now, as she began to trust Leonard and trust his love for her, she also began to open up. "In the beginning, it was very scary. Scarier than all the big things I had accomplished during my journey through cancer. As the months passed, I started first by asking him when I needed help with something," Olga explained.

For years, Leonard would watch her struggle with simple things that he could help her with, like something she did not understand on the computer. She would sit there tussling or ask everyone but him to help her. "Now she'll ask me to help her," Leonard explained. "It sounds like something small and insignificant, but it's a big step for Olga, and it was a starting point to bridging the gap between us."

All the experiences and growth she had undergone through her cancer treatment seemed to have prepared her for this new journey with Leonard. Now having the foundation of a sense of being loved by her friends, she could see clearly that this man really did love her. And having trust in herself and trusting others, she could open herself to trusting Leonard. She was lucky enough to have a husband who loved her, and she could finally begin to see it.

She thought of how far they had come from the early stages of their relationship, when she would assert out loud in front of him, "Look at that beautiful woman," beating him to the punch in order to protect herself from her fear of feeling compared to another. She had grown up watching her father explicitly admire other women in front of her and her mother. No matter how much Leonard told her how beautiful she was, she carried this fear for years of him doing the same thing to her. As they moved forward on their journey of reconnecting, she unraveled some of those old fears and began to trust him more and more.

I thought of how many women out there must have good husbands but carry these and other types of past fears and insecurities into their marriages, not knowing that it's preventing them from accessing the current love and appreciation available to them. How many times have we as women said to each other, "He's not a bad guy, but ..." Olga was not the only woman out there who did not feel loved by her man. The love was there, it was always there, but Olga simply could not access it. She had kept Leonard's love at bay in big ways like assuming he would hurt her in the same way her father had hurt her mother. She also kept him at bay in small ways, like not asking for his help when she knew he wanted to give it. She did not want to let herself need him because of the lack of trust she felt. Asking for his help was the first of many small steps she would take as she began to trust Leonard more. Eventually, with a lot of practice, she began telling him when something was bothering her. Instead of letting it fester and living in misery, they would talk it over. It took a lot of patience and practice on both their parts. "It helped that Leonard was very open. He was open to talking with me, and he learned to listen to me attentively. He was open to seeing a psychologist together. Sometimes I went alone, sometimes we went together, and there were times over the years following my cancer when we didn't go as often, but we were both committed to improving our relationship," Olga told me. "As we worked together, I saw that the love not only was there now, but had always been there.

"After a year had passed from the end of my treatment, I decided that it was very important to celebrate my clean bill of health and all that I had accomplished during the course of treatment. More importantly, I wanted to thank all the people who had supported me during the process. All the magical experiences I had along the way would not have been possible without the people who interacted with me. Their love and support had been pivotal in laying the building blocks of my new and emerging relationship with myself and with Leonard. I would not be the person I had become without the steadfast love and support abundantly given by our friends and our family," Olga said to me. "I felt it was important to celebrate being alive in a big way. The party was a huge success. We welcomed one hundred and sixty of our friends and family in our home."

Olga coordinated the entire evening, assuring that every detail was in line with the message of inspiration and gratitude she wanted to convey to her guests. She made an upbeat and moving entrance, singing to the music of Gloria Gaynor's tune "I Will Survive."

Her approach reminded me, again, that she had chosen

not to take on the victim role in this journey she was on. She did not wait for anyone to plan this celebration for her, she simply decided she wanted to celebrate life and all the wonderful people she was blessed to be surrounded by. Her triumphant commemoration got me thinking about all of the things I had accomplished and never taken the time to truly celebrate in a meaningful way. I thought of how many people accomplish great feats without taking the time to stop and celebrate themselves and all the people who journeyed with them. Olga's one-year celebration was, again, another demonstration of her giving back and of her reverence for life and all the joy it has to offer. It was also a great step toward acknowledging her own accomplishments.

"It kept me busy and focused on something positive and inspiring. It helped me battle the darkness I was still navigating my way out of," she explained to me. "The party came and went, though, and I still found myself falling back into that sense of darkness and sadness. I was still struggling, and I knew that this type of struggle was foreign to me. I decided to stop the Tamoxifen. I was off it for a month. Within just a few weeks, the change was significant. I felt like myself again. I had clarity of mind, and my mood was light and happy. Leonard noticed a huge difference as well. My next follow-up appointment with my oncologist was just around the corner, and after seeing the drastic change in how I felt, I was ready to face my doctor and tell him I was not going to continue taking the drug."

She went into that appointment ready for battle. They sat for over half an hour going back and forth. Dr. Madder calmly encouraged her to give the drug another try.

"No, Dr. Madder. There is no way I'm getting back on it. You would not want to be my husband. You would have committed a murder. I'm impossible to live with, and it's the drug. I know it's the drug because I stopped for a month, and I feel like my normal self again."

Finally, the doctor took on a more serious tone. "Olga, you do what you want, but as a physician I have to tell you the facts. You had an aggressive cancer. The fact that we discovered it in stage two and that we removed ten lymph nodes is significant. If it weren't for those lymph nodes, I would be willing to explore other alternatives, but three of those were attacked. As your physician, I have to tell you the truth: if the cancer comes back, it will most likely not be in the breast. It will manifest itself in a place that is incurable. It was a very aggressive cancer, Olga. I just want to make you aware of the consequences."

Olga sat there, just staring at him, feeling frustrated and powerless. "Four more years, Dr. Madder? That's a very long time to be feeling lousy. I just don't know if I can do it," she exclaimed.

"I understand, Olga, but my responsibility is to make you aware of the consequences. You do what you want with this information, but those are the facts."

She went home and let the discussion sink in. She had to find a way to make the side effects livable. Finally, she decided she was going to take the drug. "This is mind over matter. I am going to do this. I'm going to get through this, come hell or high water. This is not going to break me." She wasn't exactly sure what her strategy would be, but she was determined, and she knew that when she put her mind to something, she was a force to be reckoned with.

Olga's self-image was a huge part of her dynamic, and she drew a lot of it from her body image. But now the cancer had left her thirty pounds heavier than she had been before her diagnosis. Her breasts had been butchered twice, and her hair was growing back in uneven patches.

"I remember the first time I went out in public with no wig. It was back in the spring of 2007, after I had completed my treatment, so my hair was maybe an inch long. It was Mother's Day, and we were going for a family brunch. As a gift to myself, in honor of Mother's Day, I decided that would be the day I stopped wearing my wig. Jeremy was so upset that day that he refused to sit with us at the table. I knew it was awkward for him, but I knew that I was ready and that he would eventually come around.

"That day was leaps and bounds of self-love toward myself," Olga affirmed. "While in treatment, as a means of survival, I didn't allow myself to focus on the impact it was having on my body. I didn't focus on the breast scars, the weight gain, the loss of hair. I accepted it as part of the journey and kept my gaze forward. But when the treatment was done, I felt like I was coming back from war. I felt sundered and separated from my surrogate cancer family, abandoned by my friends, and like my body had been handed back to me in ruins," Olga explained. "Cancer was supposed to have changed my life in a positive way, and yet here I was feeling demolished. My breasts had been deformed by the surgery, and for me, breasts are such a pivotal symbol of femininity. Nothing about me felt feminine—not my breasts, not the lack of hair, and especially not the extra weight I was carrying. For months, I avoided letting Leonard see me naked."

I thought of the challenges women face with respect to their body image, and how it relates to their ability to love themselves. It's an easy downward spiral of self-loathing to fall into, even without deformed breasts and loss of hair. For many women, it's a constant battle to love themselves unconditionally. I could only imagine how much more difficult this could be through cancer, after cancer, and while putting the pieces of their lives and bodies back together. As cancer survivors get their bodies handed back to them looking like war casualties, they end up saying, "Who could love me this way?"

"The road back to self-love is long and narrow, but the clincher is," Olga said, "if I love myself, I can let you love me. And that's what I had to do. I had to work on myself, both inside and out.

"I started to exercise," she continued. "I had always exercised, but I got much more serious about it. Leonard took me away for a long weekend at a spa. It felt so good to be there with him. I was able to take a step back from the darkness I was in and get in touch with some feelings of hopefulness. There was a trainer there who was working with some of the patrons. As I watched him, I became inspired. I told Leonard I would spend an hour with him privately and have him teach me how to use a treadmill effectively. It was the best investment ever."

She continued, "In Leonard's eyes, I was perfect. He told me time and time again, but I had to do it for myself. I had to start somewhere and, more importantly, stick to it. This was the start of my own personal journey back to wholeness." In the same way that Leonard slowly but surely nurtured her with love as they found their way back to each other, she slowly but surely nurtured her body and mind back to health. She followed up the start of her exercise routine with a nutrition course. She laid the foundation for a strong basis of health and wellness. With diligent persistence in exercise and nutrition, and steadfast patience, she began to feel better and better. "It took a few years for my body to feel normal again," she explained. "And it took Leonard and I almost double that to renew our relationship. I look back now, five years after the end of my treatment. We've worked hard, but we are in a great place and our relationship continues to get better.

"In a way, the year and a half that I made my way through cancer magnified everything that I didn't feel good about in my life, especially with respect to my relationship with Leonard," Olga explained. "But it was also the preparation for me to be able to make myself and my life better. Especially, it helped to lay the foundation that I needed within myself to be able to move forward with Leonard."

With cancer, Olga's day-to-day routine took on a new face, and at the end of the journey, she came back to her original life as a different person. The challenge she had was to assimilate her old life with her evolved self. Where she thought that she had to throw everything out and start from scratch, she learned instead that she had to sort through things. She had to figure what aspects of herself, her life, and her relationship still served, and which were best discarded.

In her case, the common crash that happens to people after cancer opened the door for her to see everything she did not like about her life and her relationship with Leonard, almost freely, without any fear. And her experience caused her to feel as if she had nothing left to lose. This breakdown actually helped her put the pieces back together better than someone who may not have had a serious illness and the experiences and learning that she had amassed from it. Now, she had a lot more tools at her disposal. She had the seeds of self-worth planted. She had begun to love herself, and in so doing was able to open herself to receiving Leonard's love.

The side effects of the Tamoxifen forced her to work at

things routinely. She understood that there was no magic pill. Life was a day-by-day thing.

I thought of myself when I realized the colossal task I had agreed to, writing a book while maintaining a career that already required an enormous amount of commitment. Time and time again, I would walk away from a writing session disappointed that I had not finished the book in that one session. I kept expecting to finish, or get somewhere, anywhere, just because I had started off the day motivated. I eventually learned that my perspective was the problem and not the fact that I thought I was not getting enough done. I realized that a book was written line by line, paragraph by paragraph; that's how my chapters would get written. There was no magic pill. It was a day-by-day practice that eventually got me to a complete manuscript. This learning reminded me of an anecdote from one of my favorite writer's manuals, *Bird by Bird: Some Instructions on Writing and Life* by Anne Lamott. She writes, "My older brother, who was ten years old at the time, was trying to get a report on birds written that he'd had three months to write, which was due the next day. We were out at our family cabin in Bolinas, and he was at the kitchen table close to tears, surrounded by binder paper and pencils and unopened books on birds, immobilized by the hugeness of the task ahead. Then my father sat down beside him, put his arm around my brother's shoulder, and said, 'Bird by bird, buddy. Just take it bird by bird.'"

As I integrated this learning over the course of the year and a half that it took to get the first draft of the manuscript completed, I realized that this rule applied to my entire life. I was never going to get anywhere I wanted to go and be happy forever. I was not going to get the perfect job and have all my career issues resolved. I was not going to meet the perfect man and instantly be in a perfect marriage. I especially was not going to sit down to write a book and have it completed six hours later. By the time we were done writing the manuscript, I came to look at all the challenges that I encountered in this way. I understood that I was not going to resolve questions or issues in a red-hot minute. This made a world of difference in how I related to myself, to people, and to everything I wanted. This learning, in and of itself, did not come to me overnight, but rather over a period of months during the process of working with Olga and Leonard on this project.

As Leonard so poignantly annotated, "It's not very sexy, but at the end of the day, what we learned is that life is one day at a time and that's forever. There is no end. You're never done. Every day, you work on yourself a little bit. Every day, you invest in your relationship a little bit."

"And again," Olga said as I nodded in agreement, "you have to be open to it."

It's very basic, but this, for me, is the secret to the joy of life.

Epilogue

Six years after her cancer, Olga's adventure of self-discovery continues, and she is more inspired and inspiring than ever.

She continues to mentor cancer patients, now formally associated with an organization that specializes in this field. She also works with referrals from friends and family. She works with her protégés to answer their questions and appease their fears by helping them shift their perspective. Through her own process, she came to understand that cancer patients have many needs in common, but they also have different life lessons to integrate. The work they do together allows cancer patients to develop their own personal tools, skills, and resources that will allow each person to move through their own journey with more ease.

In September of 2007, Olga completed her Nutrition and Wellness Specialist certification with canfitpro—an organization that trains fitness professionals across Canada (http://www.canfitpro.com/)—because she felt that learning how to eat appropriately would accelerate her recovery from cancer and the adverse effects of the treatment. The nutrition course ignited a desire within her to learn more about how to get strong, fit, and healthy again. She went on to pursue her Personal Training Specialist certification, again with canfitpro, which she completed in August of 2011. She feels that her passion for helping others, combined with her cancer experience and her newly acquired health and fitness knowledge, is an asset in her quest to support more and more people through their own difficulties. Now having paired her nutrition certification with a personal-training certification, she not only mentors cancer patients but also works with them to design food and fitness guidelines they can follow during their treatment and, more importantly, after. She has put a strong emphasis on helping people get back into shape after their treatment, considering it a crucial aspect of the recovery process. For many cancer patients, the physical and mental recovery can feel like an overhaul. After her own long uphill recovery, Olga now understands the gradual evolution of the body's healing. Self-acceptance being one of her own paramount learnings, she encourages a bountiful attitude of love and self-acceptance among the people she works with. Her mission is to help cancer patients find the joy in their own journeys and to help survivors reclaim their health.

She continues to share her story with the masses through various speaking engagements. Her love of being in the spotlight has continued to grow, and after giving a talk at Tedx Calicon Canyon in Las Vegas in January of 2012, she decided to officially pursue her passion for public speaking. She is currently an active member of Toastmasters International, a nonprofit educational organization that teaches public speaking and leadership skills through a worldwide network of meeting locations (http://www.toastmasters.org/), and she is also taking courses on public speaking and leadership with the Dale Carnegie Institute (http://www.dalecarnegie.com/).

A contributionalist at heart, she continues to be an active fundraiser, especially for the breast-cancer cause. Her proceeds from this book will be donated in full toward finding a cure for cancer.

More importantly, she continues to inspire others to reach

beyond their limits in impromptu ways on a day-to-day basis. Olga's story and the countless testimonies I have heard are an attestation that if we are open enough, we can also experience transformation through the lives and stories of others. And this is what I found as I took on the role of torchbearer. I found my own perspective on life to be changing. Situations I had labeled as challenges now became opportunities. Things I saw as outside of my capabilities now just required a small stretch from myself, just being a little more open to the possibilities. Experiences I believed impossible now became possible. Through the process of telling Olga's story, I came to see, so clearly, my own potential for transformation. Not only had my perspective changed, my response to life had changed as well. Through telling the story of her transformation, my life was transformed. And so were the lives of many others.

Olga's good friend Beth poignantly illustrated this with the following anecdote. "I had been cycling for years, but when I cycled alone, I rarely if ever deviated from my regular training route. One day I felt moved to try something different. I thought it would be great if I could cycle my way to Olga's to drop in on her. But I clung to my feelings of fearfulness. I was used to cycling in packs and wouldn't venture into new territory without the comfort of my group. Moreover, I didn't know my way around very well. Getting myself to Olga's would mean cycling halfway through the city (and back) alone, over routes I did not know at all. I called Olga up to tell her about my dilemma. 'Oh, so great! Come on over!' Olga exclaimed with excitement, ignoring my apprehensions."

Beth explained, "Olga, of course, would not tolerate my reluctance, knowing I was holding myself back from something she knew I could do and that would greatly empower me. As I carried on with my hesitant banter, she interrupted with a firm tone of voice: 'Beth … just do it!' And there was a click. Olga had hung up. She would not entertain my nonsense. She knew I was selling myself short of what I was capable of. I had a choice to make. I could stay in my comfort zone and continue to hold myself back, as I had often done in the past, or I could take the leap I desperately wanted to.

"I thought of Olga and all that she had accomplished in that short year and half through cancer. Here was this woman who had gone out on so many ledges in a much more compromised position than I was. I had no excuse. Olga's faith and determination inspired me. Her belief in me helped me believe in myself. I put my phone away and hopped onto my bike. There was so much I still wanted to do, so many experiences I want to have. If Olga could leap, so could I."

Courage Is the Heart's Blossom:
A Note from Beth

Olga's name is courage.

Our paths crossed and our lives changed, mine because of her.
Olga is a gift and a hero unto many.
She is an avalanche of spirit, enthusiasm, energy, and joy. A force to be reckoned with! A bright and blazing meteorite.

Her energy is rivaled only by her determination to maximize every opportunity to experience and explore, to grow and evolve. She is a fountain of courage, positivity, inspiration, and motivation. She is a warrior in her absolute resolve to conquer and to go beyond cancer. She is a guiding light for myself and so many others as she creates miracles in her own life.

We say, "I can't," she says, "I'll learn."
We say, "I'm scared," she says, "I'm challenged."
What would have been an ending for most was a new beginning for her.

Her strength of human spirit, her resiliency, her vision, and her courage to say "Yes" to life, to become more than she was yesterday, to not allow negativity to pollute her life, all support her healthy disregard for the impossible!

She is all this with veracity and authenticity.
No butterfly can fly without opening its wings,
No one can love without exposing their heart.
Courage is the heart's blossom.

Thank you for more than you can imagine.
Thank you just for being you!

About the Authors

Olga Munari Assaly
Inspirationalist, Contributionalist

Diagnosed with breast cancer in April of 2006, at a point when some people's lives would have come to a halt, Olga kicked things into high gear, embarking on a momentous journey.

Driven by a heartfelt desire to help others, she values community and contribution and is compelled to use her experience as inspiration to encourage people to support each other and to boldly pursue their dreams.

Olga was born in Montreal, Canada, where she still lives. She and her husband, Leonard, have three beautiful children: Andrew, Laurie, and Jeremy.

"The best gift of all," she says, "is making a difference in just one more person's life."

You can contact Olga by e-mail at:
thejoyofcancer@gmail.com

on Facebook at:
www.facebook.com/pages/The-Joy-Of-Cancer-A-Journey-of-Self-Discovery/191014430966072

and on her website at:
thejoyofcancer.wordpress.com.

Olga's proceeds from *The Joy of Cancer* will be donated to breast cancer research. If you would like to make an additional donation towards Olga's fundraising efforts you can do so at either of the following addresses:

With the Jewish General Hospital: *Olga Assaly's Butterfly Celebrating Life Fund for the Breast Centre*

www.jgh.ca/en/makeadonation

or

With the Cedars Cancer Institute: Olga's Butterfly Fund
www.cedars.ca/

Kim Mecca

Writer, blogger, NLP practitioner, and passionate student of life, Kim Mecca is driven to share all that she learns as she journeys forward as a writer and a coach. Her writing is inspired by the opportunities for expansion and joy in the everyday moments that make up our lives. Inquisitive by nature, she loves to capture the essence of human interaction and human evolution, and she derives tremendous pleasure in using her gift as writer to express her revelations. People-watching at its best!

At a young age, Kim laid down roots as a writer of poetry and prose. Although the poet in her continues to hold a love affair with the rain, willow trees, and Sufi teachings, particularly the work of Rumi, her writing has taken a new path. In more recent years, she married her love for writing with her passion for personal growth, using her blog as a vehicle to share her work with the public.

The human connection is invaluable to her, and she warmly fosters new relationships, so feel free to connect with her by e-mail at: info@kimmecca.com
on Facebook at http://www.facebook.com/kimmecca.author
and on her blog at http://www.kimmecca.com

Bibliography

Lamott, Anne. *Bird by Bird: Some Instructions on Writing and Life*. New York: Knopf Doubleday Publishing Group, 1995.

Magee, Sherri, PhD, and Kathy Scalzo, MSOD. *Picking Up the Pieces: Moving Forward After Surviving Cancer*. Richmond, BC: Raincoast Books, 2006.

Mitchell, W. "It's Not What Happens To You, It's What You Do About It." Speech presented at The Power Within, Montreal, Quebec, July 8, 2005. Available on DVD at http://www.wmitchell.com/onlinestore.html.

"Terry Fox Run," *Wikipedia*, last modified April 28, 2012, http://en.wikipedia.org/wiki/Terry_Fox_Run.

CPSIA information can be obtained at www.ICGtesting.com
Printed in the USA
LVOW060343141212

311378LV00001B/8/P

RETURN
OF
THE TEXAN

Also by L. L. Foreman
in Large Print:

The Jayhawkers
The Mustang Trail
The Plundering Gun
Spanish Grant

This Large Print Book carries the
Seal of Approval of N.A.V.H.

RETURN
OF
THE TEXAN

L. L. FOREMAN

Thorndike Press • Waterville, Maine

Published in 2005 by arrangement with Golden West Literary Agency.

Thorndike Press® Large Print Western.

The tree indicium is a trademark of Thorndike Press.

The text of this Large Print edition is unabridged.
Other aspects of the book may vary from the original edition.

Set in 16 pt. Plantin by Liana M. Walker.

Printed in the United States on permanent paper.

Library of Congress Cataloging-in-Publication Data

Foreman, L. L. (Leonard London), 1901–
    Return of the Texan / by L. L. Foreman.
        p. cm. — (Thorndike Press large print Western)
    ISBN 0-7862-7824-2 (lg. print : hc : alk. paper)
    1. Texas — Fiction. 2. Large type books. I. Title.
II. Thorndike Press large print Western series.
PS3511.O427R45 2005
    813'.54—dc22   2005010631

RETURN
OF
THE TEXAN

As the Founder/CEO of NAVH, the only national health agency solely devoted to those who, although not totally blind, have an eye disease which could lead to serious visual impairment, I am pleased to recognize Thorndike Press* as one of the leading publishers in the large print field.

Founded in 1954 in San Francisco to prepare large print textbooks for partially seeing children, NAVH became the pioneer and standard setting agency in the preparation of large type.

Today, those publishers who meet our standards carry the prestigious "Seal of Approval" indicating high quality large print. We are delighted that Thorndike Press is one of the publishers whose titles meet these standards. We are also pleased to recognize the significant contribution Thorndike Press is making in this important and growing field.

Lorraine H. Marchi, L.H.D.
Founder/CEO
NAVH

* Thorndike Press encompasses the following imprints: Thorndike, Wheeler, Walker and Large Print Press.

1

The road looped giddily over the passes and down the canyons of that mad caprock country of the lower Texas Panhandle, and it seemed incredible that it could actually be a stage road, but the slewed ruts proved that some stageline or other used it, probably out of Travis City, the railhead town. Various trails converged upon this road, all from north and west. Southward lay the great South Plains, recently bereft of the buffalo herds and hurrying now to fill up with the White Man's cattle. The barbarous empire of the Red Man lay conquered. A white man, given energy and fierce ambition, could make a go of it there.

But here in the caprock nothing could prosper, Dain Moore mused. It was not fashioned for the prosperity of human-kind. Maybe the Devil had made it while God napped. It certainly was anything but Godlike, with its raw rock jutting blackly in swirled formations sinisterly reminis-

7

cent of nightmares of hell.

Dain Moore knew those nightmares, knew them from old. He glanced at Sid Beaugrand riding by his side, and wondered if Sid caught on. Sid was so utterly unpredictable, sometimes keen, other times seemingly dense.

"Throwed-away country," Sid said. "Wish I'd stayed where I was."

Dain Moore asked without curiosity, knowing the kind of reply to expect, "Where was that?"

He had started for what he called home, four years late — '69 — and on the slow way back he fell in with Sid Beaugrand, who wasn't bound anywhere but had the moving habit. He and Sid didn't have much in common outside of being broke, but they got along. Even here in the hungry caprock, a tough test for any loose partnership, they got along, more or less.

"Everywhere," Sid answered him. "Anywhere but here. Hell take Texas! I'm hungry. I'm thirsty. I don't like it."

Silently, Dain handed Sid his canteen. Sid eyed the canteen wryly and gulped a swig of the sun-warmed water. He grimaced, and handing the canteen back he said, "Dain, you been everywhere, I guess, like me. Did you ever see such lonesome damn country as this? Pardner, tell me truly!"

"Yes," Dain said.

"Where?"

"Where I come from," Dain said. "Where I was born."

"Where's that?"

"Where I'm going back to. Way down on the Gulf o' Mexico. Ne'er a soul in sight." He owned a two-bit kind of cattle ranch there, on Padre Island, left to him by his folks. His folks were dead. Died off during the War Between the States, in which he served as mounted trooper and scout in the Texas Brigade. He had not seen Padre Island in nearly nine years, four given to war, the remainder to futile wandering: a fighting hand for hire.

Sid said acutely, "I spotted you for a lonesome cuss when we first met up. You wasn't easy to get to know. You're like the song says —" he lifted his voice to a tuneful lilt — " 'Just a solitary stray from far awa-a-ay . . .' "

"Cut it!"

Dain Moore had a decisive bark, when he chose to use it. He was big-boned, tall, an angular kind of man who looked awkward but was not. He had a soberly thoughtful face, traced with lines of half-

9

forgotten humor, and a restrained manner that came of living too much to himself, not so much physically as in mind and spirit. The clothes he wore were all he owned: shabby range garb of an out-of-work cowpuncher. He rode a plain brown horse. People he met often took his air of gravity as a sign of wisdom. He knew they were mistaken.

Sid Beaugrand was dressed no better than Dain, and in some respects he was worse off, having lost his hat and part of his shirt to a drifting snag while fording the Canadian. He wore his bandanna tied around his head at a buccaneer slant, let his rag of shirt hang flapping, and smiled as often as ever. If his smile came sometimes unwarranted, hinting of a secret stock of sardonic reminiscences, that was nothing against him. A man down-at-heel often had to draw his humor from unlikely sources. To his credit, Sid rarely voiced complaint except in a flippant spirit. It took a more personal element than hardship to make his wicked temper pop. Dain, on the surface untroubled by misfortune, disliked any griping man. On that score he found no friction with Sid. The friction prodded elsewhere in growing irritations, sharpening to saw-edge.

They let their horses clop at a walk up a steadily ascending series of switchback humps, into another rock-walled pass where the snaking road, Sid commented, turned and bit itself. Unexpectedly, the walls spread out and the road levelled to a halting spot on the crest of the ridge. There they dismounted, stretched their legs, aired saddles.

Far on south, beyond further ridges, an orange streak underscored the blue sky. Dain nodded to Sid's inquiring gesture. "The South Plains."

Sid said, "Looks as empty as this, only a damn sight flatter. Cattle country?"

Dain nodded again, and Sid brightened up. "Got to be a town somewheres along, then. That's for me! You can have your lonesome land." He moistened his bandanna from Dain's canteen and tied it back on his head, for coolness. "First thing I better do there is get me a hat."

"Better get a job first to get the money to get the hat," Dain suggested.

Sid chuckled. "Man, when you say more'n three words at one throw you complicate talk till hell won't have it. Try whistlin'."

Early dusk caught them in the mammoth

jumble below the caprock, where granite upheavals of the Breaks crowded them on all sides. The narrow road twisted in every direction as if in a frantic craze to avoid climbing out of each crooked canyon. It still was a stage road, and for some time they heard a sound behind them that gradually sharpened into the clattering rattle of wheels and hoofs, but neither man spoke of it, knowing that the other listened to it. Dain didn't have the habit of remarking on the obvious, and Sid had fallen into one of his abstracted spells.

Only after the noise at last boomed loud enough to mention it did Sid speak. "I sure don't look forward to another hungry camp tonight. Maybe this stage comin's got some grub aboard." The chance of that wasn't worth a serious thought, but he reined in his buckskin, a snaky animal as lean as himself. "Could eat the driver if he pokes tender."

Dain pulled up with him and they sat looking back. The stage lurched abruptly around a bend, six-horse team jingling fast, and they both raised a greeting hand, expecting in return a wave and a passing word. This evidently was an express outfit making record time, the high-slung coach built light and tight like a racing rig to accommodate fast horses.

Bobbing and bowing on its leather thoroughbraces, the stage came on down the hazardous road without slackening speed, sheer cliff on one side and a canyon on the other. There was no harsh groan of brakes. The driver, a black figure in the dusk, rose up on the box and whirled his long whip and the popper cracked like a shot. He huddled back down quickly and worked the lines. Under his expert manipulation the leaders swerved slightly back and forth. The stage swayed from side to side, in and out of the worn ruts.

Dain reined his horse hard up against the cliff. He saw Sid, on the outside of the road, haul over to the edge.

The driver rode alone on the box, stiff-faced, eyes glinting, the lines slotted between his fingers, playing his maniacally dangerous trick. The coach was painted yellow, with green trim. It slammed past and even Dain's steady brown horse jerked all over and flung up its head at the close scrape.

Then it was gone and Dain heard Sid cursing in a high pinched voice, and through the backlash of dust he saw Sid yawing far over in his saddle while the buckskin clawed chunks flying from the edge of the road with its hoofs, fighting for footing. It was a long and steep drop down there into the canyon.

Sid got his horse back onto the road, missing taking the long fall by the narrowest margin. Still cursing, he stroked his gun out. The buckskin danced a circle while the racing stagecoach and team careened on around the next bend. Sid wrenched at the reins to line out the buckskin and dash in pursuit. His dark eyes glittered, slitted, the underlids quivering.

Dain said, "That buster's in a hurry."

"When I get in shot of him he'll —"

"No." Dain put his horse across to block Sid. "Lay off that."

Sid looked at him. He bent forward in the saddle with his gun rising, and all the saw-edged irritations concentrated in a flash between them. Dain shook his head then and asked quietly, "Are you that crazy?"

Checked, Sid licked his thinned lips, closed them. He lowered the gun and said, barely audibly, "Not crazy like you." He ran his stare over Dain's sober face. "Don't you *ever* get mad? He run me off the road!"

"Did me, too," Dain said. "Like to've printed me in the rock. Guess he mistook us for road agents. Knew what he was doing."

Reining out of Sid's way, he touched his horse onward. The bad moment was gone, he knew, though perhaps only postponed. Mutual liking didn't affect it much. You could like a man, and still flare into a murderous rage at him when thrown together in close company too long. Hunger, nerves, haunting memories good and bad, the loneliness for a woman, many things contributed to it. He and Sid had done well to get this far without a fight, he guessed. Be wise to part, soon as they could get off this ungodly road.

Sid muttered something and fell in alongside him, but leaving plenty of space between, for even their horses had lately taken to wall-eyeing each other. They spoke no more about it until they rode out of the rocky maze and could see the road ahead, a pale ribbon in the darkness, spiraling downward in hairpin loops before it stretched out toward the open plains. They could see, too, that the stage-driver had run into trouble.

Practically beneath their feet, but far down, the stage stood halted just beyond a steep curve that must have forced the driver to slow his team to a walk. The halt was enforced by a group of a dozen men or more who barred the way. They had lighted a fire in the road, and seemingly waited now for it to come fully ablaze so as to see what they were doing, as though they had all the time in the world and Texas. The driver stirred restlessly on the box, and firelight ran up the lines taut in his hands.

"Did I say this country's empty?" Sid drawled to Dain as they reined in to watch. "Hell, he's busted into a whole population!"

Dain folded his hands on the saddlehorn and yawned. He was perfectly willing to see the driver set back a notch. He felt uncharitable toward that road-hogging maniac.

The men were on foot, their horses bunched off the road where apparently a rough trail of some kind cut in. At that distance and angle of view they were small, foreshortened figures attached to elongated fire-shadows. About half of them wore big hats and bore the lounging stamp, distinctive and inimitable, of high-heeled range men. The others could have been anything. One of them had on riding breeches and a straight-brimmed hat, light tan or gray. He stood apart, that one, on a

16

hummock away from the fire, overlooking the proceedings, such as they were. The air of the whole group was casual and desultory, but a motionless man with the gaunt shadow of a hawk held a rifle loosely pointed up at the stage driver.

"I hope they lay his whip on him," Sid remarked, restored to better humor. "The gent off to himself is the top cockadoodle, betcha, I'm for him, spite of his fancy pants, but I wish he'd open the show and quit 'em loafin'. We ain't got all night."

"Yell a complaint."

"Wouldn't want to do that. They bulk gun-heavy, notice, an' we ain't paid for these gallery seats." Sid legged down off his horse. "Might's well get comfortable." He went to the road's edge and lay on his stomach, chin cupped in his hands, gazing down.

At a halt it was Dain's habit to loosen cinch and air the saddle. In his hands any horse found good treatment. He performed the chore now, and as he joined Sid he heard a sharp slap from below. "What was that?"

"Top cockadoodle slapped his pants for 'em to open the show," Sid answered. "Funny, I never 'member how sounds come up loud in a place like this, like out of a well. Always s'prises me. Sounds don't go down'ard much. Wonder why?"

"Far as I know, sound travels mostly on air, and air travels upward. The sound's clear according to how clear the air is. Then, too, here you get a kind of echo off all this rock."

"I might know you'd complicate it after three words."

The big-hatted men down below moved to the stagecoach team. They unsnapped the trace chains and let them fall. One of them gathered the lines and tugged at them. The driver held on, true to his type, governed by stubborn pride in his trade and position. The man swore, and another came to his aid and between them they tore the lines from the driver's hands and tossed them loose.

Dain began feeling some sympathy for the driver, who didn't appear to be a young man. He said, "In a way, we're responsible for the jam he's in. I guess he was on the lookout for trouble. He thought we were it. After he got past us, he thought he was in the clear. He didn't keep his guard up any more, and they were waiting here and caught him easy."

Sid shrugged. "There you go again, complicatin'. Any time some bullhead tries to smash me up, no matter what reason, I sure ain't gonna sob at his funeral!"

A commotion below broke out. Yelling, kicking, slapping with their hats, the men stampeded the team. The six terrified horses bolted, dragging a tangle of harness and lines. They dodged off the road and were lost in the darkness. When the racket died out, the men still stood gazing after the vanished team. The brief flurry of excitement died with the noise, and there was silence and no laughter.

A tired and heavy voice rose in the hush. "Mr. Bledsoe, you know I only work for Mr. Walker because you fired me."

It was the driver speaking from the box of the un-teamed stagecoach, each word a small explosion of sound, slow and distinct, that by the trick of acoustics seemed directed purposely upward to inform the high rocks rather than to seek to impress any human ears below.

"I'm only one of Walker's drivers, Mr. Bledsoe. He gave me orders to race to Travis City and back, empty. No shotgun guard. Just a test run."

The men who had been gazing off brought their faces slowly around. They looked at him perched there alone. By their attitudes they conveyed cold rebuke.

They didn't want to hear him.

The man standing on the hummock called curtly, "Hollen!"

"Hollen?" Sid murmured. "I knew a feller that name. Funny if he was down there. No, he wouldn't be here in Texas. Leavenworth, last I heard."

The rest happened quickly. The hawk-shadow man with the rifle nodded response to the man standing on the hummock, who then motioned to the big-hatted men.

The stage-driver's voice rose to the high rocks: "I haven't even got a gun on me! I —"

Shots hammered briskly. The driver swayed, his hands reaching for a hold as he tried to step down with some fumbling attempt at dignity. His right hand missed the baggage rail and fastened onto his leather coat, which hung folded over it. The leather coat came loose in his grasp and he pitched out. After he struck the road the man named Hollen fired his rifle twice into the body and sang out, "All right, now the coach!"

They finished swiftly, loading the dead driver into the coach and pushing the coach forward onto the fire. The leather coat was kicked to the blaze, and they mounted their horses and vanished into the rough trail branching westward. It was all over in a minute. The coach, straddling the fire, was already beginning to burn.

Sid caught at Dain's arm. "Not so fast! Give 'em time to get gone! We can't do nothin', anyhow!"

Dain shook off his hand and pushed him aside. "That poor devil might still be alive!" he growled, striding to his horse, and after tightening cinch he took off at a lope.

Dain hauled in hard and dismounted, but the brown horse feared fire and he had to run with it farther down the road before bringing it to halt. While he hastily tied the reins to the brown's forefeet, Sid flashed past. Sid got his horse turned, fought it to a standstill and tied it, and came sprinting back.

They ran to the burning stagecoach. The heat billowed at them. Sid backed off, pointing at the sign lettered across the door of the coach. The gilt paint blackening, but brightly visible in the flames, the sign proclaimed: *South Plains Stage & Express—Walker Company.*

"I reckon Mr. Walker is out one stagecoach!"

"And one driver," Dain said. It was impossible to get at the coach. If the driver hadn't been dead when they pitched him in, he was now.

Sid wagged his head, puzzled. "They didn't rob it, far's I could tell. That ain't sensible!"

Dain snapped, "Is murder ever sensible? They weren't just road agents. This was cold murder!"

"Hot, I'd call it!" Sid said. "I coulda done a neater job o' stickin' up this rig, myself. No profit in burnin' it. Say, you sure came down in a rush. That ol' nag you got ain't bad."

His restlessly observant gaze spotted the driver's leather coat smoldering close to the fire. Shielding his eyes from the heat with an upflung arm, he darted in. He looked predatory and piratical in his rakish headrag and black stubble of beard, the glaring flames enhancing the hunger hollows under his cheekbones. He whirled back trailing the smoking coat by a sleeve. A burning wheel slowly collapsed and the coach tipped ponderously as if in pursuit of him, grudging him that scrap of salvage.

Slapping the coat on the ground, he beat out the smoldering patches. A cheap wallet flew from a charred pocket. He snatched it up and shook out its contents: a few papers, no money. He examined the coat and found it too burned to wear. A search of the pockets produced nothing. With a shrug he tossed it at the blazing stagecoach and kicked the empty wallet after it.

"No luck this trip!"

The few scattered papers fluttered, sucked into the heated up-draft, winnowed like chaff in a breeze, leaving behind a piece of cardboard the size of a playing card. Dain rescued the card, frowning over Sid's careless mishandling of the dead driver's property. He turned it over and looked at a photograph. He tilted it toward the fire for better light, and kept looking at it. His face softened some. Then he was smiling, because the girl in the picture smiled up at him and that was the effect her smile had on him.

Deeply absorbed, unconscious of Sid behind him peering inquisitively over his shoulder, he came to himself with a start when Sid exclaimed softly, "A lulu! I'd go a long old way to know *her!* Whoo-ee! What I'd give!"

He swung a stare of quiet outrage at Sid. The girl resembled in no way a dancehall queen, a lulu to be had. To speak of her in that tone and fashion soiled admiration with insult. Although he knew Sid had spoken the first words that came into his head and was given to lusty reaction where women were concerned, he had to restrain a rabid urge to sling a backhand blow at Sid's mouth.

"Any idea who she is?" Sid asked, making to take the picture. "Better still, any idea *where* she is? Hell, names don't matter. I could forget my own, with her!"

"We'd maybe know, if you'd taken a little decent care with the wallet," Dain answered, tucking the picture into his shirt pocket. He couldn't keep the metallic crackle out of his voice.

Sid eyed him levelly, at once bleakly unsmiling. "You may be right. But don't jump me on it like that, hear? I'm feelin' a mite techy."

"So am I. We better split up."

"You're tootin' we better! Gimme that picture!"

"The hell I will."

They stood regarding each other, faces not more than a foot apart. The underlids of Sid's narrowed eyes began quivering.

The first shot that cracked from somewhere up the rough side-trail sent them ducking instantly out of the firelight, but Sid skated on a pebble and took a flying sprawl, and Dain passed him. Reaching the nervous horses, Dain got the brown's reins untied and he vaulted into the saddle and set off. That was enough for Sid's buckskin, never a placid brute at best. It whistled a tremendous snort, snapped its reins, and came bolting along behind.

Sid yelled once, a thin cry shredded by the hammering of gunfire. Men up the side-trail were spraying shots indiscriminately down at the road, and not always hitting it in the tricky flare of the flames, for ricochets kept whining off rocks. Dain looked back and glimpsed Sid's shape against the fire, bent over and sprinting on foot, hugging the rock-shouldered side of the road for cover. He broke gait and let the runaway buckskin come alongside and made a grab at its bridle. The buckskin shied, but on the second try Dain caught what was left of a broken rein and hung onto it. He forced both plunging animals to a tight turn-about and rode back.

Sid didn't utter a word, didn't even swear at his horse as he jumped at it and scrambled into the saddle. He was too spent, gasping for breath. In a couple of minutes, though, racing side by side with Dain, he sang out disgustedly, "Let's get away off this cussed road! It's a hoodoo!"

2

The faded sign in the window of the little cowtown café stated uncompromisingly: *Dinner 50 cents.*

Dain searched his pockets, watched morosely by Sid. Haggardly travel-worn and in need though they were, they could see nothing much about this town to arouse optimism. Its name was Flint Corners and it looked it.

It squatted at a crossroads, drab buildings baking in the sun, surrounded by plains where only dust-devils broke the flat bareness. A few listless ponies dozed droop-headed in the yard behind the smithy's shed, but the street hitchracks stood empty. Some faint sounds issued from the battered saloon on the corner nearest the café. A mongrel dog slept in the shade, stretched across the threshold of a tiny bank on the opposite corner. The bank had a single front window, barred, and had probably once been the town jail.

Nobody cared enough to come to the saloon door and look the two strangers over. Even the prospects of a free drink appeared dim in this mean town.

Dain's search dredged up a dime, surprising him. Sid, sighing, swung down off the buckskin and stretched his legs. He peered again at the café sign.

"Fifty cents, does it say? Tell me I'm wrong!" He laid a hand tenderly on his stomach. "My belly's on rubbin' terms with my backbone. That dime all you got?"

Dain nodded. He continued searching, however, hoping to surprise himself some more. When he got to his shirt pocket he automatically drew from it the small picture that he had picked up at the burning stagecoach. And once more, while he gazed at it, his face softened to a smile. Hungry and tired as he was, the smiling girl in the picture still brought that uplifting effect to him.

Sid spun slowly half around and regarded the little bank on the far corner. He moved off a few steps, idly. His shift of position put him where he was able to get a view of the bank's interior through the open door.

Despite the blazing sunshine, little light entered the bank, and a lamp burned over the teller's cage. The teller appeared to be the only busy person in town. He was counting coins from one canvas sack into another, an unconstructive occupation. Sid moved back to his horse.

He looked all around as though searching for something else to take his attention. His ranging glance whipped back to the bank, the irresistible magnet, and stayed fastened on it. A curious expression crept into his face, a look hard and reckless and at the same time comically wry, like that of a dry drunkard reaching for the drink that he knew too well would send him off once more. To Dain he murmured, "Guess I'll cash a check."

Dain nodded absently, for Sid's witticisms were obscure at times. He put the picture away and walked to the general store, where he splurged the dime on coffee beans. This being cattle country, a man could knock off a calf if the owner wasn't around, but coffee didn't grow in Texas.

When he left the general store Sid was nowhere in sight. Only then did it occur to him that Sid had not spoken in a joking tone.

Sid stepped over the sleeping dog. The

28

dog rolled its head and cocked a filmed eye upward at him, and Sid said, "Hiya, hound!" The dog twitched a languid tail and stretched more fully in the shade, groaning with comfort.

Inside the bank, Sid accomplished his business smoothly, speaking in a soft voice, his eyes like polished jet. He paced to the teller's cage and said, "Hiya, pop!"

The white-haired teller rose from his stool and came to the high counter, adjusting his steel-rimmed spectacles. He sniffed as if detecting a bad odor. "What d'you want?"

"To a blunt question," Sid said, "a blunt answer." He lifted his gun above the level of the counter.

The teller sucked in his lips. He stared into the muzzle of the gun. Without removing his eyes from it, he reached aside to a stout little canvas sack packed tight with coin and its drawn pucker-strings tied. He slid it along the counter to the pigeon-hole in the heavy wire cage.

With his free hand Sid took it. Backing to the door, he brushed past a hatrack. A hat hung there, a black beaver, flat-crowned, the type that a conservative man bought to last a lifetime. Sid transferred the sack of coin to the crook of his gun arm. He scraped the bandanna off his head and tried on the hat.

"Fits fine," he assured the teller, and stepped swiftly out over the sleeping dog.

Dain didn't at first perceive the sack cradled in Sid's arm. It was the hat that took his eye. Somebody inside the bank let out a roar of indignation that lifted the startled dog out of slumber and set it to barking crazily. The elderly teller burst forth brandishing a huge cap-and-ball pistol. He tripped over the dog, and the pistol blared a discharge that shattered the café window. The confused dog streaked off, yelping.

Noise erupted from the saloon. "Holdup!" shouted the bank teller. He loosed a second shot. "Bank robbers!"

It seemed to Dain an inordinate amount of fuss to raise over a hat. "Let's go!" Sid snapped to him, springing aboard his horse. And then Dain saw the cash sack.

He knew what it was to be out of law, but he had never yet indulged in outright robbery at gunpoint. On the other hand he could size up a bad fix pretty fast, and this one was a pip. To all appearances he had stood as outside guard. It was possible that Sid had taken his absentminded nod as assent to stick up the bank.

Flint Corners came alive, men spilling

from the saloon and everywhere else, and the teller got off another shot. A man didn't waste time on explanations in a case of this kind. Dain dashed after Sid, and together they raced out of Flint Corners. Sid threw him a look and flashed a grin; he was actually relishing the uproar that he had brought about. Its consequences only put a bright zest of challenge into his face.

The thought occurred to Dain that Sid's backtrail was most likely spotted with such lawless escapades. Hints that Sid had let drop meant that this smiling and seemingly happy-go-lucky partner of his rode Texas trails simply because he had made other places too hot to hold him. Sid was an outlaw, through and through.

In the early dawn they came upon a seep-spring and took a badly needed halt. There was some grass and a fringe of shading cottonwoods. They watered their horses and turned them loose without worry that they would drift off the patch of good grass.

Dain got a frugal fire started. He dug a blackened can out of his blanket roll, and crushed coffee beans with his gun butt while waiting for the water to boil. Bent over his task, he spoke to Sid without looking around at him.

"It was a fool stunt. The old guy might've hurt somebody. He missed me by a whisker."

Sid chuckled. He lay stretched out, arms behind his head, eyes closed, dead tired but in high spirits. "Quit pickin' on a poor young pilgrim tryin' to get along in the world. Nobody got hurt, not even the hound. The ol' coot couldn't hit a cow on the backside with a bakeboard."

"You put us out of law. We're just a couple of harum-scarum bandits, gone from bad to worse! I tell you, these Texas folks don't like strangers robbing their banks. Nor swiping their hats! That was an insult!"

Sid cast an admiring glance at the hat beside him. He had grown attached to that hat already. "We got away with it, an' that's what counts. Needed the money, didn't we?"

He sat up with a spurt of energy. "Say, let's see how much we got away with!" He went and got the cashbag, which he had slung from his saddlehorn. He had been too occupied to examine it since the get-away, what with the dust of a pursuing posse in the rear and the long ride through the night, and too saddle-worn on first ar-

32

riving here at the seep-spring. A slap at the bag now and then had reassured him that it was packed solid with hard coin.

The water bubbled to a boil, and Dain threw in the coffee and set it to simmer. While Sid's casual attitude toward lawlessness bothered him some, his chief concern for the time being hung on coffee. A jingle of coins did not entice him to turn from his task.

A brief silence intervened before another jingle. Then silence again. He guessed Sid was gloating over the loot.

Then Dain heard a searing oath. "Why, that — that scurvy ol' weasel! *Pennies!* He gave me nothin' but pennies!"

Dain straightened up and took a look. An instant of shocked disappointment hit him coldly in the stomach, and as quickly passed. Sure enough, the canvas sack contained hundreds of pennies, or had until Sid upended it. Not a single piece of gold or silver in the heap. The bank teller must have been clearing up his cash drawers and preparing to ship off an accumulation of coppers.

"That old twister short-changed me!" Sid howled.

That complaint struck Dain as funny. It crowned the whole thing with ludicrous humor: the holdup, the getaway and pursuit, desperate riding, cutting purposely erratic slants down the South Plains all night — for a sack of pennies. He roared out laughing.

He laughed uncontrollably, rocking back and forth, clasping his empty stomach. The release was a need in him, one of several hungers too long suppressed. It was the first time Sid had heard him laugh like that, but he wasn't in a mood to appreciate the rarity.

Sid was furious, craving to take out his bitter disappointment on somebody. He had been hoodwinked and gulled, made a fool of by an old codger who looked as if he didn't know enough to fool a drunk Indian. He jumped up, shaking, his temper exploding at Dain's hilarious laughter. With one full smack of his fist he hit Dain on the jaw and knocked him across the fire.

"Damn you — laugh at me, will you?"

Flat on his back, Dain lay rolling his head in a numbed effort to clear it, until the pain of fire galvanized his senses. Flame was eating through his pants, and scalding coffee soaked his shirt. He flopped a complete turn and sat up, slapping out the burn with his hands. He got to his feet.

He stared somberly at Sid, and went at him.

He landed a long-armed slam, partly a shove, that rocked Sid back on tiptoe. His follow-up, a slashing left, all but stood Sid on his head. Sid sprang up, jutting out a boot that missed Dain's groin and caught him on the right thigh. That kick set the tone. The fight slid to all-out brutality. They had ridden too long together.

Dain owned an advantage in size and reach. Sid had the edge in shifty tricks garnered from the far wandering of a dark past. Neither of them carried any fat. Lanked down to bone and muscle, nerves tight-strung, their flood of animal ferocity found release in the inflicting and taking of violent hurt. Sid caromed off a cottonwood trunk, sprawled, got slammed back down as soon as he scrambled up, and dug a hand at his holstered gun in raging exasperation. Dain promptly followed suit, taking a spread-legged stance.

Their glares met, locked. Sid thought better of it and flipped his gun aside. Dain drew and dropped his. He started for Sid, but Sid raised his bloodied face to gaze beyond him. Not to be thrown off by a stale trick, Dain made a feint to turn around and changed it into a right uppercut, all his weight behind it. Sid dodged out of reach.

Thrown forward off balance, Dain saw Sid's next move and couldn't block it. Sid's boot flashed out. The high heel fetched him a fearful crack below the knee. As his leg buckled, Sid smashed him a blow in the middle and rocked off to his horse. Wincing, Dain limped after him to even up that score.

Sid, hearing him coming, spun around and pointed northward. "Look for y'self, you crazy wampus! Is that the posse or ain't it?"

One rapid glance informed Dain that it certainly was. He picked up Sid's gun and his own. At his second look he observed how slowly the posse traveled, its members strung out, horses shuffling at a weary walk. A tired posse, too worn out to stay ahead of its own dust.

Sid caught up both horses and trotted them to the saddles. "I don't reckon they seen us yet. Sure draggin' their tails, ain't they?" He accepted his gun from Dain with a nod and holstered it. "You lamed?"

"Not too bad. I'll get over it."

"Yeah, I need new heels, that's the trouble."

They saddled up swiftly. Sid scooped the pennies into the sack, remarking that waste brought sorrow and a little something beat nothing.

They broke from the cottonwoods. At sight of them the straggling posse beat up a flurry. Sid grinned, wiping his cut and bruised face. "Maybe," he said philosophically, "it's a good thing they happened along. Our little argument there coulda got serious, y'know, if we stuck with it."

"Could have," Dain said. "Can't tell."

3

By any measure, Muleshoe Junction stood a good distance from drowsy Flint Corners. Muleshoe Junction had the bustling air of a town that never slept. An important shipping and supply center for the vast cattle country of the South Plains, its busy streets resounded to the racket of wheels and hoofs, whips and drivers' oaths, and the whoops of cowpunchers in on the loose. Near midday the broad main street was a bedlam.

Having put up their horses at one of the three livery stables, Dain and Sid came out onto the main street. They blinked gravely around in the manner of saddlemen afoot in a strange town, and as a matter of course they sought a saloon, haven for weary wayfarers. They had managed to clean themselves up a bit, using a horse trough back of the livery, but still they felt pretty rough and Dain suspected they looked it.

Sid nudged Dain, motioning at a string

of yellow-and-green Concords drawn up before a stageline office. The sign over the office repeated in larger letters those painted on the stagecoaches: *South Plains Stage & Express–Walker Company.*

In turn, Dain nodded toward a three-storey hotel, a big saloon beside it, and an office window on the other side. The Walker Hotel. Walker's Bright Chance Saloon. Milam Walker, Cattle Dealer.

"Mistah Walker," observed Sid, "walks consid'able large tracks here. Should we up and gently break the news to him what sad end came to his stagecoach back there in the caprock? Might be worth somethin' to us."

Dain shook his head. "Old news to him by now, if he keeps tab on his rolling equipment — and you can bet he does. I'd as soon not call attention to our backtrail since that Flint Corners bobble. Let's find a place to eat."

"A drink first to lay the dust?"

"Could use one."

They pushed into the Bright Chance. Sid breathed gustily, taking in the warm smell of whisky and tobacco smoke and mankind. It seemed to do something for him, uplifted him. Dain relaxed, feeling at home on the familiar ground of a crowded

39

gambling saloon, hearing the level roar of voices, slapped cards, chiming glasses, seeing the mahogany and mirrors and green baize. A well-appointed establishment, this Bright Chance. Mr. Milam Walker could be proud of it.

They shoved their way to the bar, where Dain caught the eye of a bartender and stuck up two fingers. While waiting for the drinks he idly looked the crowd over. Cowmen, stageline hands, freighters, some professional gamblers, a couple of soldiers, and the usual sprinkling of nondescripts. A touchy conglomeration.

It would be well to step light here, he mused. This town couldn't help but be like others of its kind, up-and-coming and raw with seething animosities. Cowmen invariably swaggered as if they owned the earth, and freighters stood ever ready to contest the claim. Stageline men, especially the lordly drivers, rated themselves a cut above everybody else. And of course gamblers and saloon-men entertained secret contempt for the whole caboodle. On the fringe prowled the pariahs, some of them women.

Muleshoe Junction, self-styled Queen City of the South Plains, had everything. The Bright Chance contained the essence.

This was a golden era of expansion, of exploiting a rich land recently held by the Indian, and even the drunken derelicts and ne'er-do-wells wanted in on the harvest. White men, weren't they?

The bartender, a surly bruiser, banged down a pair of thick-bottomed shot glasses and filled them. He must have come from some discourteous place east of the Mississippi, for he thumped the cork back. "Two for a buck," he growled at Dain.

Dain inclined his head at Sid. "Give *him* the bad news, not me. Mr. Beaugrand has got the money. I'm his guest."

Sid, falling into one of his impish moods, thrust both hands deep into his pockets and clinked coins. "Which pocket," he asked the bartender airily, "will you have it out of? I'm loaded down with hard cash an' crave to get shed of it."

The bartender squeezed a smile. A coterie of gamblers near by raised interested eyes and began moving in. The bartender said almost affably, "Your right pocket, friend."

"Right it is," said Sid, hauling out a fistful of pennies. And, poker faced, he started counting a hundred onto the bar. "One, two, three, four . . ."

The gamblers chilled. The bartender

grew red in the face. ". . . Fifteen, sixteen, semteen," Sid droned, stacking the pennies carefully. "Hey, this'n's bent. No good."

Dain grinned, expecting Sid to tire of his clowning before carrying it too far. To hasten its end he banged his palm on the edge of the bar, causing a tremor that upset the stack.

"Now you made me lose count," Sid reproved him. He swept the pennies up and began all over again. "One, two . . ."

In the sluggish recesses of his mind the bartender evidently concluded that he was insulted. He took a swipe at the newly begun stack. "Drag out o' here or I'll put your eyes back o' your head!"

Sid might have overlooked the gesture, but never the threat. "Think you could do it?" he inquired, and then Dain knew the thing had gone too far.

The bartender reached down and came up with a heavy brass spittoon over his fist. Sid hurled the fistful of pennies into his face and swarmed over the bar at him, upsetting bottles and glasses and the tempers of several dodging bystanders.

Bouncers and housemen came on the run, jostling roughly through the crowd. A well-intentioned cowman tripped up one of them. Two bouncers swiftly pounced on him and he sank under a striking blackjack and a brass-knuckled fist. In such a place as this, where antipathies ran strong, setting off a riot strained no talent. The job of the bouncers and their fellow workers was to nip it in the bud, even at a cost of killing or crippling a customer or two. On they came, an efficient and merciless execution squad.

The coterie of gamblers took a hand. Dain was nearest to them and they tried to jump him. He wished Sid had used some restraint. He slammed down a gambler who slashed a broken bottle at him, and a detached part of his mind wondered if he was not outgrowing this kind of violent flare-up, for he felt a little sorry for that gambler, who would breathe troublesomely through a broken nose the rest of his life.

With his fists and boots he cleared a space and set his back to the bar. He snapped out his gun. "Wup, there! Stand off!"

His gun ranged over the oncoming bouncers. He held his left hand poised above it, fingers stiffly spread. Fanning a gun was not a trick he placed much faith in, but at close quarters it could be devastating. He could do it if he had to.

"Sid, come out from behind there! Time to go!"

Tiers of glasses crashed to ruin behind the bar. Sid was struggling with the bartender for possession of the brass spittoon. The spittoon struck a backbar mirror, the mirror fell in slivers, and the two of them vanished below the bar, sending up loud thumpings.

Dain, holding the bouncers precariously at bay, said, "Damn you, Sid, come out or I leave you here to the wolves!"

"Hold the fort, mah johnny-reb friend!" Sid bobbed up, grinning like a devil in glee. "Wolves? Why, these are only —"

"That'll do!" called a voice. The speaker was a man who had just entered the saloon, a large man leaning toward heavy middle age, wearing a baggy black suit and low-heeled boots. He wore no hat, and his thick shock of iron-gray hair stuck up in all directions; but the aggressive strength of his big red face saved him entirely from appearing comic. The crowd quieted down, nobody questioning his dominating authority.

"Who started this?" he demanded. But he already had his eyes fastened on Dain and Sid. "You two?"

Dain said nothing, watching the

bouncers over his gun, but Sid responded cockily, "Who cares who started it? We finished it!"

"Who are you two?"

"Who the hell are *you?*"

Swaying up groggily behind the bar, the beefy bartender mumbled about jokers throwing pennies in his face. It was a dazed and disjointed account, adding up to little sense. To hear him tell it, the air had been full of pennies, thousands of them, all pelted at him.

The bareheaded man ran a comprehensive survey over the damage, the scattered pennies, and the mob of frustrated bouncers, housemen, gamblers. To Dain and Sid he said, "I'm Walker. Milam Walker. Come with me!" He turned to leave. At the swinging doors he looked back. His jowly red face took on the ghost of a smile, queerly benign. "I said come with me — gentlemen!"

Dain exchanged a glance with Sid, shrugged, and they paced out of the hushed Bright Chance, following Milam Walker, the man who walked considerable tracks — as Sid had understated it — in this up-and-coming town of Muleshoe Junction.

A wooden railing with a swinging gate

45

divided the front office of the South Plains Stage & Express into two parts, one private and the other for the public. The private section contained Milam Walker's sanctum — big rolltop desk and swivel armchair and several straightback chairs — and the high desk and stool of the book-keeper.

To the right of Walker's desk an open door led into a small office where a girl worked, her back to the door and her head bent over her desk. Alongside that door stood the gunrack for the shotgun guards of the stageline. The racked guns there, Dain noticed, were good weapons, but in need of better care. Sweaty finger marks should have been cleaned off before racking and the blue steel wiped with a lightly oiled rag. He could imagine the condition of the bores and breeches, and itched to take them apart. He respected tools of any kind, being instinctively a craftsman in a world where craft had come too often to mean, since the war, craftiness.

Cheap skill and slipshod work and slo-venliness exasperated him. The same with slovenly thinking. Though not smart, not very wise in the management of his own af-fairs, he had that simple gift of a sense of realism. Nor was the sense clouded by too much cynicism, at least not so far. He still remained essentially honest, to himself as well as to others, in spite of everything.

He had long borne with him the uneasy suspicion that he was in some ways an odd number. The suspicion had cemented to certainty shortly after Appomattox. A medal-covered big somebody, addressing tattered troops, made a noble speech, studded with classical allusions to the grandeur of defeat. In the holy silence following the last throbbing flight of mellifluence, Dain had found himself privately weighing the splendid words, picking at their gilded shells. He spoke his opinion of their sorry emptiness, dared to question the sincerity of the orator, barely escaped getting shot for it, and left in utter disgrace.

He learned to keep his mouth shut. Reticence became his habit, personal integrity his vague but stubborn path. Sentiment, of a kind, he could not control, but he had some mastery over his mind. So he was a square peg imbued with a distrust of neat round holes. He knew that when the heart and the head disagreed, the heart won, often to catastrophe. He fought against it.

Milam Walker seated himself at the rolltop desk. He motioned Dain and Sid to chairs. His manner made it a command. The girl in the side office did not raise her head. Nor did the bookkeeper, an emaciated old man whose green eyeshade cast a mildewed hue over his lined face.

Sid, loungingly insolent, sat down and stretched his legs out. He remarked that he and Dain had a drink coming for the one they lost in the Bright Chance ruckus. "I parted with maybe two bucks there," he said. "In pennies, but what the hell, it's money."

Dan stayed standing. Conscious of Walker's scrutiny, he met it, expecting a clash. But the hard eyes were impersonal. He realized that the man was only studying him. It occurred to him that Walker displayed no anger at the damage done to his Bright Chance Saloon, which was curious. But Walker owned so much of this town, perhaps a little destruction here and there was marked up to ordinary overhead expense.

Walker opened a drawer and took from it a silver flask and glasses. He set them on the desk and nodded to Sid. He watched Sid unscrew the stopper and fill the glasses. Dain, disliking Walker's manner, made no move, so Sid drank alone.

Walker took his eyes off Sid, flicked

them to Dain, and looked up at the ceiling. Folding his hands behind his head, he said, "You handle yourselves well. I thought I had a tough bunch in the Bright Chance, toughest in the South Plains. Are you for hire, gentlemen?"

Sid lowered the emptied glass from his lips and bent forward. "Are you hirin'? For what? The pay's gotta be high, for us. We're rollin' in cash."

Walker's faint smile returned. "That's a lie," he murmured. "You must have been broke to stick up the Flint Corners bank. You got bilked there. By old McManus. Twenty-seven dollars. In pennies!" He brought his tousled iron-gray head forward to regard them through amused, hard eyes. "The reports I receive are fast and accurate, and private. I own the bank at Flint Corners, among others. You —" he inclined his head at Sid — "did the job, according to description. You pulled a dud!"

Sid flushed. "Any man can make a mistake! D'you figure to law us or hire us, which?" His right hand was down near his holster.

Milam Walker said, "I'm always on watch for men I can use." And he watched Sid. But mostly his hard eyes rested on Dain, who did nothing, said nothing.

49

"Good men, and I mean *good,* are not too plentiful here, strange as it may seem, here in the South Plains — the new country, the new empire where you'd think every man worth his salt would head for!"

His remark was aimed at Dain, to stir him. Dain knew it, and he said slowly, "Texans have known about this South Plains country a long time. It was Indian country. Now the government at Washington has cleared out the Indians and opened it for white men. The big men who've moved in are mostly Northerners, rich, maybe with some political pull. Damyankees. You, sir, are an exception."

"You're right," Walker said. "I'm Texan. So are you, I judge. I'm related to Ben Milam who died at the storming of San Antonio. You?"

"My name's Moore," Dain said. "A Moore died at the Alamo."

Sid shifted restlessly. "Talkin' o' funerals," he said, "I rise to remark that sev'ral us Beaugrands got killed in the recent war, some for the Union, some for the Confed'racy. Soldiers o' fortune. Or misfortune! Whichever, the hell with it. History don't int'rest me. Mr. Walker, you mentioned a need o' good men. We're good — damn' good! Well?"

50

"Perhaps I expect too much," Walker said. He had changed his tone and spoke musingly, but his face could not lose its arrogant aggressiveness. He continued inspecting Dain and Sid, weighing their possible value to him. His eyes took note of their slightest actions. He probed everything and missed nothing.

He said in that same reflective tone, "I still hope to find a man able to boss this stageline. I own other businesses that require my time. The man would have to be better than good. A tough man, with brains. Able to run this stageline on strict schedules, no excuses accepted. Able to keep the whole crew up to scratch, make them all fear him, and have no friends among them. It's the only way to handle this bunch. Big job! Big pay! For a big man!"

At his desk, the silent bookkeeper moved his head furtively. Dain saw a weary grimace pass over the sunken features, and he wondered how many times the poor old drudge had heard this talk. Walker's words could be the usual bait offered to new men, to get the best out of them. The dangled carrot to the jackass.

On the other hand, it was equally probable that Walker actually did need a boss

51

for the stageline, had been on the lookout for some time, and hadn't been able to find one to suit him. His requirements were high and hard to fill, in a newly developing country where, except for the cattlemen and businessmen and speculators, most men drifted in on the make or on the run, or both. The Texas Panhandle had become a temporary haven for busted adventurers, gamblers, gunmen, outlaws and plain riffraff, untrustworthy when not shiftless. The businessmen and speculators couldn't be hired, and cowmen generally wouldn't take on any job they couldn't handle from the saddle.

Sid helped himself to Dain's untouched drink. "Look no further," he stated to Walker. "I'm your man."

"You might be. Or —" Walker sent Dain a nod — "you. I put you down as an outstandingly tough pair. But that's not enough for me to go on. Nowhere near enough. The job I've been speaking of is a one-man job. It's special, needing a special kind of man who must prove to me that he can qualify. I can pick up the other kind any day."

He drew a cigar from his vest pocket, bit off its end, and barked a single word: "Light!"

The elderly bookkeeper dropped off his high stool as though pricked. He trotted to Walker's desk, scratched a match alight, and held it for Walker's cigar. Walker puffed the cigar to an even glow, taking his time. A box of matches stood on his desk before him, but he ignored it as he did the old man, who, obviously inured to this thankless servitude, shambled back to his stool.

Walker went on talking. "I'll put you two to work and see how you do for a while. Can you handle fast teams?"

Dain nodded. Sid allowed cautiously that he had acquaintance with stages fast and slow. Dain could tell that Sid was losing interest at the prospect of having to work up to that special big-pay job. Winking at Dain, Sid poured himself a third drink and asked Walker brazenly, "Can we draw an advance on our pay?"

In the girl's little side office a chair scraped. Brisk heels tapped the floor-boards. Bearing a sheaf of waybills, the girl came into the main office. She placed the waybills on Walker's desk and stood quietly waiting. She was a smoothly efficient clerk doing her duty, presenting necessary business papers for the big boss to check through. Trimly immaculate, as quietly

53

garbed as a sparrow, she could not disguise her complete femininity. She possessed everything a woman required to take and hold fast the attention of men, whether she wished it or not. With her calm presence she brought a rich sparkle into the big drab office.

Sid, in the act of swallowing his third drink, jerked up in swift recognition, sloshing whisky down the front of his tattered shirt and not heeding it. He muttered something that sounded prayerful.

Dain stood motionless, feeling his face chill and then heat up, staring, like Sid, at the girl. She was not tall, as he had thought that she must be, and her brown eyes and hair utterly lacked flamboyance. Any man would tangle his tongue, trying to describe her, trying to explain.

She certainly was not smiling now. Her eyes held focus on a distant point, cooled by her knowledge of a couple of male roughnecks staring at her.

Dain, though, knew well her lifting smile. He carried her and that lovely smile in his shirt pocket. He took off his hat. Sid did the same, after springing to his feet. The girl, seeming oblivious of them, scanned the wall above the bookkeeper's head.

Laying down his cigar, Walker thumbed through the waybills. Soon it grew clear that the task did not wholly occupy his mind. Between inspecting each waybill his eyes slid off to rove sidelong over the girl from her waist down. Her hips and thighs owned flowing form below the slim waist, young and curved, in womanliness matching the high breasts, and any estimating eye knew that under the full skirt and petticoats her legs were long. Walker's large nostrils flared ever so slightly, and Dain thought that the girl must surely sense the virile maleness directed at her. Directed at her from three sides, for he was anything but unstirred by her, himself; yet an unreasonable rage rose in him against both Walker and Sid. Particularly against Walker. The man had no right at his age to let his thoughts run unchecked over a young girl. Old stallion!

Walker bunched the waybills together. The sleeve of his coat brushed the burning cigar and sent it rolling off the desk to the floor. He stepped on it and crushed it out. Leaning back in the swivel chair, he handed the sheaf back to the girl and drew out a fresh cigar.

"Light!"

The bookkeeper, sighing, recommenced

his subservient act, but was saved from it by the girl. She took a match from the box on the desk and struck it alight. She held it up between thumb and forefinger, not cupping it as a man would, and Walker lit his cigar unevenly from it.

Raising his heavy face, Walker showed by his expression an inordinate pleasure. He smiled up through blue smoke at the girl. "Why, thank you, Tresa!" It evidently could never occur to him that her spontaneous action was merely a kindness meant for the downtrodden old bookkeeper.

Tresa, Dain thought; *her name is Tresa.* . . . He wondered what the murdered driver of that stagecoach up in the caprock country had been to her, and if the reason for that apparently senseless crime was known to her and to Walker.

Turning back to her office, the girl nearly fell into Sid, who had sat down again and hitched his chair forward purposely to force her to notice him. Sid flashed on his best smile, spreading his arms to catch her. He won nothing for his trouble. As she whisked around him, her gaze touched Dain's for a bare instant. He had never seen a girl's eyes so expressive of angry scorn.

In his normal tone of curt command

Walker said to the bookkeeper, "Put these two men on the payroll and advance them ten dollars each."

That ended the interview. Walker had begun staring coldly at them. He was a perceptive man, Dain guessed, where other men were concerned. He had detected, without doubt, the effect of the girl upon them, and he did not like it, would not tolerate it.

They wrote their names in the pay book and drew a goldpiece each. The girl came in again, carrying an open ledger, but not even Sid could seem to think up any excuse to linger. At the front door they both looked back. She was gazing after them. Because of a rear window behind her it was impossible to know on which of them her shadowed eyes rested, and with what kind of look, but Sid tipped his black hat, smiling. She averted her face and spoke to the bookkeeper. Dain, fingering his stubbled jaw, decided extravagantly on a shave, hot bath, and new shirt.

Sid eyed him and swaggered on out. He spun his goldpiece, caught it, looked at it, frowned. "Best two out of three," he muttered, spinning it again.

Dain asked him, "You pulling out?"

"That ain't what the toss says. It says I

stick." Sid sent an elaborately casual glance back into the office. "Figure I'll win that job. And that gal, too! How 'bout givin' me that picture in your pocket?"

Dain rasped, "Nothing doing! It belongs to her. I'll give it to her, first chance I get."

"That," said Sid, "is what I figured to do. What good'll it do *you*? Man, you don't stand a shake! What with my charm an' brains, not to mention all the time I've lately wasted without a woman, all I need —"

"What you need is a bust in the mouth!"

Sid stiffened, but said banteringly, "Now, Dain, it's no use you gettin' jealous at me. Get jealous at Walker if you like. He's the big rooster, long-spurred an' high-hackled. *You* won't get nowhere there, but I'm the hot gamecock who'll take his pretty li'l chicken while —"

"A bust in the mouth!" Dain repeated.

Sid flushed darkly, the underlids of his eyes creeping upward. He opened his mouth to speak, closed it, and walked off up the bustling street.

As his anger slowly subsided, the thought came to Dain that it was some-what fantastic on the face of it for him and Sid to clash so close to the fighting point over a girl they had just seen for the first time and who had not given them a word. He looked into the stageline office.

The bookkeeper sat hunched over his high desk, in outline resembling a starved crow dismally contemplating crumbs. The girl had retreated to her office, and Walker was following her in and closing the door. Dain watched that closed door. A sense of deep and hideous wrongness disturbed him like an urgent frenzy.

A man couldn't go charging in there, though, without any other reason than a sense of wrongness. It was too easy to make a blundering fool of himself by forgetting past mistakes.

He cut quickly through the line of yellow-and-green stagecoaches drawn up before the office, crossed to the hotel, and paid for a room. Being committed to work for Walker, certainly he was expected to patronize Walker's hotel. A top boss, big man of the town, could command privileges from his hirelings.

He refused to pursue that bleak and dangerous line of thought. It led directly to the killing rage that Sid believed him free of, and which he himself knew smoldered too readily under the surface. Best to stay cool and detached, never becoming involved. Some day he would get back to his home on Padre Island, to settle down with a sigh of relief. After which, nothing would ever again disturb him, because there would be nothing but the silent dunes and the empty Gulf of Mexico all around. And the half-wild cattle to work on, horses to break in, the old ranch to repair and put back into running shape. A thousand tasks to keep a man fully occupied. Peace.

No sound had issued from the little side office after Walker closed the door. The girl had not called out for help.

Dain wished he had never picked up her smiling little picture. It burned his pocket.

4

Standing broodingly at the window of his hotel room, Dain came to instant attention on seeing the girl hurry out of the stageline office to hand waybills up to a waiting driver. To the crack of the driver's whip and a shattering whoop, the high stagecoach jolted forward, horses lunging into the collars, and rolled off down the crowded street. Another pulled up into place. The girl spoke to some waiting passengers and watched them climb aboard.

By that time Dain, almost without thought, was out of the hotel and crossing the street in long, rapid strides. Despite all, she was a magnet to him, irresistible. Her name, Tresa, kept on ringing like a silver gong in his mind. He could not rid himself of her, and, manlike, he played vaguely with the ambiguous idea that if he could win her he would thereafter be free of her. She was Walker's property. It would be a double satisfaction to play hob with Walker's enjoyment of her. That was his urge, wholly cynical at the top of his mind, unconsidered, as he circled and came behind her. A virile man on the make, stalking a strikingly attractive and desirable girl. He had done this before, and usually won his way when he really tried, when the girl met his curiously fastidious taste.

He thought her unaware of him behind her. But when she turned from seeing the second stagecoach off, she betrayed no surprise at seeing him. Instead, she took the wind out of his sails by saying evenly, "How d'you do, Mr. Moore?"

"I'm Dain Moore," he said uselessly, and found himself stammering, which irked him. "I'll be working here."

Their eyes met and held. His look and manner of sudden confusion caused her to thaw. She said, "Yes, I know. You're driver and your friend Beaugrand is to ride guard with you, on the Littlefield run. I'm Tresa Ogden. Among other duties, I fill in as express clerk and dispatcher whenever the regular man is off drunk, as he is today. I hope you'll like working here."

Dain pulled himself together. "You sound doubtful, Miss Ogden. Is it such a tough job?"

"You won't find it easy to work for Milam Walker."

"*You* work for him."

"There are few choices open for a girl who must make her own way. I have no folks left." She nodded toward the great wagon yard alongside the office. "My father drove a stage out of there and never came back. Killed. I tell you that as a warning."

"Why was he killed?" Dain asked.

She gazed at him wonderingly. "I thought everybody knew of the fight between the Walker line, here, and the Bledsoe Mail Stage Company, of Travis City. But I forgot you're a stranger. It's a fight for control of all stageline operations of the South Plains. A murderous fight. Bledsoe has hired gunmen. We're doing the same."

Me, he reflected, and Sid. Fighting hands once again. Could a man never get out of his trade? He said to Tresa Ogden, "As usual, a fight for money. It's the same everywhere, since the war."

He had to look away from her then, because her eyes became too soft, her nearness too overwhelming, and he saw Sid Beaugrand leaning out from the batwing doors of the Bright Chance, watching them. "But," he said, "I remember when Texas was a free and easy land where nobody thought of getting rich and we all trusted and respected each other. Now we've got — what? Even this Milam Walker, a Texan from way back. Out for the money, like the rest. You're Texan, too, I judge by your speech. So am I. Born on Padre Island. Trying my way back."

"Corpus Christi, me," she said, smiling up at him. "On the Gulf — as if you didn't know that."

His slow smile enveloped her, wholly. "A couple of Texas sea lions, you and me. We're rare. I bet you swim like a fish."

"I bet you do, too. We're out of our element here on the dry and dusty plains, aren't we?" Impulsively, she extended her hands. Small hands. Soft, yet firm, as he found to his ready grasp. She stirred him more than any other girl had ever done.

"For that reason," she went on, "I'm glad you're here, Dain Moore. Most of these people are not my kind of folks. So it's a selfish reason. In fairness I should warn you that the Littlefield run is a hard test. Very hard. No driver has yet made it without trouble, and never on schedule. And you'll be given harder routes — such as the one, perhaps, that cost my father his life. The caprock run north to Travis City." She shut her eyes tightly, and he felt a shudder run down to her hands. "He was burned to death!"

"No," Dain said. "He was shot dead first. Riddled. It was quick. No struggle. He didn't suffer, believe me."

She snapped open startled eyes. "Walker ordered him to make a fast test run to Travis City and back. It was to see what might be done toward winning the government mail contract away from Bledsoe. My father knew the danger. He had driven for Bledsoe. He carried no gun. It wouldn't have helped, if Bledsoe's killers caught him. As they did! You're lying to me, Dain Moore! Trying to — to rid me of nightmares!"

"No," he said. "I saw it." He released her hands. He took the little picture from his shirt pocket and gave it to her, to clinch the truth of his statement.

In a minute she raised from it a dead-white face, eyes frozen. She stared at him in bitter shrinking, as if he had become all at once sinister to her. Suddenly she whirled away from him and darted into the stageline office.

He was abruptly left alone, with a dismal but undefinable sense of having somehow pulled a terrible boner. What exactly it was, he couldn't fathom. After a while he recrossed the street, went back to the hotel, frowning.

Sid hailed him from the Bright Chance. "Hi, Lochinvar!"

Evidently Sid had not witnessed Tresa Ogden's chill-off; only her hands in Dain's and her glorious eyes turned so dazzlingly and trustingly upward. Sid was frowning, too, in his secretive way. He said, as Dain joined him, "I sure pegged you wrong. You s'prise me; you work fast, friend!"

The "friend" came tardily. Dain said, disregardingly, "That South Plains stagedriver we saw killed was her father. Her name's Tresa Ogden. She's all alone in the world."

"So am I! So are you, I take it!" With angry irony, Sid inquired, "Should we hold hands? You and her, I saw —"

"Cut it!" Dain said to him. "Cut it, dammit, you hear me?"

Sid settled back. "You're wild," he said, almost gently, taking his hands off his guns. "Wild, like me."

"How d'you mean that?"

"Wild for *her!* She's Walker's private girl. Don't glare at me — you know she is! You

66

want her. So do I. Wasn't for her, we'd push on south, wouldn't we? Me for the Mexican border, you for your lonesome land."

"I'll grant you that. Walker's not a man I'd commonly choose to work for, even though he's a Texan. And even though he's got a hold on us, 'count of your damfool penny holdup!"

Sid grimaced wickedly, showing his white teeth. "Lay off, Dain. If you and me tangle, I'm tellin' you it'll be —"

"We better not, I reckon," Dain said. "For a start, we're slated to take over a tough run, together. Let's have a drink or two, and try it out."

"All right. But we already been too long together, 'specially since seein' her, and you know it."

"That's a fact," Dain agreed. "So let's try to keep our traps shut and get along a spell longer. I need a little time here."

"Me too!" Sid said.

The tall yellow-and-green stagecoach, drawn by a splendid six-horse team, jounced over a sketchy road deeply rutted by previous wheels and gashed by dry arroyos. Rock outcrops bent the road into hairpin curves, where the coach swayed perilously and the groaning passengers inside tumbled like dice in a cup.

There was one thing to be said for Walker's South Plains Stage & Express lines. It maintained excellent equipment and fine horses. But its roads were atrocious, never repaired. It was up to the drivers to discover the best passable route at any time of year, rain or drought, mud or dust. This was so because the stageline owned none of its roads; they ran through public and private property. To that degree the South Plains Stage & Express — called by everybody the S P S & E — was a wildcat outfit, although the chief stageline of the Texas South Plains and in fact one of the most important of the whole Panhandle.

Dain drove. Sid Beaugrand rode shotgun guard. The act of driving required much more than the mere fingering of long leather lines. It demanded an acute and exquisite feel of the team, the power and authority of the leaders, and the brakes. Stage-driving was an art. It kept Dain busy, making the best possible time without wearing out the team or wrecking the coach. He had put in short stints of driving here and there. Although he didn't regard himself as expert, he had a knack for it.

Sid, inclined to boredom, drummed a heel against the locked express box behind his feet. "Any valuables in this thing, d'you know?" he inquired idly.

"Pennies, maybe," Dain responded, gentling the brakes for a blind stretch of winding downgrade.

Sid threw him a dirty look before glancing at the sun. "I guess we'll make it on schedule." He shoved the shotgun behind him, and drew out tobacco sack and papers. Balanced on the swaying seat, a match between his teeth, he commenced making a cigarette. "A tough run, this? What's so tough about it?"

They swung past the first blind bend, and suddenly there appeared before them a single horseman sitting at halt off to the side of the road. Big and dark, the man bore the marks of that close-mouthed breed of men lately trickling down through the Texas Panhandle. He was bearded and shabby, the earthy look grained into him; but he wore two guns, and a rifle was strapped beneath his left leg. And his horse was a good one, though uncared-for.

He sat motionless, a drifting outlaw obviously scanning the stagecoach for possibilities. Then his round black eyes were boring unwinkingly at Sid Beaugrand, in what Dain took to be cold recognition. The hard, corrupt face subtly altered, its expression becoming sardonic and sly. Sid was staring back, the unfinished cigarette poised in his fingers.

The incident lasted only a few seconds, for at that moment the fast-rolling wheels plunged into and out of a dry washout that scarred the road. Sid should have been prepared for the violent bump and lurch, but wasn't. It tossed him, and tobacco and papers went flying. He saved himself by a clutch at the guardrail and regained his seat beside Dain. Cursing softly, he wiped tobacco from his face. By that time they were making the next bend and had left the horseman behind.

Dain then replied to Sid's last question. "I just can't imagine what's so tough about this run," he said drily. After a pause, he added, "But you're looking a little sick, Sid."

Sid muttered, "Swallowed the goddam match, that's why!"

But Dain noticed the match lying on the floorboards. It was not a match that had so disturbed Sid, nor anything like it. Sid looked as if he had seen a ghost. A dark ghost from his past.

The dust-covered stagecoach and

sweating team rattled into Muleshoe Junction on a sunlit Thursday afternoon, Dain and Sid on the box. This now was their third run, and traffic on the main street parted for them, it being already known that they had a way of crashing through regardless of obstacles.

They unloaded at the office — express box and baggage and bruised passengers — and turned into the stageline yard. There the yard boss griped, wanting to know how they picked up so much dust and grime that had to be scoured off the stagecoach. "How in hell," he complained, "ev'ry time you two fellers —"

"Professional secret," Sid told him. Like Dain, he beat the dust off his clothes with his hat. His black-and-white horsehair hatband, an adornment which he had added to the stolen black hat, fell off and he looped it back on. While he was doing that, Dain went ahead into the stage-line office.

Walker was not at his desk, and Dain entered Tresa Ogden's little office. "Checking in, Tresa," he said to her.

She raised a rich and involuntary smile. "Hello, Dain. On time again, three in a row — a miracle!" Then again came that shrinking withdrawal, as though an ugly knowledge prodded her, and she said quite

71

stiffly, "You're building up a record."

Sid came in then, wiping his company shotgun. He threw a leg over a corner of Tresa's desk and grinned down at her. "All that tough Littlefield run needed was a good man," he drawled. "Meanin' me."

His lively gaze played over her hair, her face, and brazenly over the contours of her body, lingering on her breasts and falling slowly on down. Captivated by her complete young womanliness, her ripe femininity so overpoweringly attractive, he made no bones about it with his eyes. His manner toward her, while not actually offensive, was pushingly familiar.

She asked with quiet irony, "Who does the driving?"

A bit nettled, Sid retorted, "Aw — Dain holds the lines, but it's me ridin' shotgun that keeps him up to snuff."

She sent him a scoffing glance. Opening a ledger, she looked for her pen. Sid was sitting on it. He produced it and when she took it he caught her hand. She struggled to free it, but he held on. His face flushed very dark at the touch of her. His eyes stilled in a rigid stare.

Dain's smile faded. He said, "Let go, Sid!" Then much more sharply, "I said let go, dammit!"

Sid released Tresa's hand. He turned a dull glare on Dain and said softly, thickly, "I wish you'd get the hell out o' here!"

Tresa sent Dain an urgent and pleading look. Dain said to Sid, "Hold it! I'm here and I'm not leaving. Hold it, boy!" He held his hand close to his gun, watching Sid.

After a long pause Sid inquired with great politeness, "D'you want we should step outside and talk this over?"

"If it's what you want," Dain answered without any politeness at all, "step ahead." He stood aside for Sid to precede him.

Sid came abreast, paused again, and whispered, "Y'know I can beat you on the draw."

"You can try," Dain said, and followed him out. He saw Tresa stand, staring wide-eyed after them. Her right hand she held partly upraised in an attitude of protest, her left hand pressed to a spot just below her breasts, a gesture utterly feminine, and to her he said gently, "Don't worry, honey, this isn't your fault. It's been building up a long time." The "honey" slipped out naturally, an easy-going Texas term of affection, not disrespectful.

But it seemed fated that he and Sid were not to stake all on a shoot-out. Their time had not yet fallen due. The Bledsoe weekly mailstage from Travis City, deadly rival of the South Plains Stage & Express, was rolling into town, an event that commanded precedence over personal affairs.

5

The coach, like all those of the Bledsoe stageline, was painted dark red, with black trim. Bigger than Walker's springy Concords, this followed the mud-wagon type, burly, heavy, drawn by six tall Missouri mules. Two men grimly rode shotgun, and the driver wore a pistol and had a rifle slung within handy reach.

At the rear on top rode a fourth man, double-gunned. He had a dour hawk face, the prominent nose a straight extension jutting down from the forehead. At first glance Dain recognized him. He was unmistakably the same man Bledsoe had called Hollen, the crew boss who fired his rifle into driver Ogden's body, on the stageroad below the caprock. Hollen. Sid had remarked at the time that he had known a man by that name.

The Bledsoe mailstage met no welcome here. Carrying the mail, it was a necessity, to a limited extent sacred — mail being under the arm of the U. S. Government, the carrying of it a solemn business of contract and franchise. But deep grudges existed between the Walker and the Bledsoe lines, spreading inevitably to hatred between the two towns of Muleshoe Junction and Travis City, for Walker ruled one and Bledsoe the other, and this was the Texas Panhandle, a violent land.

It drew up before the stageline office, the team mules throwing their heads and shooting long ears distrustfully. Animals close to men always caught on instantly to men's moods. After setting his brakes, the driver fell to work throwing off the mailsacks, helped by the shotgun guards and watched by a grim gathering of Walker's stageline hands. Hollen did nothing, but his eyes ranged everywhere.

Two men, passengers, stepped out of the coach. They were without baggage of any kind. Each wore a buckskin jacket that did not wholly hide a belted gun and long knife. They looked around in the searching way of strangers, then glanced up at Hollen. The killer slid his eyes away, whereupon the pair headed over to the Bright Chance. Walker was coming out of the hotel, and in passing he sent them a short, stiff inspection.

76

Coming on, Walker nodded to Dain and Sid. Dain returned the nod, but Sid was gazing strangely after the two men in buckskin jackets. Evidently growing conscious of being watched, Sid slowly swung his head and his gaze found Hollen.

In Hollen's eyes, fixed on Sid, Dain saw the same cold recognition that he had observed in the eyes of the dark outlaw on the Littlefield trail. And he saw again on Sid's face that expression of quiet shock, as if Sid saw ghosts — old ghosts from his past, closing in on him.

Tresa came out then to sign for the mail, and Hollen's eyes switched to her and remained on her. She pretended unawareness of that steady, sinister stare, but it obviously took effort. Dain remembered that her father had once worked for the Bledsoe line, got fired, and come to Walker. Fired for what? He wondered.

He also wondered if the two shotgun guards, as well as Hollen, had been among the Bledsoe crew that waylaid and killed Ogden. It was probable. Hollen was crew boss, and men who took orders from such a cold-blooded killer were not likely to be particular as to the nature of those orders.

He sighed. Competition for trade was all very well, but here the struggle for control of the Panhandle stagelines was a murderous feud, and he hated to see the cynical like of it gain a hold in Texas. He and Sid were entangled in it. So was Tresa, a puzzling girl who greeted him warmly and then froze up on him.

The big Bledsoe stagecoach made its turn, driver's whip cracking over the mules, shotgun guards sitting like images and Hollen ranging a final mocking stare over the onlookers. It rumbled north out of Muleshoe Junction. Men moved off, cursing after it. The feud was not only based on crude principle, it was personal and virulent.

Dain said to Sid in a quiet voice shaded with meaning, "Well?"

But Sid had somewhere lost the urge to carry the quarrel on further. "Forget it," he muttered, looking blank for a moment. "Let's eat . . ."

But in the café Sid ate little of his meal, and had nothing to say. Dain finished eating and was deciding between rolling a cigarette or buying a cigar, when a shadow darkened the window behind him. Sid raised his head slowly, as if reluctant to do so. His face showed no expression. Dain twisted around in his chair in time to see the two men in buckskin jackets moving off. He knew they must have been peering in.

Presently Sid shoved his plate away and laid down half a dollar. He stood up, unsmiling and deliberate, and walked to the door. Dain spoke after him in a carefully neutral tone:

"Need any help with anything, Sid?"

He and Sid Beaugrand quarreled and snarled and fought almost to the gunning point. Yet let one of them get in a jam, the other was on tap to lend a hand. It had been that way right along, more or less. Curious, but it seemed natural somehow.

Sid paused to ask irritably, "What makes you think I might need help with anything?"

"Call it a hunch."

Sid went on out without replying. Dain shrugged, had another cup of coffee, and paid for his meal. Afterward, he strolled to the Bright Chance to buy the cigar he'd decided on. There he found Sid at a table playing blackjack with three men. Two of them were the men in the buckskin jackets. The third was the dark, bearded badman he had last seen on the Littlefield trail.

Dain swerved, going close by the table, and he asked, in that same neutral tone again, "Everything all right, Sid?"

79

All four looked up. The two in buckskin jackets, who bore some resemblance to each other as though they might be brothers, didn't know him; and he could tell that the bearded badman wasn't sure about him, having had only a brief scan at him on the trail.

Sid snapped at him edgily, "Dammit, leave me be!"

"All right." Dain paced on to the bar, where he bought his cigar and studied the four at their table.

The blackjack game wasn't right. A cover-up for talk. The cards flipped out, Sid dealing, but the four paid them scant heed and coins on the table lay unchanged. Heads lowered, the four were back in deep discussion. Sid looked strained. More than that, he looked weary and actually old.

Feeling that Sid was getting crowded, Dain examined the set-up meticulously. Their table, off to the side, stood near a row of curtained booths. He wandered idly from the bar, his cigar alight, and eased unobtrusively into the booth nearest the table. The booth was empty. Listening to the talk at the table, he could not hear much of it except for Sid's clearly recognized voice.

Sid said sharply, cutting into whatever

80

the dark man was murmuring, "Nothin' doin'!"

The dark man didn't stop. Dain caught some mention of a bank shipment northbound out of San Antonio. The pair in the buckskin jackets said nothing, heads bent, upturned eyes watching Sid with narrow intensity.

Sid shook his head. "Lay off it!"

Scanning his cards, the dark man changed the subject. He raised his voice a little, speaking of a Kansas mail-train robbery, a dead conductor, and federal warrants. "Hope nothin' happens they get tipped off you're here. Ridin' shotgun." He snorted a short laugh. "You!"

They were all four keeping their hands in sight on the table, and with his left hand Sid scraped off his black hat and used his right to wipe his forehead as if worried. The horsehair hatband fell off, unnoticed. As he replaced his hat, though, he dropped his right hand swiftly below the table.

He inquired softly then, "Want to make your meanin' clear?"

They spotted his move as soon as he made it, and it was a measure of his explosive potentialities that they grew instantly still and wary. But the dark man gave his reply.

"Sure I'll make it clear! We cut in together on this job, or your game here, whatever it is, gets all blowed to hell! Clear enough?"

Sid eyed him a long time, and abruptly got up and left. The dark man said to the other two, "He was bluffin'. He's a damn smoky proposition, but he was bluffin' that once." He saw the horsehair hatband on the floor and picked it up. "He'll come through, don't worry! He ain't got no choice!"

One of the two whispered something, and he, twirling the horsehair hatband on his forefinger, disagreed. "There's warrants out for him an' a rope ready for his neck!" he was saying as Dain left the curtained booth. "Would he take that 'stead of his cut in maybe twenty thousan' dollars? Not if I know him!"

Dain walked into the stageline office mainly to see Tresa. Walker was talking with Sid at the big desk, and he called to Dain, "I was about to send for you, Moore. Beaugrand, here, is trying to tell me he won't ride shotgun with you. Why's that?"

"Trying?" Sid said. "*Trying* to tell you? I put it plain that I won't! Ask me why, not him. I'm tired of it, that's why! When do I get the top job? I want to know!"

It was, Dain considered, one hell of a way to buck for a better job. The elderly bookkeeper took down with a coughing fit. Walker scowled darkly, pushing to his feet; but then apparently Sid's unusual approach struck a hidden sense of humor in him. He sank back, with a faint, hard smile.

"You do, do you? What makes you think you've qualified? I give Moore the credit for straightening out the Littlefield run. He does the driving."

"I've already been told that once!" Sid snapped. "Get another man to ride with him — I quit!"

A chill glaze spread over Walker's eyes. The faint smile vanished. "I rarely hire a man these days unless I've got something on him. Nobody quits me of his own accord. That includes you."

"The hell it does!"

"The hell it doesn't!" Walker's tone, previously moderate, became a harsh rasp. "Flint Corners, remember! I have no doubt the law could dig up other matters. You stick! You're a good hand, Beaugrand, the kind I want. Hang on, build a record, prove you're reliable and can be trusted, and I'll take high notice of you."

"Does that mean ridin' with —" Sid swiveled a flick of a look at Dain — "this Injun-faced buzzard?"

Walker nodded. "For a while. You two work well together." He leaned forward, expanding a little. "Next run, you pick up a sealed box at Littlefield, shipped out of San Antonio." He, a Texan, pronounced it Santone. "I can't afford to lose it. Bring it through safe. Then we'll see. Can't afford extra guards, either. That would tip off you're carrying it. Wolves of the trail would gather, and — whooeeps, you're dead!"

Sid stood up. During a long moment he stared strangely at nothing. "All right," he said, "I'll stick for one more run with this buzzard. Just one. To build a record."

He stalked out, and Walker, eyes following him, remarked to Dain, "Touchy, isn't he?"

"He's been around," Dain responded noncommittally, "and he doesn't like being pushed. He's red hell with a gun."

"Guns!" Walker said. "Gunfighters! Takes more'n powder to make money, know that? I think you do. And by the way, what I told him goes for you, too. Don't try to quit me, Moore. I'll have our Texas Rangers on your heels, pronto."

Dain said, looking at him steadily, "I'm

in your town only because I like something about it, like Sid Beaugrand. But don't push me hard. I can be a mite on the scrap, myself."

That set Walker off. He slammed both hands down on his desk, rising, and thundered, "By God, I don't take such back-talk! In my day I've eaten the likes of you and Beaugrand!"

Dain saw him then as an aging man constantly at war with the increasing years. The shock of iron-gray hair appeared slightly comic, bushed out in all directions like a clown's fright wig topping the overly aggressive face and oversize body. Forceful and ruthless, yes, this Milam Walker. Wealthy. Powerful. Able to impose his intolerant will upon a whole locality. But secretly, underneath it all, struggling bitterly against age. Striving to convince himself, as he convinced others, that he towered mightily in full prime, a bold king among men.

And, because of that, as dangerous as a treacherous kid gunman hungering for reputation.

Another slam on the desk. "D'you hear me?"

"You can be heard all over town," Dain said. He paused to choose his words.

85

"What you say may be true, but that day is long gone. Now you use money in place of a gun or your fists. I don't respect money too high. When I left Texas there wasn't much of it around to respect, and nobody gave a damn. It seems to be taking hold now, I notice since I got back. I don't see it as an improvement."

Something kindled in Walker's eyes. He bent his head like an old man brought to remembrance of past times. He said in a rumbling growl, "We didn't have much, for a fact, in those days. Wild cattle. Wild broncs. Ridin' your heart out. For beans an' maybe a dollar. I went through it all." He raised his head. "Go on. What else?"

Dan shrugged. "Any time you talk to me as you did, you can expect me to talk back. You ought to know that."

"I'll bear it in mind," Walker said drily, and then he was his same self again, assertive and dominating, his eyes as hard as glass and as totally devoid of feeling.

6

Making good time on the return trip, the Littlefield stage dragged through sand, approaching the long hump of Barrier Ridge in the near distance. Although not difficult, this was the most tedious stretch along the whole route, the slow plodding through deep sand followed by a wearying climb, hard on the horses. Dain loosened up on the lines and settled back to sweat it out. He and Sid had hardly exchanged a word this trip.

They carried no passengers. The special express box that they had picked up at Littlefield, locked and sealed, was heavy, so for balance they stowed it inside on the floor of the coach.

Dain let his thoughts dwell on a project that concerned Tresa Ogden. In Muleshoe Junction, just as they readied to pull out, a man tacked a notice up on the front of the hotel. The notice gave announcement of a dance Saturday night at Odd Fellows

87

Hall — *Gents $1, Ladies Free, Come One Come All.*

He guessed Sid's mind was speculating along the same lines as his own. Arms folded, hatbrim low over his eyes, Sid sat somberly uncommunicative. It was not the same as one of his silent spells. This was a black taciturnity, dourly oppressive.

Sid finally roused. As though catching Dain's thoughts, he turned his head and stated bluntly, "I aim to take Tresa to the dance."

"So do I," Dain said.

"I aim to be first to ask her!"

"So do I!"

They exchanged looks, deadpan and stiff. Sid said, "We got to settle this." He slipped a penny from his pocket. "Toss for it, or any way you fancy."

Without a word, Dain took the penny and inspected both sides of it. He poised it on thumb and forefinger, looking at Sid inquiringly, and Sid said, "Head."

Dain spun it up high. "Catch!"

Sid caught it, showed it, and his face lit up as though he had won a fortune. "Head it is! I take her!"

He chuckled at Dain's expression. Dain was cursing himself for letting the fate of such a thing rest on the spin of a coin. Sid shoved his shotgun under the seat, and in a high humor that was almost feverish he sang jeeringly, *"Just a solitary stray from far away — far, far awa-ay . . ."*

They pulled out of the clogging sand at last, and began ascending the steep grade slanting up Barrier Ridge. Dain kept the horses to a steady walk, watching the off-wheeler, a smart animal with a slick trick of slacking off in the hard haul. He still was wrathy about the toss, and when he caught the off-wheeler doing its trick he paid it attention with the whip.

Sid had fallen serious again. More than serious. He sat hunched forward, hands on his knees, his eyes searching ahead. His face looked leaner than ever, sharpened by a fierce impatience.

They neared the top. Here the road cut and bent through jutting tiers of rimrock and around fallen boulders. As soon as it leveled out on the crest, Dain drew up to give the horses their breather. He set the brakes.

Sid took his hat off and put it back on.

Three men rose, breaking cover from separate positions among the rocks. They had guns out, but wore no masks, no disguises of any kind. One was the dark and bearded badman. The others were the two in buckskin jackets. All three fixed their eyes on Dain, not on Sid, who made no move to get at his shotgun under the seat. They were not looking at him at all.

Dain held the lines in his hands, seeing the three close in, seeing the deadly and passionless purpose in their eyes. In a flash it all connected in his mind. The scraps of conversation he had overheard between them and Sid in the Bright Chance. *This* was the bank shipment. Sid's tense mood. Sid's odd action in taking off his hat and replacing it — a pre-arranged signal, letting them know everything was right.

What he saw in their eyes was murder. They would kill him, for their future safety. Part of the plan. And here sat Sid beside him, very calm now, very cool, waiting.

"Buster," the dark man said to Dain, "put up your hands an' step off!"

"I'll need my hands to climb down," Dain said. His holster had worked partly under his leg. He needed a trick to get at it.

"Nothin' doin'! You'll take it there, then! Makes small odds —"

A gun roared, startling the team so that the lines tugged at Dain's hands. Sid's gun.

The dark man staggered back a step, staring-eyed, his face all astonished and indignant. The second shot, delivered instantly after the first, took the man on his right, who husked a heavy sigh in falling. The last of the trio cried out a name and ran backward on his toes, dodging, shooting unsteadily. Dain freed his right hand and got his gun out and dropped that one.

On his knees, fumbling with his gun, the dark man glared wildly up at Sid. "You double —"

That was as far as he got. Sid smashed two more bullets into him, and he pitched forward onto his face.

It was all over in less than half a minute, and only the team retained excitement. Dain spoke to them, horse talk, profane and soothing. He tied the lines and stepped down. Sid stepped down on his own side of the coach.

Dain bent over the dark man and picked up his fallen hat. He stripped from it a black-and-white horsehair hatband and offered it wordlessly to Sid. Sid eyed him, shrugged, and took it. Dain gazed broodingly at the sprawled body of the man he had shot down, regretting the deed, the necessity of it. He was no killer and he never could be. Even in the war, four fighting years of it, he had regretted every time. . . .

He said then, "We'll find their horses somewhere near. Let's take your friends in. Wouldn't want to leave 'em laying here."

Sid turned a narrow stare on him. The aftermath was roiling inside him, Dain knew. Even the killers had it, some of them, and sometimes the worst of them. "They wasn't no friends o' mine! I won't ever hear it said they was!"

Sid made only one more remark on the way in to Muleshoe Junction, the bodies of the three bandits roped over their horses behind the stagecoach. He made it apparently to himself, but aloud. He said, "This trip I build me a record!" And he smiled.

The town turned out in full force. Everybody wished to shake hands and buy a drink. The deputy sheriff took his notes importantly and turned the bodies over to the undertaker. There were public expressions of appreciation from Milam Walker, who would have had to make good the loss of the bank shipment, the value of which was rumored to be tremendous.

Dain didn't wish to talk about it. Shaking his head to all questions, he pushed through the crowd and took refuge in the stageline office. From in there he heard Sid expounding outside, making a big thing of it, using a kind of faked modesty — "So I just pulled my gun an' let the bastards have it, that's all, nothin' to it." Sid was in his glory.

Dain sat slumped and alone when Tresa came in. She asked him rather coolly, "Why aren't you out there?"

He didn't look up. He said, "Today I killed a man." Then he said with sluggish violence, shutting his eyes, "Dammit, when a man kills a man, d'you expect him to celebrate? Are you like all the rest? Is Texas gone entirely to hell? Every damn soul in it? In so short a time? God, find me another country! Another country that I'll love as I've loved Texas!"

He felt her hands on his head, and he caught them in his and looked up at her. "I'm glad to hear you say that," she said, and now her voice was warm and tremulous, and he saw tears spring to her eyes. "Oh, so deeply glad! It proves to me that I must be wrong — about something that has been haunting me — something that has tortured me!"

"What? What was it?"

"The picture. Mine. That my father carried in his wallet. My father was burned in the stagecoach. His clothes, his old leather coat, his wallet — all burned. By the men who murdered him. And you had the picture. You gave it to me. Don't you see?"

He shook his head blankly, and her tears welled out. "I didn't want to think it, please believe me!" she said. "I hated the thought! But it was there — it kept coming back — that horrible question. *How did he get hold of that picture of me, unless —?*"

"My God!" He shot to his feet. "You wondered how I got it unless I was with the men who killed him! I could've told you when I gave it to you, but you didn't give me a chance."

"You don't need to now, Dain."

"But I want to. When Sid and I got there everything was ablaze but your father's coat, and that was scorching near the fire. All we saved from it was your picture. Then we were fired on and had to scoot fast."

Drying her eyes, she turned her back to him. He was worried that perhaps he had left out too much and by so doing had raised the question in her mind again, but presently she said muffledly, "There's a dance Saturday night at Odd Fellows Hall."

He opened his mouth to frame the suitable and happy response, but just then

94

Sid's voice jogged his memory unpleasantly. Through the windows he had a view of Sid idly spinning and catching a coin while talking to a group of listeners. Sid had won the toss, won the right to be first to ask her. Dain ached to welsh on it, but he knew he wouldn't. The choice was Tresa's, anyway. She wasn't bound to accept Sid's bid, and the odds were against it.

Evading the issue by pretending not to have heard her, Dain asked, "Why did Bledsoe fire your father, and hate him so much as to have him killed? He was there, with Hollen and a mixed crew. Sid and I saw them. Not much good going to the law, though, because it would be only my word against theirs. I doubt Sid would want to get mixed up in it," he ended carefully.

She answered with some hesitancy, "Bledsoe tried to — to force his attentions on me. My father objected violently. Bledsoe not only fired him, but was preparing to have him thrown in jail on some trumped-up charge. We slipped out of Travis City in the night, leaving everything behind, and fled here. Walker hired us both. I'm certain Bledsoe never forgave us. He's anything but a forgiving kind of man."

So, he reflected, the crime he and Sid witnessed that night in the Breaks had been committed partly out of revenge. The atrocious revenge of a lusting man against a father who protected his daughter. And partly a blow at Walker, the rival stageline operator, the enemy. A vicious warning to stay off the Bledsoe routes. Yes, the wolves and jackals had moved into Texas. . . .

Tresa repeated in an even smaller voice than before, "There's a dance Saturday night. . . ."

She gazed at him wonderingly when he didn't reply, searching him for hitherto un-suspected signs of awkward shyness. He surely gave no impression of being a bashful man. He stood at ease, though frowning slightly over this problem, a tall, big-boned man of thoughtfully restrained manner and a rare smile.

She must have perceived something in his face to give her the courage to do what she did. Or sensed that he was in a quan-dary of some kind. Her color, already rich, deepened as she said, "Well, it makes a shameless hussy of me, but I guess I've got to come right out and ask you. Will you take me to the dance? There!"

He smiled into her shining eyes. "Could you imagine I'd refuse, Tresa of Corpus

Christi? Could you?"

"Could you imagine I would, Dain of Padre Island — if you'd asked me?"

"Consider it I asked. I planned to, at what I thought was the right time."

Sid Beaugrand invaded the stageline office like a returning conqueror accepting laurels and a diadem to boot. He was as bright as a new sword. He glowed all over. The admiration and compliments showered on him outside in the street, transient and half false as they were — and he was wise enough to know they were — had wrought upon him a glitter, a self-assured polish.

He racked his company shotgun. Meticulously, first wiping it with an oiled rag. Toward good firearms he, like Dain, gave respect. He touched Dain with a sardonic glance, then strolled with large confidence to Tresa and spoke his piece.

"Hello, honey-gal!" He put his hands toward her. "D'you know what knight in shinin' armor is takin' you to the dance Sa'd'y night? I mean at Odd Fellows Hall, all nice and genteel, nothin' rough, ladies an' gents. D'you know?"

"Yes," Tresa said, instantly and positively, avoiding his outstretched hands.

"Dain Moore is taking me."

A sharply indrawn breath, sucked through the teeth, hissingly, as from hurt; an utter silence; then she was backing away from the startling transformation of Sid's face. She edged instinctively to Dain for protection, and Dain took her arm and drew her farther back and placed himself before her. Sid's eyes slanted, satanic. The underlids jumped.

Milam Walker came in and, normally self-centered and obtuse, sensed nothing amiss. He declared with more than usual heartiness, marching to his desk, "You boys are good! I swear you make it hard for me to choose between you for top screw of this stageline, I mortally do!"

This was old Texas talk, that he indulged in comfortably when expansive. He ordinarily spoke without the drawl and the long pauses that strangers often found irritating in Texans, for he was in business and had mixed around.

He dug out a cigar and snipped off the end. "Light!"

Dain and Sid paid him no heed. Neither did Tresa. It got through to him then, and he eyed all three shrewdly while the old bookkeeper performed his servile chore. When two men, friends, fell out so drasti-

98

cally in the presence of a highly attractive girl, only one answer was possible. Walker's eyes glinted a savage jealousy. He motioned for Tresa to retire into her office, and she obeyed reluctantly. When he spoke to Dain and Sid, however, he pitched his voice to a moderate and confidential key.

"Bids are open for the government mail contract. Right now, Bledsoe holds it. This stageline has got to have it, to operate at a decent profit. There's a lot of prestige, too, in running the mail. We can make the run down from Travis City twice as fast as Bledsoe. We've got to prove we can, to win the contract! You understand me?"

They looked at him vaguely, each occupied with his own devil, and he went on, "Some time ago I sent a stagecoach up to Travis City, on a test run. The driver happened to be Ogden — Tresa's father." His eyes shaded, masked. "Ogden never got back."

"We know about it," Dain said. "We saw it. He didn't have a chance after they trapped him. He was a dead duck right there."

Walker raised his eyebrows. "You saw it, eh? A most tragic affair. Very sad." His voice lacked genuine feeling. He cut a glance at Tresa's office and back to Dain and Sid. "I take it you know the road to Travis City, then?"

"Some of it."

"D'you two men think you could make it? That is, pick up mail at Travis City and deliver it here in quick time, for the record. To show proof that we can do it faster than Bledsoe. Well?"

"Not together!" Sid grated. "I wouldn't ride with a damn welshing son-of-a-bitch if you gave me the stageline!"

Dain let that go by, seeing Sid's side of it. He said to Walker, "That's a rough road through the Breaks and the caprock. A long run. And no change of teams. You've got no relay stations up there, of course. It's Bledsoe's route. I don't think your stages are built to stand it. Too light. And it'd mean nearly killing the horses, the round trip."

Walker turned to Sid. "What do *you* think?"

Sid said recklessly, "I'll take a stage over any cock-eyed road and to hell with the opposition! But not with him! And I want to know first what's in it for me!"

Walker, puffing at his cigar, bowed his head. "I see. All right, you can each take a stage out. Tomorrow morning. To Travis City and back. The one who makes the best showing positively gets the job of top screw of this stageline. He gets a free hand to run the outfit, and forty-percent interest in it to encourage him. How's that?"

Sid said starkly to Dain, "We'll take rifles and we'll go double-gunned — huh?"

"A duel?"

"Call it that!" Walker smiled.

That evening, Dain had a talk with Tresa. He caught her as she left the stageline office, late. She worked long hours. "I reckon I won't be here for the Saturday–night dance," he told her ruefully. "The Travis City run. Sure sorry."

She almost clung to him. "Don't go!" she pleaded. "Don't make the run to Travis City! My father lost his life on it, and I'm afraid! Bledsoe and his men will do anything to stop you. *Anything!*"

"It's not that bad," he said, "or Walker wouldn't want Sid and me to try it."

"Are you sure he wouldn't?" she asked strangely. "He sent my father out, knowing that he hardly had a chance. He pretends a fatherly affection for me — but his interest in me is anything but fatherly! A girl can tell. And I saw the look in his eyes today when he realized you and Sid were on the edge of fighting over me. Milam Walker

101

would sacrifice anybody, if it suited him!"

"That may be so," Dain granted. "But this now is a personal thing between Sid and me. You understand that, Tresa, don't you?"

She answered bitterly, withdrawing from him, "I understand it's a man thing — a pride thing. You'll be fighting each other as well as the Bledsoe crowd. It's sheer madness, yet because of this crazy masculine pride you'll try it!"

Pride he did not believe was his master, as it was to Walker, to Sid Beaugrand, and, from all accounts, to Bledsoe. He wanted to tell Tresa that, to explain that this thing involved a good deal more than pride, at least for him, but she parted from him while he still sought the words. Frowning, he crossed to the hotel.

Sid sat in the lobby, and Walker was talking to the clerk at the desk. Dain would have passed on through to his room without speaking, but Sid accosted him, saying stonily, "Get this straight. I'll spill you any way I can. I aim to come back winner."

"Maybe we'll both come back," Dain said.

Sid shook his head. "I don't see it as likely. If so, Walker names the winner.

Then the other'n quits an' pulls out, flat. That won't be me."

"I take your warning and give you the same."

They spoke in undertones, two men discussing a private affair without display of rancor. Walker came over and broke in on them. His eyes shifting from one to the other, he declared, "Make no mistake about it, nothing's more important to me than winning the mail contract away from Bledsoe. It would set my stage company up high on its feet, and cut the ground out from under his. So I've just decided to raise the stakes for you boys. The one who gets back —"

"Meaning the one who gets back first?" Dain interrupted.

Walker only looked at him, and did not commit himself on that score. "On top of what I've already offered," he said, "I'll add a cash bonus of a thousand dollars. . . ."

Generous, Dain ruminated. He thought of what Tresa had said, that Walker would sacrifice anybody if it suited him. The thousand dollars stood as an extra inducement for him and Sid to strain all out to do their best. Or it could be intended as one more incentive for two men in contest to destroy each other, or one the other.

The truth lay in between, he supposed. If he or Sid should succeed, and Walker won the mail contract, well and good. If they both failed, Walker was rid of two pushy roughnecks who had proved too difficult to handle and were becoming troublesome to his peace of mind, especially in regard to Tresa. Walker certainly must have noticed a difference in Tresa lately; a rising independence and an aversion toward him. Dain glanced at Walker for some betraying sign of corroboration.

Walker's eyes, meeting his, were chill and dominant, and ominous. The man seemed without emotion, immensely alert behind a ponderous inertness.

7

The singing issued from many unseen voices, loud, not unmusical in the clear air, accompanied by a throbbing guitar and a blasting trumpet. Mexican music. Each phrase paused on a falling note. Each verse died a lingering death, attended by sympathetic sobs of the guitar and vagrant wails from the brass trumpet. Mexicans in general loved putting plenty of tragedy and heartbreak into their singing, even if the song happened to be about the love-life of a rooster.

Listening to it while driving his stage, Dain Moore got homesick all over again for South Texas — which to him meant the border country below Laredo, the Rio Grande Valley and the Coastal Plain, and particularly the lower Texas Gulf Coast where a man could fish, swim, sail, and work his cattle the same day. Where the folks, the few who had the sense to live there and with whom he had grown up, enjoyed life for itself and took an easygoing attitude toward money.

Driving a short distance in the lead, Sid was unaffected by the voices, except as from curiosity. He had no roots anywhere. His recollections, as far as Dain had ever discovered, had to do only with women and whisky and wild scrapes. He never mentioned homeplace or family.

Although Sid held the lead over Dain, and had done so from the start since leaving Muleshoe Junction, the race was not yet on in earnest. That would come later. For the present they saved their teams. They had a long way still to go, and then the long dash back. No relay stations for them on this route.

They rounded a low hill, following the road, which then dipped at the cutaway bank of a dry riverbed. Here a bizarre sight confronted them. An ancient coach was pulling up out of the riverbed. It was held together by baling wire and rawhide, and packed inside and out with a collection of singing humanity. A hitched-on springless wagon trundled behind. Shaggy burros in rope harness, led by a wrinkle-browed yellow ox, did duty as team.

Dain grinned. He had seen the ramshackle outfit before, in Muleshoe Junc-

106

tion. This was the *Grand Express,* owned and operated by Porfirio Valdez, a fat man of magnificent dignity. Its home base was San Rosario, a Mexican settlement somewhere up near the caprock, off the regular road.

As Dain had heard it, once in a while the spirit moved Porfirio Valdez to announce that he would hazard the formidable journey, with a payload, to Muleshoe Junction. A payload consisted of everybody in San Rosario who could climb aboard for the joyride. Most of them were related to him, and of course they rode free. Others paid him in dabs of produce — corn, squash, chilis — which he sold or traded in Muleshoe Junction. It took him forever to make the trip, and they had to camp out wherever nightfall found the coach, but the whole arrangement appeared satisfactory and everybody was happy. They sang all the way.

To Dain it was the kind of serenely madcap caper with which he was nostalgically familiar. He stood in favor of it. What the money-grubbing world needed was more radiantly goodhumored nonsense, more clowning, less of the grim marching to drearily logical destinations. . . .

The yellow ox came plodding placidly

onward, splayfooted. The double string of fourteen burros shuffled their little hoofs and flapped their ears. The ancient coach lurched in the ruts, its incredible load of riders loudly harmonizing, the guitarist — it was a giant *guitarron* that he banged — practically sitting on top of the driver, Porfirio Valdez, and the battered trumpet rising from the throng to peal off at the blazing blue sky.

Porfirio Valdez raised soft brown eyes in a moon-round face to see the two splendid yellow-and-green stagecoaches rushing down the road at him, their haughty horse-teams high-stepping, and he tried to pull out. A small boy, squatted at his feet, threw stones frantically from a box at the yellow ox.

Sid made no attempt to slack down. Dain saw his whip rise and fall, saw his stagecoach swerve scarcely enough to clear, and he foresaw the outcome and hauled hard on his own lines. There came to him the click of hub striking hub, one tightly steelbound, the other cracked and bound with wire, and the sound of a breaking wheel. Sid slammed by, swung back into the ruts, and drove on.

The *Grand Express* tipped drunkenly to halt, spilling men, women and children in a yelling shower all over the road. The burros brayed and the ox groaned. Dain got his stagecoach halted barely in time to avoid a disastrous pileup.

He set the brakes, tied the lines, and jumped down. Women cried, kids bawled. Swarthy men flashed their eyes, plainly wishful to commit homicide in reprisal for this outrageous indignity. These were an even-tempered people on the whole, content to enjoy small pleasures of their own making and able to get more out of life than any bunch of millionaires. But, jostled, they blew up. And no *paisano* went to town lacking a weapon of some kind about him, in case of possible eventualities. They eyed Dain like a pack of panthers with their backs up. He was a cursed Walker stagedriver like that other one, and his brace of holstered guns didn't scare them a bit.

Dain said to Porfirio Valdez, in Spanish that was border Mexican, *pocho*-Spanish, "Don Porfirio, I offer a thousand apologies for the discourtesy of my fellow driver. He is an *extranjero* that knows no better." He scooped up a tumbled little tyke that was squawling its head off, and rocked it in his big arms, crooning and making comical faces down at it. "How now, *niñocito?* What uproar is this? Surely with the gift of such a thundering voice you shall grow to become a giant whose shout will make the very mountains tremble! I fear for my ears! Desist, I beg you, little giant-to-be!"

The *niñocito* gulped back a fresh howl and solemnly inspected his face. With his bandanna Dain gently dabbed its tear-streaks.

Porfirio Valdez, placed squarely upon punctilio, swept off his straw sombrero and bowed. He wore rawhide sandals, no socks, and his cotton shirt and pants wouldn't have fetched four cents for scrub rags. Yet he had grandeur.

He assured Dain that he could in no imaginable fashion be held responsible for the upset. In fact, he said, the upset was of small consequence except that it blocked the road and most unfortunately detained Dain from proceeding on his way. For that he and his people owed apologies — a million apologies, topping Dan's a thousandfold. They would clear the road for him, *muy pronto.*

A barefooted young woman came forward to reclaim her child. She smiled shyly at Dain, taking it, then less shyly when he smiled back. "He truly has a voice," she admitted proudly. *"Magnifico!"*

They were all smiling at Dain now, recognizing by his speech and manner toward them that he was a way-South Tejano. The flash of the angry panther drained out of the men's eyes, replaced by the warm pleasure of meeting somebody from home — the borderland from which by some chance they had long wandered — somebody who was at ease and on equal terms with them. They threw anger off and forgot it. Porfirio Valdez, having the title of *don* conferred upon him, shook hands with Dain ceremomiously. Then everybody shook his hand. The official stone-throwing boy informed Dain that he had decided to become a stagedriver when he grew up.

"Be rather a cattleman," Dain advised him gravely, man to man. "The railroads are building. You may live to see little use for stagecoaches. For beef there will always be need."

To Don Porfirio he said, "We could rip a plank from the bed of that wagon, perhaps? And lash it aslant under your coach axle here, as a drag, to replace the broken wheel? It would not get you to Muleshoe Junction, I fear, but it would get you home if you are careful not to put much weight on it — which means that your passengers must walk. The children could ride in the wagon behind."

They fell to in rollicking good humor, Dain helping, and rigged the drag, lashing it firmly in place with a couple of spare lines that Dain carried in the stage boot. Don Porfirio tested it by driving up past the Walker stage, preparatory to turning back. It worked well, everybody cheered, and Dain swung up onto the box and sorted out his lines. He had spent a precious hour here while Sid raced on.

He shook up the team and raised his hat, followed by a chorus of thanks and good wishes. At first they had addressed him as *señor* and used the polite *usted* third-person form. Now it was, *"Gracias, amigo — mil gracias! Vaya con Dios! . . ."* He had toiled alongside and joked with them. *Un compadre.*

An hour lost, but he felt good. It was fine to be again among warm-hearted philosophers who didn't give a damn for time or money, people who had learned centuries ago that life was not given to waste in scrabbling pursuits, but meant to be enjoyed at leisure, with grace and gratitude.

Anyhow, his team had got a resting spell out of it. He could afford now to stretch them out in a spanking run.

The heavy red-and-black stagecoach stood halted along the road through the Breaks, its six tall Missouri mules motionless in that attitude of mindless unconcern typical of their kind. The driver and two shotgun guards were talking and gazing rearward, so occupied that Dain drove up within a hundred yards before they showed any awareness of the noise of his approach.

"Here goes!" Dain muttered, as their heads whipped around. Clearly, Sid had recently passed them, and they had pulled up to discuss it, and now here came another Walker stage bowling along the forbidden road. It was enough to make them think they were seeing delayed double, or a mirage.

Dain hitched his holsters forward where he could get at them fast. Those shotguns at close quarters could cut him to rags if he gave the guards a chance to bring them into play. Sid had had the advantage of surprise over them, passing them unexpectedly on the road. For him, little of that advantage existed. Their vigilance was sharpened, their reactions quickened.

Pulling a bluff, he let them see him reach back and slam his right hand three times down on the roof of the coach as though drumming a signal to men inside. The driver and guards atop the Bledsoe stage hunched lower and froze, and stayed so during the moment it took for him to close the distance.

He whirled by with scant inches to spare, flicking a calm glance over their staring faces. Not making the mistake of weakening his show of arrogant confidence, he didn't look back, although his spine prickled in half — expectation of a double shotgun blast of buckshot. He heard one of them exclaim, "Two! *Two* of 'em! What the hell?"

Then he was beyond range and sight, cutting curves in the crooked road, and he let out his breath and tipped his hat for the sweat to trickle down. "Woosh! Any fool wants this job can have it! Henceforth mine's a saddle, so help me! Get along, team, get along! Haa-aa, there — haa-aa-aa . . ."

He sighted Sid when well up into the caprock country. A glimpse, that was all, on an open stretch of the tortuous road. Sid had either pushed his horses too fast without rest and dulled their edge, or else he thought the competition was removed. Dain gave his team a breather, after which he settled down to steady driving, keeping the horses up to their best notch while not overtaxing them. He was gaining on Sid. That was good enough. He would not court disaster with any whirlwind dash over these up-and-down rough roads. The spent hour could be recovered best by soundly considered driving, leaving always a margin for the dash if necessary. Dain tooled his team along, whistling to them, talking horse-talk to them, always encouraging, never forcing them.

These six good horses represented his passage back to Muleshoe Junction, to Tresa. He would not sacrifice them.

The Bledsoe stage that he had passed back there, he surmised, was doubtless carrying the mail to Muleshoe Junction. In that event, this was a dry haul for him, and for Sid. What then? Would Walker pay out if they returned empty — *if* they returned, either one? He would not be bound to, according to his stipulations. Bringing back the mail was his main object. Walker had a couple of government postal inspectors on tap, whose sole interest lay in finding the fastest and most efficient means of carrying mail. Bidding for the coveted contract, that mainstay most essential to the profit and success of a major stageline,

115

Walker promised twice-weekly delivery, fast, safe, reliable. Later, daily delivery.

Without mail under the box, a record run could hardly impress officially the government postal men. They were hard-headed and they dealt in solid facts. No mail, no record. No contract.

Sid Beaugrand, rattling into Travis City, made a wide turn in the central square and drew up before the railroad station. While building a cigarette he ranged a suspicious look around. His fleetly searching glance stilled on the office of the Bledsoe Mail Stage Company, across the square from the station. He looked straight into the eyes of Hollen, standing in the doorway.

Hollen's eyes were slits in his hawk face. He sent Sid a slow nod. In that nod could be read several reactions. Formal recognition of an old side-kick. Grudging appreciation of high daring. A tinge of regret for what must be done, and a deeply thoughtful regard to the means of doing it. In this town a yellow stagecoach was anathema, a red rag to fifty bulls, and Hollen was top bull. He spoke over his shoulder to somebody within the office.

Bledsoe came up behind him, and Hollen moved aside to give him first place.

116

Together they stared across the square at Sid and the yellow stagecoach.

In his whipcord riding breeches and straight-brimmed hat, Vince Bledsoe resembled what Texans called a buggy boss, a visiting ranch owner, generally from the East, who made his inspection tour in a buggy simply because of horseback ineptness. Bledsoe was, in fact, an Easterner, rumored to be of prominent family. But he could ride anything with hair on it. He was a crack shot, rifle or pistol. He could outwork and outwear any of his hired men, and beat them all in sheer nerve.

His eyes were a kind of semi-transparent black, like obsidian, strangely alive, set in a pale-skinned face whose ascetic cast was marred by the developed lines of the satyr and the voluptuary. He had been everywhere, done everything, and spent a couple of inherited fortunes on specialized vice. Now he was forced into the position of having to make his own way. He was making it very well, for he possessed keen intelligence and no scruples whatever. The main fault he found in Texas was the quaint Texan veneration of good womanhood, sternly applied. It handicapped him in the exercise of his chief pleasure, his everlasting want, and the exquisite brutality

117

of that specialized and esoteric vice that was his. He wanted, craved Tresa Ogden; she would be a rare prize, one to shade all the others.

A mounting ruckus south drew Sid's head around, and all astonished he watched Dain Moore drive into town. He grinned at first, wagging his head. That Dain. That Texan, coming home. Then he scowled. That double-crossing behind-the-back son-of-a-bitch! Sid jumped down and stalked into the railroad station, for the mail.

Dain made the turn in the square and drew up behind Sid's stage. He saw Bledsoe and Hollen, but gave them only a cursory look. To hell with their amazement, their venomous glaring. Bledsoe ran this town, yes indeed, but it wasn't likely he'd pull anything raw, such as a gang-kill, right here in front of everybody. There were limits beyond which even a town-master could not step with impunity. The attack, the stark peril, would come on the open road, the road back. No witnesses.

In the station he came upon Sid arguing with the mail clerk, a self-important little man in paper cuffs and green eyeshade. "The Muleshoe Junction mail went out on the Bledsoe stage 'smornin'," stated the mail clerk. "Today's train run late an' they didn't wait for it. It only brung in a dab, anyhow."

He slung a mailbag over the counter, from long practice landing it accurately on a bench near the door. "An' you can't take it without you got a signed order!"

"I got one!" Sid declared, searching through his pockets. He had, too, as did Dain, but it was like him to mislay it. A slip of paper didn't mean much to Sid unless it was currency. He cut a sidelong look at Dain, who had his order ready in his hand. "Thought I shook you off back there, welsher."

"You hoped." Dain nodded. "Mind what you call me, you friend-killing Judas."

The mail clerk, pricking up his ears, hinted darkly that Sid was an impostor brazenly attempting to collect mail without authority. Dain agreed with him that Sid was doubtless a mail robber trying out a new wrinkle. "I'd sheriff him, was I you," he said, poker-faced. "Bet he's on more than one reward dodger. Looks mean enough."

Sid Beaugrand did look mean at that moment. "Damn paper must be in my jacket," he muttered, stalking out.

119

Shaking his head, the mail clerk turned to his pigeon-holes. "Here's a coupla letters for Milam Walker," he said, laying them before Dain. "Ain't much to show for the trip. Like I told you, that Bledsoe stage took —"

"Just so it's mail, the amount don't matter."

A stickler for form, the mail clerk examined Dain's order meticulously. Dain heard a jingle of trace chains outside, followed a moment later by the sounds of Sid driving off, which surprised him. The mail clerk at length gave the written order his official approval. "Seems all right." His eyes, happening to glance past Dain, bulged. He let out a yelp. "Hey! That damn thief — !"

Dain spun around. The bench near the door was empty. Sid had hooked the mailbag on his way out. Dain left the mail clerk waving his arms and shouting for the sheriff. "I'll catch up with the tricky twister, if I kill the horses doing it!" he vowed to himself, running to his stagecoach.

On the box he gathered up the lines, cracked the whip, and uttered his starting yell. The team lunged forward, but the coach didn't move an inch. With the leather lines wrapped around his strong fingers and over his wrist, Dain got jerked off the box in a flying header. He slapped flat on the ground with a force that knocked the breath out of him and nearly broke his neck. The horses dragged him, all spraddled out, the length of the town square before he could get them to halt. By then he dazedly had the cause of it figured out.

Sid had uncoupled the trace chains, correctly assuming that he wouldn't notice in his hurry.

Scratched and bruised, smothered with dust, he got up and steered the team back, picking up his hat, whip, and a fallen gun enroute. Roars of laughter at his mishap sounded all around the square. He was too disgusted and out of breath to swear. As he got the team sorted out and hooked up, he paid a look to Bledsoe's stageline office and experienced a chill feeling.

Bledsoe and Hollen and a squad of gun-slung men stood there. They were not laughing. They gazed at him, their hard features impassive, their eyes expressive only of estimating consideration, as wranglers might study a bronc to be broken.

Soon as I clear town they'll be on their horses and hot after me, he thought, and then he felt lonely, as lonely as ever in his life.

8

Over the caprock road Dain maintained a hard pace that fetched flecks of mouth-foam from the sweat-drenched horses and on sharp curves strained the underbraces of the swaying coach. This was the long dash for which he had carefully conserved the team, and he was spending everything on it. He could not afford sympathy for the hungry, thirsting animals. They had to give their best, and he forced it from them with the whip.

Twice he had to halt, to roll rocks off the road — rocks that he knew Sid had tumbled down to wreck him or at least to delay him. He grinned leanly and sardonically at that trick. In taking the time to pull it, Sid was also delaying himself, gaining from it no advantage. Sometimes Sid, obsessed with crafty wiles, forgot sound common sense.

Bledsoe's gun-crew still did not show up behind. Yet Dain felt the sense of certain knowledge that they rode on his tracks. Perhaps they had not expected him to hit such a fast pace. Perhaps they timed him wrong, too accustomed to the Bledsoe stages' mule-drawn rate of travel. They were saddlemen, not stageliners, and hardly knew the vast difference between a cumbersome mud wagon and a swift Concord. The gun-proud ignorance of brush outlaws could often be amazing. A prickly breed, quick on the kill. Devilish clever in their own element. Cow-dumb and jackass-stubborn when facing anything unfamiliar.

Dain caught up with Sid in the Breaks, south of the caprock, where the stageroad widened while cuddling a four-mile ledge, a towering black cliff on one side, a steep rock-strewn drop-off on the other. He made to pass, on the drop-off side.

Sid swung over, hogging the middle of the road, blocking him. His horses ran somewhat heavily, for he had not handled them well, not being an expert driver, nor one possessing the instinctive touch and rapport with his team, through the lines.

But on a hairpin bend Sid's horses took the inside track, close to the towering cliff, of their own wise accord. And there Dain roared into the open space and forged abreast, on the outside.

Dain twirled his whip over his head and cracked the double snap. The first snap exploded under Sid's nose and the second over Dain's racing team.

Sid panted curses in a high, tinny voice. He cut loose, his temper flaring. He slashed at Dain, and the braided lash of his whip laid a bleeding welt across Dain's face.

Dain flinched involuntarily at the stinging pain of it, and immediately struck back. It had to be back-hand for him. He drove on Sid's left. He flung the long lash out and drew it in with a savage jerk, and it bit like steel wire into Sid's neck. Shuddering, Sid continued hitting, and Dain stood up and matched him strike for strike.

The horses went frantic, both teams thundering neck-and-neck together down the crooked, precarious road, in their flat-laid ears the repetitious crack and slap of the lashes. The careening coaches banged into each other, scraped hubs screaming. In the grip of insensate rage, neither man would fall to the rear and give way. Sid flayed at Dain, and Dain took whistling cuts back at him.

Their faces became criss-crossed with

red welts, rips in their shirts showed blood, and still the mad duel went on. Feinting a blow, Sid whirled his whip twice and lapped the lash around Dain's neck, choking off his breath. He tried to haul Dain down off between the two coaches under the wheels. Dain, hanging on, clawed himself around and got it uncurled, and hit fast before Sid could stretch his whip out.

Part of his lash must have smacked Sid across the eyes, for Sid threw up his forearm and pressed it against them, his shoulders twitching. When he uncovered them his eyes were streaming. His next whack at Dain was clumsy and went to waste. Giving it up as a bad job, he crouched low and hit at his horses, seeking to pull ahead, to get out of the punishment due a loser.

Dain could then have flogged him to ribbons, but to rub defeat into a partially blinded man went against the grain. Instead, he gave belated heed to his driving and to the road ahead.

"Watch out!" he shouted.

Whether or not Sid even saw the Bledsoe stagecoach blocking the road until too late, Dain never knew. Sid went on whipping his team blindly like a maniac, the lines all bunched anyhow in his left hand, not measured to a nicety between his fingers as they should for good driving.

Dain tried to pull back. He and Sid were bearing down on the Bledsoe stagecoach at top speed, and there was nowhere near enough room to pass abreast. His maddened horses refused to obey the lines. Their blood was up and they raced out of control. The brakes squealed and smoked without effect. The distance rapidly shortened, and then it was too late.

The heavy Bledsoe stagecoach, the same one that Dain had passed earlier when going up, and that Sid also had passed before him, stood square in the middle of the road, brakes set, wheels blocked, the mules at rest. The driver and two shotgun guards were waiting, close to the coach for cover, their weapons ready.

With that formidable set-up they had every reason to believe that they could command the road, force the Walker stages to halt, and hold them for disposal by the undoubtedly pursuing gun-squad from Travis City. They had no reason to foresee two suicidal madmen slamming their rigs down abreast at them, a heart-stopping sight, not to be turned off by gunshots, any more than an avalanche.

They scrambled out of the way.

At almost the last instant the two bolting teams took it into their heads to split apart for the standing obstacle directly before them. Of their own accord Sid's leaders swerved aside and plunged into the gap between the Bledsoe stage and the rising bank alongside the road. It didn't seem possible to make it. There was a rending crash, half the jutting rear boot of the Bledsoe coach was torn off, and Sid's coach skidded and tilted from side to side, banging like a runaway cattle car off the tracks. Somehow it battered through, Sid's whip never at rest.

No such luck attended Dain. He had the outside of the road and he was stuck with it. He swung to pass on the near side of the Bledsoe stage, knowing too well that he couldn't make it, that the space was too narrow.

His near front wheel sank, sliding over the road's edge, and the coach canted drastically, then the rear wheel crunched over and threw the coach entirely off its center of gravity. At that, he might have skinned by, the team running so strong, but the front wheel hit a half-buried boulder that kicked the jouncing rig high. That was too much, even for a well-built Concord.

Dain heard the braces crack. For just an instant he saw his team bolt on with the bare under-carriage, following Sid. Dain's coach, torn from its pinnings, tumbled down the rock-strewn slope, Dain with it. It fell like an empty great crate, rolling, sliding, gathering impetus, then with sudden finality smashing itself among a jumble of rocks part way down the slope. Dain lay dazed and gasping, his mind fumbling over the thought of catastrophe.

For a while, he did not know how long, images recurred to him of the old home-ranch and of Padre Island, of the Texas that he had known and grown up in before he left to fight on the side of the Lost Cause that, right or wrong, to him had represented too simply the defense of his homeland. The lean, tough, singing vaqueros. His family had never owned slaves, not one. The slow pounding of the blue-white surf on the beach, the cockle-shell boats built, the saddles owned, the wiry ponies and hard riding after half-wild cattle, roundup and branding times. . . .

He heard the loud popping of exploding shells, the tearing impact of bullets near him, and he fought out of his dreaming haze. They, the Bledsoe stagedriver and shotgun guards, were shooting down at him from the road above, riddling the smashed coach in vengeful desire to finish him off because Sid had scraped clear and was gone on his way with mail, to Muleshoe Junction.

Dain dragged himself out of the wreckage and took cover behind a rock. His rifle was missing, flung out somewhere from its hanging scabbard beside the box, but upon touching his belted holsters he found that he still had his sixguns. He drew one, lined it over the sheltering rock, and fired up at a poking head and long steel barrel.

His eyes were not yet right. His bullet kicked dirt below the level of the road. Squinting, he tried again. The long barrel puffed, then another, and another, one on each side. They were using rifles. Confederate Jamestown Mendenhalls, by the sound and smoke of them. A Minie ball scored a light gray streak in the rock and screeched off. Dain huddled down, blinking at the sixgun in his hand.

They've got me staked out, he thought, pulling his mind together. *All they need to do is hold me here till the others . . .*

A gun from farther off blared heavily. Wondering, he listened to a howl following that mammoth report, and to voices raised in shocked dismay. He risked a look around the rock. Seeing at first nothing moving, he cocked his sixgun and waited. A moment later he came within a hair's pressure of firing at a straw sombrero high up above the road.

Beneath the straw sombrero peered the round face of Don Porfirio Valdez. The fat compadre was finishing the reloading of an exceptionally long flintlock rifle, tamping the charge home with a wooden ramrod. Beside him his small grandson appeared, scowling portentously.

Don Porfirio set the lock, took aim and let fly, and black smoke screened him until he shifted to a new position. Evidently he was shooting bits of cut nails and odd scrap-iron. The lethal load whistled, spattered the road like hail, and a second anguished howl arose. Somewhere up the road oncoming hoofs hammered out a quickening cadence. Dain quit the rock and began climbing, after slinging a shot at a head that bobbed up and hastily withdrew. Don Porfirio, good and watchful friend-in-need, himself needed time again to reload that outdated but terrible firearm.

Dan reached the road, making a lot more noise in doing so than he intended.

He was in a hurry. The Bledsoe riders from Travis City were coming fast and it meant sure death for him to be caught sprawled out in plain sight on the slope. As he peeped above the edge of the road, a small rock dislodged underfoot and in falling brought on a minor landslide.

His quick glimpse revealed both shotgun guards lying face down in the road, dying or badly hurt, fingers and toes scratching at the dirt. The driver, flat against the near side of the Bledsoe stagecoach for cover, and trying for a shot up at Don Porfirio, spun around at the sounds Dain made. Glaring, frantic, he got his Henry seven-shot leveled, dropping his spare, the big-caliber Mendenhall. Bledsoe's men went armed to the teeth, ready to meet any and all eventualities.

Dain shot the driver directly in the chest, first. Then in the heart. He watched the driver fall, knowing that the manner of it would be slow to leave him, that he would be slow to forget it.

My gun, he thought, *is so cruelly accurate. I never miss when I am myself. I am fast and sure, able to take on the best of them. God guide me!*

Above, Don Porfirio threw up an arm, urgently waving him onward. The gunmen from Travis City were coming. They came roaring down the road, in full sight, primed for the kill. Don Porfirio's long muzzle-loader blared at them, spreading broadly at that range its scrap-iron charge so that they all, men and horses, suffered the bite and stinging hurt of it. The horses stalled and swapped ends, scampering back, and the riders raised arms to shield their faces. A confused set-to ensued, during which Don Porfirio waved again to Dain, and his grandson waved, to tell Dain to get out of there while they held the riders off. Knowing the country as they did, at the right time they could scurry off through the broken hills without the riders catching a glimpse of them. The fat man hastily reloaded.

Dain hauled up onto the road and ran to the Bledsoe stagecoach. He kicked the chocks out from under the wheels, mounted the box, and freed the brakes while sorting the lines.

"Get going, mules! Haa-aagh! . . ."

Sid Beaugrand, strutting high in Muleshoe Junction, smilingly accepted the congratulations of Milam Walker and the postal officials for a record-breaking trip. The fact that he had half-killed his horses

132

didn't stand against him. He left it to the stablemen to care for the drooping animals, and tossed his mailbag to Tresa on the loading platform.

"There's the mail, honey!"

When Dain's team cantered in, pulling only the bare undercarriage, Sid remarked with genuine regret, "Poor ol' Dain!" He looked at the white-faced Tresa, ignoring all others. For a minute his face came youthful, clean and aspiring. "We won't be seein' Dain no more. Dammit, they got him! Girl, get it! He's gone!" The spasm of regret creased his mouth, making him appear old and worn. He muttered sullenly, "The hell with him!" — seeing the look of tragic loss in her eyes and on her face.

Then Dain rattled into the main street, driving the Bledsoe stage. Thick traffic parted for him. The Missouri mules stolidly trotted past riders and rigs, to draw up at the twitch of lines before the home office of the South Plains Stage & Express, right smart.

"Delivering the mail from Travis City!" he sang out to Tresa, exchanging a brilliant smile with her. To Walker he said, "Had to swap coaches. Mine got wrecked. Borrowed this'n. It's full o' mail."

For once Walker loosed a hearty guffaw, turning to the two postal officials and declaring, "That's my man! They can't stop that kind of man, gentlemen!"

Tresa stood transfixed, her face aglow. Looking hard at her, then at Walker, Sid rasped, "All right, so I miscalled the score on him!" He wiped his bloodshot, swollen eyes. "He was a dead duck, far's I could tell. All right, dammit, he's here alive, but who got back first? Tell me that? Him or me?"

"Why, you did, Mr. Beaugrand," Tresa responded clearly and coolly. "With a sack of mail for Fort Worth — not for us! You apparently picked up the wrong sack, Mr. Beaugrand."

Walker pealed out again his loud guffaw, then abruptly cut it off, seeing the wickedly dangerous cast of Sid's face. He said to Dain, "The top job's yours, and the thousand-dollar bonus. You may yet have to earn it," he added, looking at Sid.

But Sid spun on his heel and strode off, not giving back a word. Dain watched him stalk straight to the livery stable for his horse. Sid was getting out of town. He was all through here. Dain looked for him to glance back. Sid didn't, and Dain felt a real pang of regret. After all, he and Sid were sidekicks, after a fashion.

Walker raised his voice so that all the crowded main street could hear him. "Another dance party next Saturday night, folks, with all the trimmings — and it's on me!" he bellowed. "In honor of the occasion! We're a cinch to win the government mail contract, and I've got the right man to handle it — Dain Moore! I tell y'all, Muleshoe Junction's the up-and-coming town! Travis City will wither away from here on!"

Sid Beaugrand then looked back, from the door of the livery barn. His expression reflected intolerable disappointment, searing regret. Milam Walker, watching him, could read that look. There, but for a small slip, a hasty error, Sid Beaugrand could have stood in the place of Dain Moore, accepting the congratulations of men crowding about him. Taking hearty handshakes. Being made much of. And Tresa, gazing at him with that glistening light in her eyes. Tresa, most of all. Tresa. . . .

Walker, too, looked at Tresa and perceived with definite finality how it was. There was no mistake about it at that moment; he could not hoodwink himself. In spite of all he had done for her — protecting her in this rough town, carefully hiding his true feelings, showering a fatherly kindness upon her and waiting for it to work into her affection — she had fallen in love with another man, a younger man who could do comparatively nothing for her. A damned drifter whom he, Walker, had hired merely to use as a tool as he had used others in his climb to wealth and power.

Sid Beaugrand's bitter rage he could understand. He had it himself, tenfold. Shortly, Sid emerged from the livery, mounted on his buckskin. Walker watched him ride out of town. Northward. Walker could guess what that portended. It crossed his mind to send some riders out after Sid. He dismissed the thought.

No. Let him go. Let that young-faced, old-eyed killer go. Walker nodded to himself. There were always methods of getting rid of an unwanted man without yourself being involved. When you found you had a pair of unmanageable men, the trick lay in setting them against each other. A contest, to serve a purpose to your benefit. If they both got through alive, then lavish rewards on one and let the other go empty-handed.

The rewarded winner wasn't like to last

long, his defeated rival being vengefully on the prowl.

Walker smiled benevolently and blandly at Dain and at Tresa, letting them see his smile, in his mind's eye seeing Dain lying dead and bloodstained under the crash of a smoking gun. Seeing Tresa in his own arms some day, willingly, as he wanted her.

He had never yet lost anything that he wanted badly enough. He would not lose now.

9

Sid Beaugrand waited until it was well into morning before riding into Travis City. He rode openly into the town square, something he would not have cared to do by night, and reined in at the Bledsoe Mail Stage Company office. Without dismounting, he stretched his arms and yawned, and while stretched he called into the open door of the stageline office, "Hollen, you in there?"

He had shivered some while waiting out the dark hours, and hung over his thoughts, bitter and bleak and not free of anxiety. In running over the score he felt that Dain Moore had crossed him up all the way. That tongue-tied solitary stray was heaps trickier than he looked or acted. It had to be so. Look at the score. Dain now had everything, sitting on top of the world, while he had nothing. Yet he had worked as hard as Dain, tried as hard — or harder.

Sid let none of that show on him at this

minute, waiting for Hollen to come out, waiting to see what the gun ramrod would do. A few men along the loading platform kept looking at him, trying to recall where they had seen him recently, he guessed. They couldn't have got much of a look at him when he drove in the stage, all that dust; any man looked different on horseback.

Then suddenly Hollen stood in the door of the stageline office, silently regarding him. Sid brought his stretched arms down, slowly. He folded his arms over the saddlehorn and said pleasantly, " 'Lo, John," and sat with muscles tense. His future, his very life, hung on Hollen's response.

The hawk-faced man presented a negative front. His long arms dangled flat down his sides, hands loose, offering no indication of any forthcoming activity. But his feet were spread a little apart, and Sid knew that stance. The prominent nose jutted straight at Sid like the blade of a slanted hatchet. The repellent eyes clawed coldly over Sid's face, while a minute dragged by and the watching men on the loading platform made small movements that concerted to arrange them in a half-circle, single line.

And Sid sat smiling, outdoing him in nerve-stretched patience. This man was his contact here. This minute was his tall gamble, and he had weighed its chances and taken them. Through this man he might capture what was lost. He would not quit this country broke and defeated, after the efforts he'd spent and the high dreams he'd had. The mere thought of it twisted his insides. No, no, anything but that. He'd forever despise himself. There would be opportunities to manipulate the Bledsoe-Walker feud to his own advantage. He'd figure them out. He was smart. And if in the end he couldn't win what he wanted, then he'd grab what he could and smash the rest. Satisfaction in that. Restore his self-esteem. Damn John Hollen, he was taking a long time. . . .

Hollen murmured, "Three good men we lost with that mail stage on the caprock road."

Sid shrugged lightly, not removing his folded hands from the saddlehorn. "Not my doin'. I skun by 'thout firin' a shot, near blind from Moore's whip. It was him knocked 'em over and took your stage."

"With help from somebody up in the rocks. You don't deny it was dirty pool you played," Hollen pursued, "on Big Blackie an' the Coffin boys."

"Nobody puts the squeeze on me," Sid said. "I guess you know that, John. If it was you sent 'em, they oughta told me. They'd still be alive."

"So?" Hollen's tone was neither believing nor disbelieving. "What you after here?"

"I've broke with Walker."

"What of that?"

"You and me have worked together, John. I'll take a chance on Bledsoe. He could use me, the way I see it."

The brazen effrontery brought a thin twitch to Hollen's upper lip. "You ain't changed, Beau! Heard you got killed in Kansas."

"Heard *you* got Leavenworth."

"We both heard wrong, then." Hollen stepped back. "Tie up and c'mon in."

He was at a closed door at the rear of the big main office when Sid went in, and after Sid got his eyes adjusted to the shade and joined him there, he tapped on the door, a rapid tattoo with his fingernails. A voice beyond called irritably, "Come in!"

Hollen opened the door and stood aside for Sid to precede him, a rare politeness in that man, if that was what it was. Sid walked in and confronted Bledsoe across a flat desk. Behind him Hollen drawled, "This is the Beau, Mr. Bledsoe. From Muleshoe Junction, on his own. You know — the son-of-a-bitch who drove the first o' them two Walker stages up here. Shall I let him have it now, or d'you want to talk to him first?"

This was a comfortably furnished room, having upholstered chairs and a couch, and rugs on the floor. Touches of prettiness, not quite masculine, were lent to it by a Spanish shawl thrown over the couch, the brocade and tassels on the upholstery, the pinkish hue of the flowered rugs. A large oil painting hung on the wall nearest the desk, depicting a nude reclining on a bed of roses. The roses looked so real they could be smelled. Sid sniffed, and realized that he was smelling perfume.

Vince Bledsoe had shot to his feet at Hollen's announcement. The lines of the voluptuary vanished from his pale-skinned face, stretched smooth by muscles of anger, so that the features took on more of the ascetic cast, but it was a cruel asceticism, inhuman like the stone head of an ancient Toltec idol.

Not turning, knowing that Hollen held a drawn gun inches from his spine, Sid said, "Speakin' o' dirty pool, John, you ain't doin' well by a trustin' old pal!"

Hollen breathed a laugh. Bledsoe, his strange black eyes fixed on Sid, demanded, "Why are you here?"

" 'Cause I can be useful to you," replied Sid. "For one thing, I'm better and faster on the gun than any man you got. That takes in Hollen. He knows it, or he wouldn'ta waited till he could get at my back. For another —"

Hollen broke in, "This way is easier, is all, Beau! The Coffin boys, too, was pals o' mine, and I trusted 'em more'n I ever did you!" He asked Bledsoe again, "Shall I let him have it?"

Bledsoe made a slight, restraining gesture. "The rugs."

"I'll be careful."

Sid said, "For another thing, I can show you how to bust Walker. He claims he's got the mail contract tied up, and I guess he has, 'cause a coupla gov'ment fellers he had there sure looked agreeable to it. That'd put you in the hole, huh?"

Although he said nothing to that, Bledsoe let his lips compress. His pale face whitened, bloodless as paper, and he gave a muffled sound, swallowing. Loss of the mail contract meant the loss of the mainstay of his business. His stageline, now growing fast, would shrink and fall lower and lower to the bottom level, a shabby, poverty-cursed outfit scrabbling for trade, any kind of trade. When a big stageline started down, beaten by a successful rival, it soon fell to pieces of its own weight, unable to meet the heavy overhead, let alone repair and replace equipment.

The stageline was his main resource. Unlike Walker, he owned no others that were as yet profitable or even self-sustaining. From its revenues they, his other enterprises, still drew nourishment for their growth. He had spread out so far. If the stageline went, everything else went with it. Gone, the expensive crew of armed hardcases who maintained his supremacy and his right to dictate terms. Gone, the heady vision of eventually controlling the entire Panhandle and South Plains.

A has-been, a broken-down adventurer, ex-gentleman, ex-this and ex-that. . . . He had seen the type all over the world, cadging drinks, mumbling of past glories, and looked on them with scathing contempt. Yet they must have been debonairly dangerous in their time, at least some of them, before they came their final cropper.

Sid perceived the unspoken answer to his question, and he murmured, "Yeah, it sure would. I can show you how to put old Walker out o' business, though. Good thing I come here, huh? For you, I mean." In background, breeding, culture, in everything he was poles apart from Bledsoe, but he could judge the inside workings of Bledsoe pretty thoroughly. A crook was a crook, in whatever pedigree and however he operated. No difference. It was the honest men who raddled up all your calculations.

The streets of city slums had bred Sid Beaugrand, and dog-eat-dog was his culture. A knifing had sent him dodging westward before he was twelve, from police to whom he was already well known. The West offered him a larger field and he learned to harvest it. The change of tools and methods didn't throw him off for long. He was adaptable, a quick learner able to surpass his teachers, and over the years he initiated some improvements of technique. His early culture stayed with him.

And Vince Bledsoe could judge him, too, for what he was. He eyed him with concealed distaste. Guttersnipe. Illiterate mongrel. Dangerous. With a dash of class showing through, injected into his bastard line somewhere at some time by some gentleman amusing himself with a scullery maid. This type he had met often, too, and loathed them more than he did the out-and-out trash. Too assertive and inclined to step out of their place.

But useful, yes. Invaluable hirelings. This swashbuckling gunman had the nerve of the devil. "And how," Bledsoe inquired urbanely, "would you handle this matter, Mr. — er — ?"

√"Just call me Beau." Sid hooked his thumbs in his belts. He could afford now to ignore Hollen and the gun at his back. Hollen had served his purpose as contact. "Listen. Old Walker's throwin' a big dance party Sa'd'y night. It's in honor o' Dain Moore and on account o' the contract. He's givin' Moore full charge of his stageline. Ev'rybody'll be there, natchally. Nobody wears his gun to a party like that, right? Okay, so . . ."

As Sid talked on, Bledsoe at first shook his head, rejecting the plan. Then the sheer boldness of it intrigued him and he saw its practical aspect and potential side-issues. He sat down and leaned forward, motioning for Sid to draw up a chair. After that Hollen came around, sliding his gun away, and he too listened intently, nodding.

They had hung red-white-blue bunting

and Texas flags in Odd Fellows Hall, and many lanterns and colored paper streamers? They had set up tables loaded with fried chicken and pit-roasted beef and all the fixings, and pitchers of lemonade. The jugs and bottles of stronger refreshment were kept discreetly outside, out of respect to the ladies, who knew perfectly well of their existence but indulged the men in the pretense of innocence.

Everybody of any consequence in town and from the outlying ranches attended. The hall was packed, the streets crowded with horses and rigs of every description, from a transplanted Britisher's handsome surrey to a new settler's covered wagon. Most newcomers quickly took to the Texan fondness for dances and blowouts, and a party such as this was not to be missed. Whatever Walker did, he did on a lavish scale, freehanded, nothing left out.

He had imported a band all the way from San Antonio for the occasion, rushing two special stagecoaches to carry musicians and instruments, arranging for fresh teams every forty miles. After rendering *Dixie* and *Bonny Blue Flag*, and having a go at the national anthem as a passing nod to the Union, the band got down to business, opening with a waltz.

Later on, when the kids were bedded down and the crowd thinned a bit, there would be good old rollicking reels and square dances, but right now there wasn't room to bow. Kick up a heel and you cracked somebody in the shin.

Dain, squiring Tresa, claimed the first dance. Walker had advanced him money to buy new clothes, and he felt stiff and stalky in an unaccustomed town suit, white shirt and cravat. Tresa, though, assured him that he was very handsome. He wasn't. In turn, he told her that she was lovely, wonderful, *muy bonita*. She was.

She wore the popular basque, a close-fitting bodice, in blue silk, cut low, and a full skirt in contrasting apricot gauze. Her hair she had arranged in a modified waterfall fashion, heavy at the back, the ends caught up underneath, airily curled at the sides, and she had dared the coquettish item of a single curl in the middle of her forehead. But mostly it was her inner quality of complete femininity, glowing through. And her smile. Her shining eyes. She was the girl in the little picture, gloriously real and alive.

"Bit tight here," Dain commented, apologizing for bumping a stout woman who promptly bumped him back, without rancor.

"Do you want to stop dancing?" Tresa asked demurely.

"No!" He grinned, tightening his arm around her.

He noticed Walker up on the stage in front of the band, still wearing his baggy everyday suit, his gray hair an untidy mop as usual, big face red as fire. Walker was playing the genial host, smiling benevolently out over the crowd, but his eyes never strayed off Tresa for long. He had never before seen her look like this. Once, Dain caught Walker's eyes resting on him. The eyes were hard and ruthless, anything but amiable.

"Let's see if we can work through to the front end of the hall," he suggested to Tresa.

Perhaps she had got a glimpse of Walker's eyes. She nodded.

It was fun, sort of, working through the jampacked dancers. A game of skill. A small gap, edge quickly into it, and a gain of perhaps a yard. A firmly pressing shoulder, somebody gave way, and another yard. Tresa began laughing, admiring Dain's crafty skill and complimenting him on their progress. Dain told her modestly that it was simple when you got the hang of it.

He hardly knew when it was that he first became conscious of a growing hush set-tling over the hall. The low thunder of dancing feet faltered and slowly subsided. Some kind of mass communication swept mutely through the crowd and soon there was no dancing, no talking, everybody staring curiously at the windows all around. The band went on playing, but it sounded wrong, out of tune. It degener-ated from bad to worse, the musicians ex-changing questioning, nervous glances, and gave up altogether, leaving the stage in murmuring silence. The band leader spoke to Walker, still on the stage under the brightest lights. Walker shook his head worriedly.

And now noises could be heard outside. Brief commotions, as of muffled fighting, and deep voices. Some of the men had slipped out for drinks. It sounded to Dain as though they had already fallen into a scrap, but were trying to keep it quiet. He reasoned that the people nearest the win-dows had heard it first, and so begun the questioning hush that spread throughout the hall.

The windows were all open to the cooling night air. The heads and shoulders of men loomed up to appear at them, looking in. And guns, glinting in the light of the lanterns, poking into the crowded hall.

A voice called harshly, "Deputy sheriff and posse from Travis City! Stand still, everybody, we don't want to hurt any of the women an' kids!"

Disregarding the command, Walker jumped down into the crowd, out from under the bright lights where he made so easy a target. A gun blared at him, missed. The bass drum bonged. The unhappy musicians went into a scramble.

The deputy sheriff from Travis City put a stop to that with a roar. "Don't move! We're here to recover a coach and team stolen from the Bledsoe Mail Stage Comp'ny — and to arrest one Dain Moore for highway robbery an' murder! Which one is he, Beau?"

Sid sang out, "That tall joker near the front, dressed to the nines, with the lady in the blue-an'-yaller dress!" He laughed as he said it, high-pitched. "Step right up, Dain, and get fitted for cuffs!"

10

A town woman burst into hysterics. Half the wide-eyed children exploded screams, some more women caught the infection, and in seconds the place was a bedlam. The whole crowd began milling, angry men seeking out their frightened wives, mothers frantically rushing to find their wailing children. Walker's men, mostly single, rubbed their empty hip pockets and cursed under their breath the rigid etiquette that dictated that no man went armed to a purely social affair. Nobody had dreamed that Bledsoe would pull anything like this.

The shouted warnings of Bledsoe's gunmen and the deputy sheriff went largely unheeded. Walker came thrusting roughly through the crowd to Dain.

"That's your friend!" he grated. "That's your precious friend who's behind this! It's his work! I can smell it!"

"So can I," Dain said, "better than you."

"What're you going to do about it? My

God, all these women and kids — they'll panic in a minute! We'll have hell here!" There was genuine horror in Walker's tone. "The devils! The damned devils!" He asked again, "What're you going to do about it? It's your name they called!"

Dain said very soberly, "I'm on my way out to them." Tresa clutched his arm. "No! You can't!" Her voice pitched close to a scream. "You mustn't! I won't let you!" He put a hand to her hair in a quick, brief stroke, saying, "You don't mean that, honey. You know we can't let all these folks get smashed up on my account by those —"

"They won't open fire into the crowd!" She shook her head wildly. "They wouldn't dare!"

"Look, they're killers. I saw them murder your father in cold blood and burn his body." He gently pried her fingers from his arm. "I don't say they'd shoot women and children. But some of them are bound to be that dumb they'll trigger off a few shots, thinking it'll scare them quiet. Then we'll have panic, people trampling on one another. Then Bledsoe's men will get rattled, and they'll shoot some more because that's all they know to do, and — well, far's I know there isn't a gun in our crowd."

He pushed off, calling, "I'm coming out, Sid!" He looked back and found Tresa following him, fighting off Walker who tried to hold her back, and he said, "No, honey — no! You can't come with me!" But on she came, and he hurried faster to the door, shoving people aside much as Walker had done. He heard her cry after him as she struggled in his wake, "Dain, wait for me! Don't leave me!" And he almost turned back.

Voices set up a shouting, in which the dominant word, echoed repeatedly, rang the alarm most dreaded by human ears: *"Fire!"* What they shouted was, "They've set fire to the stageline buildings! The office is on fire! It's on fire!"

Fire was the word most clearly distinguished by most of the trapped crowd. It was a word piercing the brain, carrying with it the atavistic terror of burning alive. A woman screamed out that the hall was on fire — she saw the flames.

She saw flames reflected on a corner window, and so did others, but nobody paused to consider that explanation. The panic broke. They rushed at the door and the open windows, men with their wives, women with children, single men, single women, a mob of humans out of control.

154

Guns could not hold them back had the muzzles blasted into their faces. It was mass hysteria. The gunmen from Travis City gave way, themselves shaken by the fear and the pandemonium.

Dain, close to the door, spun around to go to Tresa's aid. Now that it had happened, the thing that he and Walker had dreaded, and with such suddenness, his first thought was for her. He glimpsed her hair, her upturned face. That was all. She was pressed in and being borne toward him in a solid pack of humanity surging to the door.

He could not, for all his strength, make headway against the crowd. It buffeted him onward like a moving wall, propelled him on out through the door, and there it blocked until those foremost clawed free and loosened the jam.

Somebody waiting in the outside darkness struck at Dain with a gunbarrel. Dain jerked his head away from the descending stroke, catching a flickering glint of the metal, and threw up his left arm. The gunbarrel thudded onto his upper arm near the shoulder. He sucked a hiss of pain, stumbling back. The man came at him like a tiger, lean and lithe and as sleekly coördinated: Sid

Beaugrand. Sid in a merciless hurry.

Dain dived at his legs, striving to topple him down, but his numbed left arm wouldn't work for him and his right wasn't enough. He heard Hollen call out, "Beau!" Sid hit him again, cursing softly. Dain hung on with his right arm, trying to shield his head against Sid's legs and still trying to bring Sid down.

The people rushing out of the hall swarmed over and around him, stepping on his outspread legs, his back, heedless of the punishment they unwittingly inflicted on him. Sid tore loose and stood against the stampede. Dain looked up, making another grab at him.

This time he failed to dodge the stroke of the gunbarrel.

A stage driver, Lankford by name, restored Dain to consciousness by turning him over and pouring a bucket of cold water onto his face, nearly drowning him. Dain rolled his head away from the deluge, gulping and snuffling. Encouraged, Lankford went off for more water.

Dain wiped his eyes clear and got them open, wincing with every pounding throb of his head. He lay where he had fallen just outside the hall, which now was empty, although still lighted up. There was much noise somewhere close at hand in the town, but none right here. He tried to recall the events of the evening. They ran jumbled through his dazed mind, like segments of a half-forgotten dream. He touched fingers to the back of his head, where it hurt the most. It was sticky, and the fumbling contact of his fingers there caused him to wince again.

Pushing to sit up, he used his left arm and it gave way under him and let him down. Pain there, too. That brought on a lucid thought: Sid must have hit a nerve or something with that first swipe. The second had smacked across his back, leaving soreness that was part of the general soreness left by trampling feet. It was the third that had laid him out, buffaloed him. That much he got straight.

Using his right arm, he struggled to his feet. A sick wave of dizziness came over him, and he stood teetering on widespread feet, head hanging. Something of enormous importance prodded at him. He couldn't place it. There was only the feel and sense of it, of its atrocious wrongness and its insistence on becoming known to him.

Lankford, returning, watched him for a

while. He set down his filled bucket and offered him a pint bottle from his pocket, first uncorking it for him. Dain took a long swallow, gasped, and handed it back. "What," he asked when he regained his breath — "what happened?"

Lankford tipped the bottle for a short one and brought it down. "They're gone. Took that Bledsoe stage along, with two teams o' Walker's horses hitched to it. Sure sailed out fast. They set ev'rything afire — office, barns, coaches, ev'rything! We've about got it under control now, but Walker stands a pretty bad loss. I guess the stageline won't be runnin' for some time. Not at full schedule, anyhow."

"Hell with that," Dain muttered. He placed his right hand against the doorframe of the hall to steady himself. "Our folks? That panic —"

"Some broken bones, cuts, bruises. Nobody killed. Doc's got 'em. Ev'rybody's lendin' a hand. It coulda been worse. Quite a party, wasn't it? The dirty, stinkin' bastards!"

Dain nodded heavily. "Quite a . . ."

The insistent thing burst like a bomb in his mind. "Tresa!" he said. "Tresa!" She had been behind him in the panicked crowd, but that she had panicked was unbelievable. Her fierce struggle was not to escape any peril, but to be with him. She had emerged with the rest, seeking only him, and so seeking she could not have avoided seeing him lying there senseless on the ground. Why, then, was she not with him now? "Where is she? Hurt?"

Lankford moved his head from side to side, slowly, gazing at the bottle in his hand. "She ain't among the hurt. I looked. Walker sent me lookin' for her. That's how I found you here. She ain't home, nor at the fire. I can't find her." He raised somber eyes to Dain's.

"Swede Voll," he said, "that ol' drunk, y'know — he was layin' out on the edge o' town when the bunch rode in. He says there was a woman in that Bledsoe coach when it left town. An' Bledsoe, himself. I dunno, you can't hardly tell what ol' Voll's sayin' half the time, an' the other half he's off his head, so . . ."

His voice trailed off. Dain was pacing away, seemingly, aimless and not very steady on his legs.

Many of the horses and rigs still lined the streets. Dain borrowed a dun horse whose owner was elsewhere. He chose it mainly because the stirrup straps were let out for a long-legged man and he didn't feel up to adjusting any stirrups with one hand. He tightened the single cinch and untied the reins and legged aboard. Despite his compelling urgency, it felt good to be in the saddle again where he belonged. He heeled the dun, and the animal responded, and he turned into the main street.

Smoke poured from the buildings of Walker's South Plains Stage & Express Co. In the flicker of flames here and there men ran carrying buckets of water and sand. The horses had been run out and gathered in the street, where they snorted and whinnied, stamping, flinging their heads high. Over all roared a medley of commands, exhortations, and the babble of onlookers. Milam Walker, somewhere in the thick of it, was not visible. Dain rode on past, lifting the dun to a lope.

He was an hour out when his head cleared and he was able to think coherently. A tingling down his numbed arm told him that he would be regaining the use of it soon. He sent his thoughts forward into the near future. Not much chance of catching up with the Bledsoe raiders, they having such a long start. No good if he did, one against the bunch.

It occurred to him that they would post

rearguards along the road home to Travis City, to beat off possible pursuit. Most likely in the caprock, best terrain for snipers. Don Porfirio, with an old rifle in the Breaks, had held off twenty men on the road, and then made a clean getaway. He would have to take off on one of those side trails and try working in to Travis City.

Finally, and most belatedly, it came to him that he didn't have a gun.

In the darkness, the Mexican settlement of San Rosario was a little collection of mud boxes, equipped with doors front and rear, no windows, and each having its useless-looking little fence around it. The fence, however, was necessary in the matter of courtship etiquette, playing an important part. These people from the border drew their customs from Mexico, which had drawn them from Old Spain.

Don Porfirio, proud bearer of the ancient conquistador name of Valdez, stood in the tiny plaza vowing that he and his sons and cousins and nephews would march with Dain to Travis City. What they would do when they got there, only God knew, but go they would.

"No, Don Porfirio," Dain said. "They would later come and wipe out this fine

161

town of yours. They are evil men. You have your women and children. And already I owe you much." They spoke in Spanish, a gracious and ceremonious language even on the border.

"Among friends are no debts of favors."

"True. I am in need of a weapon."

"I give you my splendid rifle!"

Dain considered that muzzle-loading flintlock, yard-and-half long and probably weighing fourteen pounds, that had to be charged after each shot. "If they caught me with it, they might know it is yours, and kill you for giving it to me. A pistol?"

Don Porfirio shook his head, his moon face and soft brown eyes sad. Not a pistol in the settlement. "My knife?"

"*Gracias.*" Dain slipped it in his belt, under his coat. "I am in need of a guide to Travis City, not by the road."

"My grandson. You know him. A good boy, who throws the stones with extreme expertness and artistry." Don Porfirio clapped his hands. "Ho, Francisco Jesus!"

Francisco Jesus promptly darted out of a door, giving rise to the suspicion that he had been listening. No doubt all the people of the settlement had their ears cocked behind doors. They would not intrude upon a private conversation. Dain had ridden in and called Don Porfirio's name, and they sensed that he was in trouble.

The small boy stood straight before Dain, gazing steadily up at the tall man. Don Porfirio said, "By his friends he is called Pancho."

"May I?" Dain asked the boy.

"Si, señor."

"Will you guide me to Travis City, Pancho, not by the road?"

"Si, señor!"

"Get your burro," said Don Porfirio.

The dun horse followed the burro, its hoofs beating out a cumbersome rhythm compared to the neat and light *tuka-tuka-tuka* of the smaller ones. By what means the boy struck his course Dain couldn't tell. No landmarks existed in the pitch-black moonless night. He could barely make out the figure of the boy on the gray little burro.

The boy never looked around, never spoke, taking his task gravely. Great upheavals of rock slid slowly by, and sometimes they scraped through gaps and crevices only inches wider than the dun horse. Dain knew that he could never find his way back over this tortuous route.

At last a cluster of lights moved up into view ahead and below, seeming to hang suspended in a pool of blackness. "Travis City?" Dain asked the boy.

"*Si, señor.*"

"I shall go on alone from here, then. Thank you, Pancho. You have the eyes of a panther and the heart of an eagle. *Adios!*"

"*Adios, señor,*" said the boy quietly.

A strange kind of kid, Dain thought, going on. Most Mexicans enjoyed conversation, plenty of it. Seldom did you find one to whom taciturnity came natural, and practically never a silent one among the children. Then he put the boy out of his mind, for he was approaching Travis City and his thoughts crowded forward to Tresa.

11

Dain moved in on Travis City in the manner of a cavalry scout of the Texas Brigade reconnoitering an enemy stronghold. He did not use the main road, nor any sidestreet. Urgency pricked him to hasten, to take any risks. Common sense and training imposed sound caution. Nothing good could be served by throwing his life away. So he circled to west of town and moved in across vacant land to within a stone's throw of the town's outskirts.

There he dismounted. Whether or not the dun horse would stand ground-hitched was a question. It was not his horse. The centerfire rig on it indicated, though not positively, that its owner might be from northern states and territories where such refined training was not always considered essential. They had trees up there to tie to. Thinking it over, he decided not to risk it. Come the getaway, if it came, he would need to know exactly where the horse stood. First rule for a cavalry scout of the Texas Brigade: When you got to go on afoot, know where your horse is at.

He led onward at a curtailed walk, searching for something to tie the dun to, and fetched up against a backyard fence before he knew it. He tied the dun to the fence. The reins were long, and he made a double lap and a bow-tie knot. That horse would stay right there unless badly spooked.

"It ain't I don't trust you," he murmured, finger-nailing the dun's roach, "but you and me ain't full acquainted yet, see?"

He had a way with animals. The dun horse snorted slobber all over his fancy town suit, affectionately. Dain gave it an appreciative thump in the neck, and prowled on.

The front office of the Bledsoe Mail Stage Company presented a closed and darkened exterior to the town square. The hour — it was between three and four in the morning — could account for it. On the other hand, a stageline of such size and wide-spread consequence usually stayed more or less open around the clock to accommodate the late-throughs and the specials. Some light seeped out from a draped window at the rear of the office.

Dain crept to that window. It hung open. The drapes were of velvet, and faintly he smelled perfume, which astounded him. For a minute he thought that he had gone askew and was scouting the rear of some madam's joint, until he heard men's voices beyond the velvet drapes and pinned one of the voices down to Sid Beaugrand.

In Vince Bledsoe's private office Sid said coolly to Hollen, "We only took the dep'ty sheriff along for the show of it, you know that. The loudmouth slid out soon's it got bad. And you? Where was you? It was me planned the whole damn thing and put it through! Don't talk like you done it!"

"Jeez!" Hollen breathed, staring at him. "You got the gall of a drunk Injun!"

"I ain't no Injun," Sid grated. "Call me a liar, I'll step outside with you! Guns, knives, fists or whatever! I'm the Beau, don't forget!"

"Now, now, men!" Bledsoe interjected from behind the desk. "No quarreling among ourselves. You both did a good job. We have the girl in my house, where she is —"

"I've got the key to her room!" Sid broke in.

Bledsoe inclined his head. "True. But

look at the facts, Beau. I'm not afraid of you. I'm fast and sure on the shoot, too, don't think I'm not! I can break in her door —"

"Would you, though? Would you want her that way?"

Bledsoe's smile wiped all of the esthetic quality from his face. He grinned like a devil in gleeful conquest. "Apparently," he said to Sid, "you have robbed yourself of certain delights. To tie and rape a virgin — life holds nothing better."

"You're a queer one," Sid observed bluntly. "Here in the West you could damn near get strung up just for what's in your mind! Specially here in Texas!" His face had lost color, becoming gray and ugly. "I don't advise you break in her door! She's mine, nobody else's! Hands off!"

Hollen cut a glance at Bledsoe and said placatingly, "We only brung the girl to draw Walker here, remember. He's hot for her, the old buzzard. The road's tapped, I seen to that. The Beau's right, Mr. Bledsoe, in a way. In Texas you just don't mess with any girl, less'n she asks for it, an' not always then. Women are scarce here, and the men give 'em respect. This ain't the East, sir."

"I'm well aware of it," Bledsoe snapped,

eyes gleaming. "I'm damned sick of this antique Texas chivalry! The girl came to my house of her own accord — we've all got to swear to that, or we all hang. We're all involved. You'll get your reward, don't worry. I'm an honest man. After I'm done with her, you can have her. It's merely my prerogative, as your employer to take her first!"

"A gentleman," murmured Sid. His diction became precise, his tone mocking. "I saw a gentleman lynched in Kansas once. He had all the airs of a lord. They lynched him for — well, something men don't talk about in mixed company."

Bledsoe, touched on the raw, came to his feet. "Is that an insinuation? Are you daring to hint that I'm degenerate?"

"What that means I don't know," Sid said. He lapsed into his common speech. "Them two-dollar words throw me. I do know it was me conked Dain Moore and hustled the gal out o' that screamin' mob. She's locked in your house 'cause it was the best place to put her, that's all. It don't follow she's yours. You may be the big boss here, but —"

"I certainly am! And I don't know why I waste my time here arguing, when I could be asserting my rights!"

169

"She ain't your property!"

Two men at odds over possession of a forcibly seized girl. In Texas. It was incredible.

The atrociousness of it evidently touched Hollen. "She ain't rightly property at all," he put in. As steeped in crime as he was, he recognized a few sins as lying beyond the pale. Although he had no more scruples or pity in him than a rabid wolf, he shied away from committing acts which were condemned even by his own kind. They just weren't worth the risk. But in a flare-up he would side with Bledsoe against Sid, any time.

He said remindingly to Bledsoe, "We're waiting here to see if Walker's bunch shows up. I got the whole town loaded, as well as the road. A damn cat couldn't slip in unbeknownst. It's better we stay together so we don't get crossed up in the ruckus if they come."

He paused, and added carefully, watching the diabolic hunger mounting in Bledsoe's face. "We kidnaped that girl. Didn't strike me at the time, but that's what we did, plain. Nobody can prove it. But if *she* ever gets to tell of it, we're sunk. There'll be lawmen an' posses out by the hundred. No place we could hide in all the

West where they wouldn't smoke us out. We're too well known."

"What're you gettin' at?" Sid asked.

Hollen laid a blank stare on him. "You know! Quit this fussin' about her. Forget it. Too damn chancy. You know, well as I do, when you're in too deep to pull back only thing to do is get rid o' the evidence. Sooner the better." It went in line with his nature to shy from one crime and cold-bloodedly propose murder in its stead. Murder and a secret burial, nobody the wiser. "Ain't that reason'ble?" he demanded.

"I guess it is," said Sid, uncertainly.

In the darkness outside, a voice behind Dain rasped, "Turn round slow with your hands up an' let's see who y'are!"

There were five of them. The one who had uttered the command scratched a match alight. He was the deputy sheriff. He examined Dain's face briefly in the flare of the match, and stepped back in surprise.

"Well!" he said, letting out a breath. "Damned if it ain't him! Moore — Walker's new stageline boss! The man I want! How'd *you* get here? Keep those hands up! Watch him, fellers! Keep him covered!"

"You talk too much," growled one of them contemptuously. "We see who he is. Knock on the door an' tell Bledsoe. We'll bring him. All right, Moore, follow the dep'ty."

The deputy hastened around to the side and banged. After an exchange of words, bolts clicked and a door swung open into Bledsoe's office, letting a fan of light spread out.

"I just caught Moore!" the deputy announced. "He was skulkin' right out here, an' I —"

The four men brushed roughly past him, shoving Dain on into the lighted office, their guns at his back. The one who had spoken before said warningly, "We ain't searched him yet."

Bledsoe and Hollen stood staring at Dain. Sid only wagged his head and sat down, eyeing him in sardonic contemplation. Sourly, Hollen said to Sid, "Thought you said you caved his skull in!"

Sid shrugged. "A mulehead like he's got, you could bust a gun on it. At that, I didn't do it any good. Look at it! Thought I broke his arm, too. Damn fool's made of iron, I tell you!"

"We'll see!" Hollen searched Dain thoroughly. He found the knife and tossed it on the desk. "That cleans him. No gun. Can you beat that! How'd you get in town, Moore, past our men?"

"Flew," Dain said.

Hollen hit him in the mouth, a vicious backhand blow that staggered him. "Who's with you?" he demanded. Dain gave no reply, and Hollen hit him again. "Iron, huh? We'll see!"

Sid looked on thoughtfully. Rising, he reached over the desk and picked up the knife and inspected it, turning it in his hand. He put it back on the desk and returned to his seat, resuming his position of interested spectator.

Bledsoe spoke to Hollen. "No blood on the rugs. If you can help it." To two of the men who had come in with the deputy sheriff, he said, "Go and spread the word to keep extra sharp watch. I've an idea Moore got in alone, but we can't be sure. The rest of you stay. Hollen may need some help."

"Not me!" Hollen grunted, lunging a fist at Dain's middle. Dain twisted aside and escaped taking most of its force, and let his hands down. The deputy sheriff slammed the door shut and put his back to it, and the two men remaining with him clenched their fists ready for a smash at Dain.

Hollen snarled, "Keep your hands up, Moore!"

"Go to hell!" Dain said through split and bleeding lips. He was one against six, counting Bledsoe and Sid, and unarmed, but he would not take a crippling beating without a fight. He lashed out and sent Hollen stumbling, and put his back to the desk. Bledsoe had come from behind it when he spoke to the two men, and now he was tugging on a pair of heavy leather gloves to take a hand.

Dain whipped over a hard, driving right, with follow-through from shoulder and torso, and this time he connected squarely with Hollen's jaw. The hawk-faced ramrod went flying backward and crashed into the wall. He bounced off and fell to his knees, where he hung balanced for an instant, eyes upraised and showing the yellowish whites. He swayed forward onto his outspread hands, slowly sinking and shaking his head.

The deputy sheriff at the door reached to his holster. He wore the trigger-happy look of a windy bully who liked to shoot as long as no shooting came back.

Through carelessness or deliberate purpose Sid had placed the knife on the front edge of the desk. Dain reflected swiftly that

174

it was not Sid's way to be careless in such a matter. Purpose, then. *He wants me to snatch it up!* Dain thought. *He wants that excuse to shoot me!*

But the deputy was pulling his gun, anyway. He was going to shoot. His eager, brutish eyes proclaimed so.

And Sid sat there, not moving, elbows resting on the arms of his chair, dangling fingertips almost touching his gun butts. Watching Dain.

Dain snatched up the knife by the tip of its blade and let fly, and this was a trick that he had learned as well as any kid on the border. Judging distance for the spin, and letting the arm come down smoothly, were the two main secrets of it. Aim came natural. You never lost the knack, once you mastered it. Border kids threw for score, at seven long paces splitting straws.

The deputy gurgled once, horrified astonishment on his face, the handle of the knife protruding from his neck.

Sid's guns roared.

There had been numerous occasions, in war and out of it, when Dain felt the very breath of death on him. And later, a moment or an hour, known the genuine surprise of finding that he still was living. Sometimes it was a shock, requiring a little time in which to adjust to the realization that his number had not come up yet. After accepting death, it disconcerted when it by-passed. If not reasoned out, it was apt to arouse a dangerously irresponsible belief in sheer fatalism. Or, worse, superstition.

But this came so fast, like near lightning and its instantly accompanying clap of thunder. The quiet *thunk* of his knife, and the immediate explosions from Sid's guns. And his heart still beating. He first had the mad notion that Sid had missed — mad, because Sid couldn't miss at ten times that range.

Then he saw Bledsoe dive behind the big desk, clawing off his heavy leather gloves so that he could get at his pocket pistols. He saw the two gunmen, one on each side of the falling deputy sheriff, falling with him. He spun around.

Sid pushed at him, muttering, "Grab a gun and get out, quick! They'll be all over us!"

Dain scooped up the deputy's gun and wrenched open the door. His eyes were full of light, and when he jumped out he couldn't see a thing, but he heard men running toward him, talking. Sid, coming out and shooting back at Bledsoe, who had got his pocket pistols into play and was firing over the desk, yanked the door shut, cutting off the light.

Sid had cat-eyes in the dark, besides an acute sense of hearing and direction, probably developed in his city slum days of dodging the police, heightened by outlaw years on the range. He fired, a running man cursed painedly, and he called a low-voiced "This way, Dain!"

Dain followed him. They didn't run. They trotted on their toes, making as little noise as possible. Dain came up abreast of Sid and asked, "Where? In case you're killed — where?"

Sid laughed shortly, without mirth. "You'd like to see me killed, I bet."

"For what you did to Tresa, I'd kill you now, myself, if I didn't need you!"

"Straight talk! You better shoot as straight, you solitary stray bastard! I need you, too, tonight. Bledsoe's house, the one with the lights on, there ahead. It's guarded. Slow down. Leave the talkin' to me."

They walked toward Bledsoe's house. It was new brick, two floors, square, built on the style that in New England passed for austere elegance, that Westerners called plain, and that the Spanish Southwest termed prisonlike and uninviting. But the West and the Southwest had not caught up with East Coast sophistication. Never would.

The ground floor was lighted, narrow windows all around breaking the darkness outside. And two door-lamps showed a kind of unroofed tiled porch in front. A portico, Southwesterners would call it.

Two men stepped around the house onto the portico. They carried rifles. They craned their heads to listen, and one called out, "Who's that comin' here?"

"The Beau," Sid called back, walking on to them. "We hit a little rumpus. You hear it?"

"Sure we did."

"Hear this!" Sid said. He fired twice. While the two guards folded and collapsed, he said to Dain, "I bet they didn't hear that! C'mon on!"

They ran to the house. They passed between the two dead guards and opened the front door and burst into some kind of large living room decorated in pink. In the light Sid's lean face looked worn and bloodless, ghastly, eyes burning, lips pressed in a thin smile.

A Mexican butler in a spotless white jacket bustled forward. "Hello!" Sid said, and shot him in the chest. "Up those stairs," he said to Dain, "and to the right. Second door. Here's the key. You only got one gun? Hell, man, you need two and a rifle and plenty shells to get out o' this jackpot! Wait, I know where he keeps 'em. I need shells m'self." He struck across the living room to a door. "They're comin'!" he said. "I hear 'em comin'! This is a lousy mess, know it? But here we go, dammit, here we go!"

12

Dain bounded up the stairs. The second door on the right was locked. He fitted the key in and unlocked it, but the door held fast against his pushing. "Tresa!" he called in an agony of suspense. "Tresa, are you in there? It's me — Dain! I'll smash the door —"

"Wait!"

He listened to her remove a chair that she had jammed under the doorknob. She flung open the door. "Dain! Oh, Dain! . . ."

The room behind her was in darkness, but by the reflected light from the floor below he saw that she had been supplied with a nightgown, over which she had donned a thin wrap. Her feet were clad in slippers, pink silk, and the wrap was pink and so was the nightgown. Pink of one shade or another was obviously Bledsoe's favorite color. In an untroubledly rational corner of his mind Dain thought, *It would be!*

She said, "I heard shots, and Sid's voice! It was Sid who caught me, pulled me away from the crowd! I was frantic to find you, Dain! Then Hollen and the others —"

"It's Sid who's helping me try to get you out of this now. Maybe he's gone off his head, I don't know. He looks queer. Come on, we've got to go!"

"I —" She took a step toward him, and retreated. "I must get dressed!" she said, and the onlooking corner of his mind brought up the thought, *To a woman no emergency exists greater than her modesty. God help us!* A man would sensibly run out stark naked where a woman would chance her life finding something to cover herself with. Women were strange creatures, as often exasperating as lovable. Especially the good ones.

"No time! You're dressed enough!"

An outburst of gunfire thundered below. Perhaps, Dain thought, Sid had gone altogether shooting-mad and was killing the aroused servants of the house, as he had wantonly killed the Mexican butler.

Sid came leaping up the stairs. Near the top he turned and fired again, then came on. He said very calmly to Dain, "One gun and some shells was all I could find. Here, take 'em." Tresa he totally ignored, although it was noticeable that he kept his speech clean of its usual profanity.

"They're comin' in," he mentioned, motioning downward with a smoking gun. He did look strange, like a sick man, in fever. "Bledsoe's with 'em, and Hollen — that jigger's got a stone jaw, I coulda told you. I oughta gunned him, but Bledsoe was crackin' off over that desk at us. . . ." He cleared his throat noisily. "We're sure in a hole, feller. Two against the town. This is Bledsoe's town. We ain't got no friends here, none at all."

"We'll get out of it," Dain said. "We got to, Sid!"

"That's you. Long's you don't talk too much, you make sense." Sid peered down over the bannister. He fired. Somebody tumbled down the stairs. "Yeah, what sense?" he demanded in an abrupt surge of savage anger. "They're swarmin' in! You got a horse?" For the first time he paid recognition to Tresa, though only indirectly. "You got a horse for *her?*"

"I've got a fairly good horse for her, tied west of town."

"We better get her to it! How, I dunno. I ain't been long here, not long enough to know all the slants an' alleys o' this antigodlin' town. It's up to you —" Sid

182

forgot his acquired manners — "you, you solitary stray bastard, to get her out o' this!"

It was odd that he should talk in that manner, by turn quietly rational and then cursingly violent. Something was riding him. The impression he gave was that of a man driven inexorably by a force outside of himself, at which he rebelled. Sheer nonsense for him to claim ignorance of the topography of Travis City. Give him one look at any town of any size and he knew all of its ins and outs by instinct.

That a devil lurked in him he gave no doubt by leaning again over the bannister and rapidly blazing a gun empty into the big living room below, although nobody there was trying for the stairs at the moment.

The resultant outcries and scrambling commotion seemed to soothe him. He said, quietly once more, "Anybody who takes Bledsoe's pay is rotten all through. Anybody. Old Walker's a connivin' son-of-a-bitch, but this Bledsoe is a — what was that two-dollar word? Oh, yeah — degenerate. That's him, all right!"

"You worked for him, you should know," Dain observed. "You kidnaped Tresa for him!"

Sid gave no retort to that. He was reloading his emptied gun. "Better move
back with her," he advised. "They're
readin' to rush us. Listen to Hollen
yappin' at 'em. Old John's scared. Get rid
o' the evidence, he said. Meanin' her. He'll
do it, too, give him a straight shot at her,
don't think he won't! And bury her himself
way off in the hills. Got ice in his veins,
that'n!"

"He'd have to kill us as well."

"Natchally."

Hollen could be clearly heard downstairs
exhorting the gun crew. "Only two of 'em!
Only two! What's the matter with you hair-
pins? Hell, one charge and it's all over!"

"I don't want the girl injured!" Bledsoe
put in sternly. "Be careful you don't shoot
her!"

There was a short silence. Then Hollen
declared, "I'm goin' up them damn stairs!
C'mon!"

They advanced, clattering up the stairs,
shooting as they came, and for a minute it
was a toss-up whether they would make it
or not. Had the stairs been wider, allowing
them more frontage for the attack, there
could have been no stopping them. But the
stairs gave room only for two abreast, and
their own crowding worked against accu-

racy in their shooting.

Dain joined Sid at the overlooking bannister, and there they crouched low, firing through the rungs. Gunsmoke choked the stairwell, stinging eyes and nostrils, and the terrific racket in the enclosed space made the ears ring. Twice the gun crew stormed up the stairs and almost gained to the top, incited by Hollen's relentless determination, before an edge of prudence prodded them and they fell back into the living room.

Hollen cursed ferociously, threatening to burn the house down if he couldn't get at the defenders in any other way. That drew a sharp objection from Bledsoe, who vowed to let nobody burn his house. "Besides," Bledsoe added, "as I said before, the girl is not to be —"

"Yeah, yeah — I know!" Hollen snarled. "All the same, she's likely to get us harmed plenty, happen she gets away from us!"

"She can't get away!"

"While she's got them two fightin' for her, anything can happen! What in the hell knocked the Beau off his rocker like that, anyhow?"

Listening from the second floor while he reloaded his guns, Dain murmured, "It'll be breaking light soon."

Sid pressed his face close against the rungs, looking down. "Four dead on the stairs. And we punctured some o' the others. Not a bad score, but not good enough to stop 'em from tryin' again. We oughta got Hollen. When it gets light they'll see better to pick us off as they come."

Dain turned his head and looked at Tresa. He said, "We've got to get out of here before then. Any backstairs to this house?"

Sid shook his head. "If there was, they'd be up 'em. No, they got us up a stump here. They know that much."

"A window?"

"Jump an' break a leg? This is a tall house. To make it worse, the ground slopes sheer off at the back — and I guess you don't figure we're gonna float out a side window for anybody to see us bust when we hit, downstairs lighted up like it is. Tresa couldn't make the jump, anyhow. Nor me. Bledsoe, damn him, lodged a slug in my back as we scooted from his office."

"I didn't know that."

"Well, you know it now," Sid grunted, and Dain realized what was riding him, making him so changeable in temper, giving him that worn and ghastly look.

186

Spasms of pain. "If we just had a rope, maybe —"

"That's it!" Dain exclaimed. "Bedsheets! Can you hold the fort a spell?"

"If I can't you'll soon know!"

Dain stuck his guns into his belt and hurried into Tresa's room, to find her already there ripping the sheets off the bed. She threw them to him, and while he knotted them together she darted off to obtain others.

Gunfire broke out again. Dain stepped into the corridor and found Sid at the same post, dueling it out with some of the gunmen who, crouched alongside the lower stairs for cover, were snap-shooting up at him. Sid appeared to be holding his own. Tresa brought a double armful of bedsheets, and Dain went on with his task of making an emergency escape rope.

Suddenly drowning out all other noise, a large-bore shotgun discharged a fearful blast that shredded the rungs of the bannister. Sid jerked back, a hand to his head.

"Jeez!" he spat. "A sawed-off scattergun! They'll come up behind it now!" He raised his head, taking his hand away. A pellet had gashed his forehead, and blood was running into his eyes. "Man, get her out if you can, right away!"

"Come on, you too!"

"I'll be along. Got to hold 'em off."

Dragging his impromptu rope, Dain sprinted down the corridor to a rear window at the end. Tresa behind him. He opened the window and peered out. The sky was graying along the eastern rim. He looked down at the slope below. It sheered off close to the house, leaving a narrow margin of ledge. He gained the impression that it was neither very steep nor long. If any watchers waited down there, he couldn't detect them. No lights showed in that direction, for the house stood outside the town, facing it.

Finding something to fasten the knotted string of bed-sheets to posed a problem. He solved it by tying onto the knob of the nearest door. It robbed the rope of some of its length, which couldn't be helped. When he lowered the loose end out the window he saw that it didn't extend much more than halfway to the ground. There was no time to add to it. The shotgun blared once more, followed by the sharper explosions of sixgun shells. Sid was having a rough time there.

Dain went first down the rope, to test its strength and to be on hand to catch Tresa. His feet landed on the narrow ledge and he caught his balance. He braced his back against the wall of the house, spreading his legs and holding his arms out.

"Let go when you reach the end," he whispered up to Tresa. "Fall sort of sitting, your legs out level, if you can, and I'll catch you all right."

She proved herself extremely agile, like an athletic boy gifted with perfect muscular coördination. Modesty being impossible to achieve in that situation, she discarded any attempt at it. Nor did she betray the slightest sign of nervous hesitation in following Dain's bidding. Dain remembered that she was a Texas sea lion from the Gulf Coast, a swimmer and no doubt a fearless rider of half-broken ponies. Femininity such as hers was deceptive.

She dropped, slim legs horizontal, and when he caught her in his arms he found that she had not made the mistake of stiffening, but had let her body go slack, much in the manner of an experienced range rider pitched off his horse. He set her on her slippered feet beside him, where she straightened out her nightgown and wrap and brushed her loosened hair back from her face.

"Sid?" Dain whispered twice before Sid

189

poked his head out of the window.

"Comin'!" Sid peered down. "This thing ain't nowheres near long enough for me!"

"It's got to be! Come on, I'll break your fall." Dain braced himself and Tresa stood clear.

Sid clambered laboriously over the window sill and lowered himself hand over hand, cursing softly. The shooting inside the house continued, and from town rose the noises of a stirred-up community.

At the end of the rope Sid looked down dubiously, his face a pale mask. "Here I come!"

Catching him was a different matter. Either he wouldn't or couldn't raise his legs. One of his high bootheels scored Dain in the chest, narrowly shaving his chin, and tore his shirt to the waist. There it hooked into Dain's belt. Sid began taking a header and Dain grabbed him. At that angle the sudden weight, increased by the fall and complicated by Sid's bootheel, yanked Dain off balance. He held onto Sid's middle, Sid got a hold on his hair, and over they went, rolling down the slope, locked together, Tresa sliding after them.

At the bottom they released each other. Dain helped Sid get up. "You'd break my fall!" Sid muttered, smearing blood from his eyes. "Break my damn neck!"

"You came near caving in my chest and scalping me!" Dain growled at him. "Well, we came down fast, anyhow. You all right, Tresa?"

"Yes. You dropped one of your guns. I picked it up."

"Better hold onto it."

The shooting in the house abruptly ended, replaced by an uproar of voices and clumping feet. Their escape was already discovered. "Which way's that horse?" Sid asked.

"Let's keep on along this hollow and work wide around town," Dain said. "It's on the west side, tied to a backyard fence."

They set off, Tresa between them. Behind them a banging of upstairs doors told of a rapid search. Soon, somebody gave a yell, evidently finding the knotted bedsheets dangling out the open window.

In the self-conscious way that he had taken on toward Tresa, avoiding speaking directly to her, Sid commented, "Them slippers won't last her long. She'll need that horse."

"Maybe you'll need it as bad," Dan said, noticing his stumbling walk.

"I'm all right."

"Hope so. They're scattering out now on

the hunt. And Hollen already had guards posted round town, eh? They'll be on watch for us, too. If we do slip by 'em, we'll still be in for a running fight. Only a question of time before they spot our course. It'll be tough. And my first concern is Tresa."

"Mine too," said Sid wearily.

13

Under the silver-gray light seeping into the eastern sky, darkness began to lift so that the outlines of objects grew visible at fifteen or twenty paces, and the air took on the quality of stillness peculiar to that early hour when the slightest sound became acutely distinct. It would be full light before the morning breeze sprang up, to last only the short time until the sun rose high enough to lay its daily blanket of heat over the land.

The hunted three, never knowing where guards and pursuers might be watching, skirted around well clear of town, until Dain judged that they were close to the spot where he had left the borrowed dun.

"Wait here," he murmured to Tresa and Sid, and walked with care toward the town.

The shape of a small house came up before him. It had a fenced backyard. He had estimated correctly its location. He saw the horse, tied to the fence. A man stood by it.

Dain paused, drawing his gun from his belt.

The man said something, low-toned, and another stepped around the horse, carrying something in his hand. A rifle. He, too, spoke in a mutter, and there was further movement and then there were four men, faces all turned in Dain's direction.

Dain stood motionless. They had heard him, perhaps glimpsed him moving in the thinning darkness. But they were not sure. So they waited, ears and eyes alert.

One of them, the man with the rifle, sent forth a query into the gloom. "Who's that? Who're you?"

Dain uttered no response. A voice from near by inquired, "Who you talkin' to, Slim? Us?"

"Hell, no! Didn't you hear somebody walkin', or are you asleep?"

"Heard somethin'. Thought it was you!"

"It wasn't us!"

The four crept forward. The one with the rifle snapped the weapon to his right shoulder, crying out, "I see him! There he is!"

Dain lashed a bullet at him. He whirled and fled, knowing that the dun horse was beyond hope of recovery. Shots spurted blindly after him, the four and the others promiscuously spending shells.

He sprinted back to Tresa and Sid, waiting on the open ground. "They found it!" he told them. "I think they came on it only a minute or two before I got there, damn the luck!"

Tresa took the bad news stoically. Sid said, "Them shots are bringin' the whole bunch! What now?"

"We cut on round to the south and get into the caprock," Dain answered him. "A kid guided me in through the caprock from San Rosario. Maybe I can find that track." He privately thought it a miracle if he could. "Can you manage to run?"

"Whadda you care? She's your first concern, ain't she?"

"Right, and I need your help, like you need mine! That's all I care about you!"

"And that goes for me! If you get dropped, the hell with you! Guess I can do as well for her as you can!"

"Last night you proved you could do as bad for her as anybody!"

They were running while they snarled at each other. Two men in desperate straits, a girl between them. The town, a turmoil of sounds, spilled out men searching the graying darkness for them. Without a doubt, the searchers had orders to kill on sight. Some, mounted, could be heard quartering the open ground outside of town like roundup hands cutting a wild herd.

The caprock loomed up, black and still, like immense sea-waves frozen, rounded, seemingly impenetrable. "Which way?" Sid panted.

Dain didn't know. He couldn't locate the place from which he had emerged into the open bowl that formed the site of Travis City. It all looked the same, fold upon fold of solid caprock spewed up by some earth-shaking convulsion aeons ago.

A small figure moved into sight from behind an upthrust shoulder of rock, and Sid instantly leveled a gun at it. Dain batted the gun aside with a long sweep of his arm. "Don't shoot, Sid! Pancho? That you?"

"*Si, señor,*" said little Pancho calmly. "This way. You have trouble? I expected so."

"Plenty trouble, *amigo* — plenty! I thank you for following after me and waiting for me." You never discarded courtesy when speaking to one of the old Spanish strain, whatever the circumstances. "You have your excellent burro?"

"My humble Gertrudis is here, *si.*"

196

"The lady's slippers are thin. She left hurriedly without opportunity to dress." Dain spoke as to a grown man.

"Gertrudis will be proud to carry the lady," Pancho said.

Behind the rock, entrance to the tortuous track, Dain lifted Tresa and mounted her on Gertrudis, who evinced no particular pride outside of twitching a long ear. The boy led the way, his burro plodding behind him with Tresa, Dain and Sid in the rear.

Sid inquired, "Is this kid on the square? How d'you know he ain't givin' us the go-round? Meskins'll sell you out for a dime!" He had little actual experience with Mexicans. His knowledge of them came from hearsay, from mean-natured men who probably had writhed inwardly from being out-ridden and out-roped by *vaqueros* of the border.

"Pancho's my friend," Dain said tersely, and Pancho turned his head and nodded.

Sid subsided, shrugging, but he was in a suspicious and ugly mood. He kept one gun out in his hand, cocked and ready, trudging with Dain. He watched the boy closely, while flinging an occasional glance to the rear.

"You're rattled," Dain accused him

bluntly. "Sure, they're combing this caprock. I hear 'em. They ain't found this track yet, though, that's certain, or we'd have 'em on our necks. Quit being so nervous! It ain't like you."

"I'm nervous for *her*," Sid muttered, "not for me. Nor for you! They'll find this track soon or later, you know they will."

"Meantime we're making distance."

"They got horses!"

"All right, this track, most of it, they can only come at us one or two at a time. We'll handle 'em! But so far they're still searching. We'll get to San Rosario, Sid. I'm telling you!"

"How far?"

"Just a few miles."

"Lord in heaven!" Sid complained. "Somehow I figured it was just a step!" He stumbled badly, and recovering himself he said, "I ain't never got used to these distances. In the West, I mean. I was brung up in the gutter, y'know. Gutter rat, that's me. Ten blocks was a long ways off. Big-rich bastards lived there, in their big-rich bastard houses. We only sneaked there at night, to steal somethin', see?"

He was trying to explain himself. He was losing his head. His wound gave him fever.

Tresa said to him gently, "You've come a

198

long way since then, Sid. You've left it behind you."

"Thanks," Sid said, reeling, stare-eyed,
"but I ain't. Once a gutter rat, always a
gutter rat, underneath! I know! I'm one!
And I'm a fightin' son-of-a-bitch, I'll have
you know! I'm the boyo who out-gunned
Marshal Frank Weep at Dodge! Where are
them Bledsoe buckaroos? Bring 'em on!"

"Take it easy, Sid!" Dain said to him.

Sid tramped on, mumbling, half delir-
ious. Occasionally he narrowed a glare at
Dain as if believing him to be the enemy.
He was apt to go crazed and shoot, and
Dan kept an eye on him. At that point it
was a question whether Sid was a help or a
hindrance. He was a dangerous ally,
treacherous in that black mood, for in his
mind he had slipped back to the dog-eat-
dog viciousness of urchin days in the gut-
ters of city slums.

The sun came up. Barefoot, the boy set
the pace. They were climbing. On one upward stretch they had to crouch low, Dain
sighting pursuers on horseback combing
the caprock behind them. Hollen would
never give up the hunt, nor Bledsoe.

They topped a rise and met a furious
gust of wind from the south. The wind
whipped off Tresa's loose wrap and sailed

it high in the air before dropping it somewhere back among rocks. Dain gave a sigh.
That flying, flimsy garment was as good as
a smoke signal to the keen-eyed hunters.

He took off his coat and handed it to
Tresa. She flushed, whether from embarrassment over her thin nightgown or regret
for the tell-tale pink wrap, he didn't know.
What he did know, hearing distant shouts,
was that their course was spotted.

Sid, whose eyes and ears were usually so
acute, paid no attention. Head hanging,
sunk in a dream-world, he lagged behind.
He had not even noticed the flight of
Tresa's wrap. A few minutes later, however, a tardy percipience struck him. He
brought his head up and asked in a fairly
normal voice, "They after us?"

"They're trailing us," Dain said. "We
can't outrun them. They don't see us yet.
Maybe this antigodlin' trail throws 'em off
so they have to keep combing around. In
any case, there isn't a thing we can do
more'n we're doing, till they come up to
us, right?"

"Right," Sid agreed, and caught his toe
in a rock and fell headlong. He stayed
there, breathing heavily.

This was bad. Dain picked Sid up and
steadied him. "Hang your arm around my

neck," he told him. "You're too heavy for me to carry, but I can —"

"You go on with her," Sid said. "I'll be along after I rest a spell."

"You couldn't follow the trail. I couldn't, without Pancho, and I've been over it once. Fact is, I don't think he follows any trail. Just finds his way by instinct or something."

"So can I."

"The hell you say, city boy! Hang on, or I'll drag you, and that wouldn't be any easier on you than me! These rock chips are sharp as glass!"

The silent Pancho had stopped his steady pacing when Dain stopped to help Sid, whereupon Gertrudis the burro halted with Tresa and stood waiting, endowed with all the endless patience of its ancient tribe.

Tresa urged Gertrudis to turn, saying to Dain, "Let Sid ride! I can walk!"

"No, on these sharp rocks your thin slippers —"

The shot, from a distant rifle, cracked into the argument and put an end to it. They ducked their heads. The coming whine told Dain at once that the marksman had made a long drop-shot. Had the range been shorter and the trajectory flatter, that venomous sound would not be heard at this end until the bullet passed.

The smack of the slowing bullet, crunching into flesh and bone, made another ugly sound among them, of unmistakable significance. Dain had heard its like before, too often, and seen the results.

They stared at one another. Their eyes contained the terrible question of which one of them was hit. It seemed an eternity of waiting. Waiting for one of their little group to fall. They could not move, it was physically impossible until they knew, though the marksman might well be sighting in for another long shot, nursing his carefully rested rifle wherever he lay.

Gertrudis collapsed with a kind of sigh, and didn't kick at all after she fell. Dain pulled Tresa clear. Tresa was bruised and cut about the legs and hands. Dan said bleakly, "That bullet was aimed at you! The range was too far, and it dropped. Hollen, I'd guess. Let's get on before he tries again."

Pancho gazed down at his dead burro. His dark eyes moistened over. "Gertrudis —" he murmured, then turned quickly and took up his walking. Dain pushed Tresa onward and helped Sid along, and they followed after him without a word. There wasn't much to be said to a boy who had just lost his very own faithful steed.

Pancho was barefoot. But his feet were tough, never yet shod in boots. He could walk on nails. Not so Tresa. Within half a mile her little silk slippers shredded. She limped, trying not to show it, picking each step, head up and arms bravely swinging. And at each step she left a smear of wet blood. She couldn't hide that. Her feet were tender and easily cut. They couldn't possibly bear her to San Rosario, for the trail got worse in the Breaks. This was nothing by comparison.

Sid, leaning heavily on Dain's shoulder and stumbling unsurely like a drunken man, began a mumble in Dain's ear. He was light-headed, a liar and cheat in ordinary circumstances, and nothing he said could be given full credence. Still, he sounded truthful and sincere now, and in a rambling sort of way he carried conviction.

"Get this straight. I didn't catch her for Bledsoe. Man, I hurt . . . I hurt all through. My back . . . No, I didn't catch her for him. Not me! I heard him an' Hollen, see? Talkin' 'bout her. Gonna grab her. Take her back to Travis City, for Bledsoe to . . . He's a goddam degenerate. He'd —"

"Shut up!" Dain told him. Tresa was too close. She could overhear.

"Uh-huh," Sid said vaguely. "So I figgered to grab her first, ahead of 'em, see? Take her way off. Not I'd harm her. Jeez! I love her! But I wanted her mine, natchally. I dunno, I didn't think much. Jeez, my back! Is it broke?"

"No, or you couldn't walk at all. Got you in the kidney or somewhere, I guess. You'll get over it, Sid. You got over others."

"Got me some place it hurts, that's sure! What was I sayin'? Oh, yeah. So I conked you. You was in my way, y'know. Stubborn big bastard! Coulda crushed your skull easy. Didn't wanta shoot, all them scared folks. Wimmen and kids, y'know. I jest tapped you on the noggin an' laid you out."

"Hell of a tap!"

Sid said, "Hollen and the crew came up about then. Get this straight! I was pullin' her out o' the crowd. Hollen wasn't thinkin' right, either. They pitched in and helped. First I knew, they had her. They rushed her into that Bledsoe coach and cleared out. All I could do was follow along. I got the key to her room, anyhow, and made pretty sure she was safe, when we got there."

"You're a queer duck!"

"Nothin' queer 'bout me! I'm a natchal man! I love her, that's all there is to it, what more?"

"Shut up!" Dain said, lugging him along.

Sid began explaining all over again, claiming that he had tried to carry Tresa off only in order to save her from Bledsoe. He insisted that he had intended her no harm, but where he had meant taking her to he didn't say, nor how long he would have held her against her will. Perhaps his thoughts had not extended that far. In any case, Hollen and the raiding crew came up and took the matter out of his hands. Quite likely Hollen had not planned on seizing Tresa, regardless of Bledsoe's orders, but seeing Sid struggling with her outside the hall, he had jumped to the hasty conclusion that the act was being accomplished for Bledsoe. And so, with the crew, Hollen had done his part and more, which he since regretted to the point where he would commit murder to wipe out testimony and evidence of it.

With repetition Sid's tale grew rambling. His mind wandered, and he connected the pursuers with some posse or other that had once given him a tough chase. "They'll never take me alive!" he muttered.

"You're right they won't," said Dain grimly. "They've got no intention of it!"

14

Tresa could walk no further. Her feet were a mass of cuts, the slippers reduced to bloody remnants giving no protection. She sank down with a little moan and sat huddled over, pitiful in the flimsy nightgown and Dain's coat. Dain had not realized the plight she was in, she voicing no complaint and he occupied in helping Sid.

He lowered Sid to the ground. He took off his shirt and tore it in two, and used it to pad and bind up Tresa's feet. While he knelt before her to the task, she studied his bent head and unusually gentle hands.

"Sid's getting worse," she said, "isn't he? And I don't think I'm able to walk much more. How are we ever going to reach San Rosario?"

He took a moment to answer, giving the question thoughtful consideration. The condition of Tresa's feet certainly ruled out much walking. And Sid was indeed worsening, becoming weak from effort and probable internal bleeding.

"I can help you both along, one on each side of me." He glanced at the boy. "Not far now, Pancho, is it?"

The boy shook his head. Tresa asked, "What will we do if we get there?"

"Might be horses there." The boy again shook his head, and Dan said, "A couple of burros, then. Maybe we can hide out in reach of the settlement. The folks there are friendly. Pancho's an example."

"Do you never give up, Dain?"

"Not while there's a chance."

"What chance?" Sid muttered, but Dain didn't reply to him, being unable to make any chance sound plausible to a man in Sid's dire straits.

He helped Tresa up and did the same for Sid, and ranged them one on each side of him, his arms around their waists and theirs around his shoulders. He was glad that he was a big man, muscular and solidly built, packing a vast store of stamina. For a moment he listened to increasing sounds raised by the searchers, and surmised that they still were going constantly astray and losing time. He nodded to the waiting boy.

"Lead on, Pancho!" he said, and set off after him, Tresa hobbling painfully on one side, Sid lurching on the other.

In the Breaks the going was as bad as it could be, giant boulders impeding their way so that they had to twist and turn to get through the jumble. But their cover was better. Nowhere could a rifleman pick them out.

Their halts grew more frequent, mostly on Sid's account. Tresa, with help, could manage. Dain could have carried her and made faster time, although trembling with fatigue. Sid was losing the use of his legs, though, and Dain had to bear most of his weight. While at halt they listened to the searchers calling back and forth, sometimes far off, sometimes comparatively near. There was no let-up. Bledsoe would not abandon the hunt. Neither would Hollen. Those two were prodding on the crew.

The boulders thinned out, and the ground sloped down. Dain had no idea of how far San Rosario might yet be. He felt that he had tramped at least twice the distance that he had ridden in the night. He was worn out from bearing up Tresa and Sid. The sun beat down harshly on his blood-caked head, flies tormented his punch-puffed lips, and his feet were blistered raw inside their tightly fitted boots.

He kept on going by telling himself that it couldn't possibly be much farther.

The boy ahead spoke over his shoulder to him. "*Señor,* we are nearly home now!" Dain didn't know how bad he looked, until the boy added sympathetically, "You need rest very much."

Sid got his legs helplessly tangled, and down he went, falling against Dain and causing him to stumble and let go of Tresa. The boy darted to save Tresa from falling, and with his aid she succeeded in staying upright, but she obviously could not walk alone. The padded bandages on her feet were bloody rags. She stood swaying wretchedly from one foot to the other, deep shadows under her eyes, the trembling of her lips a sign that she was reaching the end of endurance.

It took extreme effort for Dain to hoist Sid up and get him propped. "Come on, we're nearly there," he urged him.

"Bring the damn place here," Sid groaned. "I can't go one more step. I'm ruint an' you might's well leave me an' take her on."

There was that temptation. . . .

Sid had, after all, been instrumental in securing the capture of Tresa by Bledsoe, even if unintentionally. He had turned traitor and brought about the disastrous raid on Muleshoe Junction, with its burnings and panic and violence. For those monstrous acts he surely deserved to be left behind for Hollen to murder.

Dain couldn't bring himself to do it. On the credit side Sid had gone a long way toward wiping out his mischief and rectifying himself, at high cost. Sid was Sid, gutter rat with a dash of class, unpredictably swinging from one extreme to the other. Good traveling companion. Good sidekick. Mortal enemy.

With Sid leaning on him for support on his right side, Dain drew Tresa close to him with his left arm and said to them both, "We'll make it. We got to!"

Sid chuckled crazily. "That's you, pal, that's you! The solitary stray," he tried to sing, "from far away . . ." His voice cracked and broke. He forgot the rest and why he sang it.

Tresa sighed on Dain's left shoulder. She was quietly crying from pain and exhaustion and most of all from helpless compassion for him in his heartbreaking struggle to save them.

In that fashion they shambled slowly into the little Mexican settlement of San Rosario, where the dogs barked at them and the people ran out to them.

Don Porfirio Valdez vowed extravagantly, gesturing about him at his rock-
and-mud hut, "Here in my house you are
safe! I would defend my house from any
man!"

He had not yet got over his shocked distress at sight of the three shambling fugitives hanging onto one another, Dain
battered and bare to the waist, Sid a hag-
gard wreck, and — more distressing still —
Tresa, in soiled and torn nightgown and
man's coat, her feet bound in those blood-
clotted rags, moving like a dazed old
woman. He and other villagers helped
them into his house. They brought them
water and fresh goat's milk and some
homemade brandy. The three were too
done up to eat.

From the people crowding at the two
doors, front and rear, rose a murmur of
agreement with Don Porfirio's sentiments.
It was not clear how he or they would go
about defending anything from a crew of
hardcased gunmen. They possessed few
firearms of any value, and the women and
children outnumbered the men by about
six to one. In the heat of excitement and
brave intentions they ignored the details.

"No," Dain said, "we're not letting you get your heads shot off for us. What good? Much better if you can supply us with some means to travel on."

Don Porfirio, who didn't recognize in Sid the driver who had clipped a wheel off his Grand Express coach along the road, asked Dain, "Could your wounded friend ride? Could he even hold himself on a steady burro?" He shook his round head, answering his own question.

"We'll tie him on."

"And —" Don Porfirio motioned delicately — "the young lady?"

"I'll hold her on."

"For how far, *amigo?*"

"Till we find some place where we can hide out. Is there any such place near here?"

"Yes," the fat man nodded. "I will take you there." To his people he snapped, "Get burros for us, quick!"

They bustled out to do his bidding. Those at the rear door fell suddenly silent, standing stockstill after they turned to go. Their air of blank dismay stabbed a cold premonition into Dain, and Tresa, lying on a cot, raised her head as though he had spoken to her. Sid, sitting slumped against a wall, chin on his chest, didn't stir, too far gone for perception to touch him. He needed a doctor's care, and wasn't likely to get any very soon.

Don Porfirio went to the open rear door and joined the silently staring group there. When he looked back in at Dain, his moon face contained less color. "They are coming!" he announced somberly. "So many! I did not know there are so many of them!"

"Too late for the burros, eh?" Dain asked, and got a sad nod for answer. "All right, you folks clear out and get to cover! Don't take any part in this! If they catch you — if they give you a chance to talk — swear we took possession of this house at gunpoint, understand?"

"We understand," Don Porfirio sighed, and the doorway cleared and Dain saw the oncoming Bledsoe riders.

They came threading down through the scattering of boulders on the long slope, at a walk, in no great hurry now that their prey was at last run to earth. They had witnessed the village people break away from one particular house, and knew perfectly well the meaning of that. The foremost of them, steadily shortening the distance, slid out saddle guns, talking and nodding back and forth. Perhaps they meant to take a few potshots at the running villagers, as punishment for helping the fugitives. Perhaps not, but it did look like it.

Dain opened fire from the doorway. The range was pretty long and his first two shots dropped short and kicked up dirt. He tilted his gun up a bit more. On his third shot a rider bent forward, wheeled his horse slowly, and walked it back.

They all halted, seeming to regard the low-roofed little house with deep cogitation. Dain caught a flash of pink among them, an incongruous touch. One of them had picked up Tresa's lost wrap and brought it along.

Tresa left the cot and knelt on the other side of the door from Dain. She had held onto the gun that Dain had lent her, and after watching the angle of Dain's gun she sighted off a shot. A horse tossed its head and shied, and the rider got down to inspect it for damage.

The shooting brought Sid to with a start. After blinking groggily about him, he started crawling along the floor to Dain and Tresa. Dain said to him, "No room here, Sid — take the front door! They'll hit us on all sides when they come! Right now they're only casing the set-up, finding out what we've got."

Some of the foremost riders raised their saddle guns and took to firing, and bullets smacked the outside walls and zipped through the open door.

Sid, crawling on hands and knees toward the front door, stopped when halfway there and exclaimed, "Who's under the bunk?"

Dain thought Sid was off his head again and seeing things that weren't there. But a bumping and scraping sounded underneath the cot, and out wriggled the boy, Pancho, dragging his grandfather's long flintlock rifle. They all three stared at him. He bobbed his head to them politely, one after the other, and with his free hand concealed a purely bogus yawn, trying to give the impression that he had been taking a nap.

"Pancho!" Dain rasped at him. "You shouldn't be here!"

"But, *señor*, I am here," replied the boy composedly.

Dain swore under his breath, and Tresa looked ready to weep over the boy. Sid, however, said, "He's for a fact here! Let him run out now and he's got a lively chance o' catchin' one o' them slugs! Well, it ain't my worry." He crawled on to the front door and settled himself there, stretched out flat, the only position that

216

appeared to give him any relief from pain.

Pancho patted the flintlock. "I hope to shoot the man who killed my good Gertrudis!" he confided.

"You ever squeeze off that old cannon, boy," Dain warned him, "its kick'll cartwheel you into the next county! Put it away and hunch down!"

The gunfire increased, more of the riders putting their saddle guns into play. Hollen was nowhere in sight, but Dain could see Vince Bledsoe's conspicuously garbed figure sitting a halted horse up the slope beyond the crew, far out of pistol range. Bledsoe was personally directing operations in the style of a campaign commander, making it obvious that Hollen was absent for some reason.

Raising an arm, Bledsoe called down to his men, and the firing subsided and they all looked back up at him. Three of them detached themselves from the rest and rode to him, and he spoke to them at some length, motioning with his arm and outlining to them what he wanted done, as well as how to do it. Dain guessed he was in his glory, and only hoped he would pull a few bad boners and destroy the confidence of his men. It would not take a whole lot to do that. Bledsoe, supercilious and condescending, held himself apart from his men. He was not one of them, as was Hollen.

Bledsoe soon proved that he was no fool at conducting an attack in the open. A man who had been everywhere and done everything, he had picked up abilities along with vices. The three men nodded to him and trotted back down to the bunch, where they passed his orders on.

Bledsoe evidently had appointed the three his lieutenants on the spot, for after some consultation they split the bunch up and each took charge of a group. The middle group left their horses and moved forward on foot. They moved cautiously and not at all fast, watching the little house intently.

"They're keeping in mind the taste we gave them on the stairs of Bledsoe's house," Dain said to Tresa, wishing to keep her courage up. "We'll do it again!"

The two remaining groups parted company, one riding to the left, one to the right, wide around. The design of Bledsoe's strategy then came clear, and Dain scrambled to Sid. "Watch it, Sid, they're closing in!"

"Why don't they just shoot the hell out o' this shack?" Sid muttered. "Oh, yeah,

Bledsoe said, 'I don't want the girl injured! Be careful you don't shoot *her!*' The bastard!"

"Yes," Dain said. "Bledsoe's running this show. Running it damned cagey, too. You can manage, you reckon?"

"Can manage a gun!"

15

From his vantage point up the slope Bledsoe used a system of plain signals to control the flanking parties — a flat downbeat of his hand for one to slow, a circling motion for the other to move faster. Concerted action. Strike on all sides at the same moment.

He was using a pair of fieldglasses now, and through them he must have got a glimpse of Tresa in the rear door when she looked out to shoot. He raised a shout and sent a signal. The middle group ceased firing, but continued working forward in a strung-out line, warily eyeing the door.

Tresa poked her gun out. They promptly flopped, and she drew the gun back. When they rose crouching, Dain got off two fast shots from his side that sent them down again, one rolling, another clutching his thigh, and the rest baffled.

"That'll hold 'em for a while!"

Pancho, watching Dain reload, observed wistfully, "If I could shoot my grandfather's fine rifle only once, *señor* —"

"No!" said the *señor*. "You stay out of sight."

Sid's guns suddenly exploded a double series of discharges. He was firing to the left, into the tiny plaza. Dain, knowing Bledsoe's carefully timed pattern of attack, yelled, "Another bunch on the right, Sid!"

Sid paid a glance that way, and pulled back like a stung pup. A bullet whacked into the doorframe where his head had been. "Damn if there ain't!"

The feel of kicking guns in his hands seemed to restore much of Sid's strength and dull his pain. That, or the red rage of a triggerman at bay. He rocked from side to side on the floor, head and forearms over the threshold, blasting shots left and right.

Dain left Tresa to guard the rear, and went to give Sid a hand. He leaned out and fired over him, giving him a chance to fill his guns. The two flanking parties, quitting their horses outside the settlement, had spread out and were now converging simultaneously upon the hut from all directions. There was not a building on the plaza that did not shelter armed men. They darted from one corner to the next and around the little boxlike houses, pausing to shoot, disappearing, then turning up elsewhere, closer.

Dain slung a shot across the plaza. His bullet cut fine, chipping plaster from a corner, but went on through. The man he aimed at stepped forth, dropping a carbine and covering his face with both hands. Somebody hauled him back, and Dain switched to a man running behind a fence and brought him down. He heard Tresa's gun rap twice, then once, and no more, but he couldn't yet leave Sid to handle the front alone.

A bullet flayed a furrow under his armpit, and an instant later a second nicked his ear. From the angle of the shots he located the fast shooter, one of Bledsoe's lieutenants, behind the raised planks of the settlement's open well. On speculation he drove a shot into the planks. The lieutenant jumped, and Dain got him left of center, high.

Sid opened up again. He was spending shells at a reckless clip and not doing his best shooting, but he had already raised havoc among the attackers. Among men sprawled around the plaza, one was dragging himself toward cover, shaking his head. Sid lashed a shot that stilled him. It was an unnecessary waste of a bullet, Dain thought. Sid, venting his rage on anything that moved, didn't have his mind on conserving ammunition.

Pancho tapped Dain on the back. "Shells for the lady!" he said, and Dain slipped back to Tresa.

She was trying to hold off the line of men back there with an empty gun, and they suspected it was empty. Some of them stood up, alert, mutely daring her to shoot, and when she didn't they spoke to the rest and came on.

Dain gave her a handful from the supply in his pocket.

She moved back to load, and this time he took her side of the door. He poked his gun out, as she had done, and at the same level. The advancing men barely looked at it. He triggered twice, emptying the gun, and nailed two of them squarely and created a scramble.

"Give 'em something to think about," he said, exchanging his gun for Tresa's filled one.

"Give a hand here, Dain!" Sid called, and when Dain joined him he said, "Hold 'em while I load up. Hey, I'm runnin' low!"

"I don't wonder, the rate you burn 'em up!"

"How can I shoot without usin' shells?"

"You got me there."

223

It went on like that, Dain working both doors as needed, Sid blazing like a madman, Tresa holding her station at the rear. There was no breeze. The sun beat down on the low roof of the smoke-filled hut, and inside it grew stifling. Their eyes streamed, it became a torture to the throat to breathe, and their hands were red and swollen from the heat of their guns. Little Pancho coughed in a corner, shaken by the fury of the fight.

Soon, Dain had to feed Sid shells.

"Gimme shells, dammit, gimme shells!"

And Tresa: "Some shells, Dain!"

And his own gun, too, eating them up. He thought of the wry joke made by an Irish sergeant to a squad of green recruits: *Never let the enemy know when you run out of ammunition, just keep on shooting and fool 'em.* A bad old joke, he considered it now.

"Shells — quick!"

"Dain . . ."

He gave Sid five and Tresa three, and that emptied his pocket.

Sid loaded and rattled off his shots. When he quit, no return fire answered except a single report. Dain looked to see if he had got hit, he was so quiet. Sid lay glaring out the door. He waved his empty guns in a flapping, listless way, and choked up and emitted blood from his mouth. Nothing moved in the plaza. There was nobody to reply to his attempt at a challenging gesture.

At the rear, something set Tresa to firing, and then she said wonderingly, "They're moving off! Dain! Is it all over?"

That was hardly credible. Dain looked for a trick in it. From east and west of San Rosario trooped the two flanking parties, helping their injured and leading riderless horses. Seeing them, the middle group had begun a withdrawal, too, and it was their movement that had caused Tresa to spend her shells.

Bledsoe signaled them to go back, his motions angry. They slowed, conversing together. The two remaining lieutenants rode on ahead to him. By the actions of their hands they obviously argued with him, and Dain surmised how their argument ran. They felt the need of a respite. Guns to be cleaned. Injured to be given some attention. No, they were not pulling out.

At length Bledsoe nodded. He kept the two by him. Presently he prodded a finger at them. He pointed down at the hut. He made some kind of positive statement and brushed his hands together.

Dain said to Tresa, "It might be all over if it wasn't for Bledsoe. There's nothing in it for them, outside their pay and maybe a little extra. He knows what guns we've got. He has a fair idea how much ammunition we started with. He's telling them we can't hold out."

"He's right," she said. "Sid's guns are empty. So is mine. Yours?"

He checked. "Two shells!"

Either that last shot in the plaza had got Sid, or else the aftermath of the fighting let him down from his temporary upsurge of strength. He lay face down, his head pillowed on his folded arms, empty guns slack in his hands. Dain went and pulled him farther into the hut. He would have shut the front door, but he wanted the smoke to go out.

Coming back to Tresa, he repeated, "If it wasn't for Bledsoe! If only . . ." He stopped, and sent Pancho such a look that the boy's eyes widened. "What's that old rifle loaded with?"

"A fine big bullet of lead, *señor!* My grandfather made it himself, and loaded —"

"Does it shoot straight?"

"As straight as truth!"

"Far?"

"Farther than any eye can see!"

"Let me have it! I'll shoot it for you!"

He pulled the cot up close to the door, for he felt that he required a gunrest for this brute of a weapon. He set the percussion cap and the lock, and got down behind the cot and laid the rifle aslant the edge. Butt to shoulder, he took slow and careful aim. Ordinarily, he didn't hold with taking such slow aim, because in four or five seconds a slight shake was apt to develop. He had the cot for steadiness, however, so the heavy barrel was no drag on his arms. He drew an imaginary line down from Bledsoe's chin, another horizontally from shoulder to shoulder, and made the intersection his mark, dead center.

He fired.

The long barrel spurted a tremendous roar. It packed a vicious kick, as he expected. He heard the rattling scream of the slug. Smoke from the discharge hazed his vision, and he asked Tresa at the door, "Hit or miss?" She had her hands over her ears, and he had to ask a second time.

"Tresa! Hit or miss?"

"Why, it — it knocked him right off his horse!"

The smoke thinned and he saw the horse, the two dumbfounded lieutenants on either side, the men below all staring.

The roar disturbed Sid. He rolled his head half over and mumbled unintelligibly. Dain said to him, "Bledsoe's dead, Sid."

Sid's lips silently formed the word, *Good!*

Dain handed Pancho back the old flint-lock. "A fine gun. An excellent gun, Pancho."

"I told you it was," said the boy proudly, and added a mild compliment. "You shoot it pretty good. Not as good as my grandfather, but nearly."

They watched the men congregate around the two lieutenants. The bunch kept staring down at the hut uncertainly, not knowing what next to expect from it. Dain's glance lifted on up beyond them at something slowly moving.

"Damn!" he grated. "Damn him to hell! Pancho! Any more powder and lead for that gun?"

"No, *señor*, not here."

Dain said, "Sid! Hollen's coming! D'you understand me, Sid? Hollen!"

Sid only sighed, and didn't stir.

Riding a badly lamed horse, at that distance Hollen was a cumbersome object inching into sight down the boulder-strewn slope. His horse more than limped; it hopped, in agony. Hollen could have walked faster. He was just the kind of man,

Dain thought savagely, who would subscribe to the opinion that three legs were better than two. That was a ruined, suffering horse.

One of the men rode up and met him with a spare horse and Hollen transferred over. He descended at a trot. The bunch opened for him. For a while he looked down at Bledsoe's body. He shrugged, cutting his eyes to the hut, and put a question to the men around him. They pressed in, all talking. Unlike Bledsoe, he listened, taking it all in. Now that he was here, another attack was a certainty.

Dain laid his gun on the cot. "There are guns in the plaza," he mentioned, hitching up his pants and drawing his belt a notch. "Dead men's guns. And shells."

Tresa shook her head in protest, her eyes fearful. "From where they are they overlook most of the plaza! And they have rifles!"

"I know," he said, and sprinted out the front door, her cry following him.

Strong sunlight had long burned the bare and dusty ground the color of cinnamon, and in full day no shadow touched it. He raced toward the man he had dropped behind the open well, the nearest source of ammunition.

Within seconds of his leaving the shelter of the hut, rifles on the slope cracked, their volume rapidly rising to a fusillade. Bullets whipped the earth, and a tug at his leg threw him off stride. It was no use. Even if he reached the well he could not hope to return. What he had done to that man, they could do to him. He whirled and raced back to the hut.

"Can't make it!"

Wordlessly, Tresa tore off a sleeve of her nightgown, and he used it to bind up his leg. He took up his gun. And now, he reflected bleakly, Hollen will know for sure we're low on shells, in a bad way.

Hollen knew. He made it evident by leading the bunch down to seventy-yard range. From there he sent forward a hail. "Come out! Your irons are cold!"

A bitter-eyed man who carried one arm in a bandanna sling drawled, "Don't bet on it, John! They're full o' tricks!"

"Gun tricks are the Beau's specialty," Hollen granted, and scanned the men's faces. They manifested little interest in going on to get stung again. Bledsoe had positively assured them that the defenders in the hut lacked rifles and must be practically without ammunition — and next minute a whacking big rifle bullet laid him low. With Bledsoe's death went their pay. No, there was nothing in it for them and they'd gone cold on the deal.

Hollen didn't attempt to harangue them. He held a brief palaver with two of those nearest to him. They glanced searchingly over the settlement, nodded, and Hollen slapped them on the back. The pair broke away from the bunch and rode eastward. Dain watched them drop south, circling in.

"Now what?" he muttered.

Time lagged. The bunch on the slope smoked, loafing in their saddles, Hollen among them keeping eye on whatever maneuver it was that he had set in motion. And always they watched the hut, rifles ready to pour in a volley at sight of the slightest movement.

Dain listened, thinking he heard horse sounds. The bunch stirred and a few of them cracked off some shots. To make him and Tresa keep their heads in. Or to smother those sounds. Or both. He went to the front door and looked out. The plaza remained empty of life, but a sound near at hand was not to be entirely smothered by another farther off, and he heard it again.

A goat shed stood a few paces to the left of the hut. He asked Pancho, "Any goats in that shed?"

"No, *señor,* they are turned out on the range all day."

"Well, something's there!"

He eased on out the door. He could not be spied from the slope as long as he stayed close to the front of the hut. He slipped along the front wall and gradually craned his head to put an eye past the corner.

One of the two men was tip-toeing out from around the goat shed, into the bare space between it and the hut. From somebody's woodpile he had chosen a chunk of pitch pine, heavy with resin, and he had set it well alight. He held it back, poised to toss underhand onto the flat roof of the hut. His left hand gripped a cocked gun.

His eyes darted everywhere. He spotted Dain instantly, and because he was nervously on edge he twisted violently, slinging up his gun, and Dain's bullet only tagged across the side of his chest. It jerked the man up and he missed his shot, Dain revealing only an eye and an arm.

And because of that flaring chunk of pitch pine Dain spent his last shell.

Dain started forward to get the sagging man's gun and belt of shells, chancing the rifles of the bunch. The man clung to his gun, fumbling to cock the hammer. A rope snaked out from the goat shed and the loop snapped around him. Hoofbeats started up and he was plucked out of sight.

Dain backed into the hut. "Tried to burn us out," he told Tresa. At her questioning look he nodded. "Had to shoot twice." He thought about it and said, "Maybe Hollen figured that in. Draw our fire. Make us spend our shells."

Rather calmly, Tresa asked him, "We don't have a chance now, do we, Dain?"

"I don't know, honey," he said. But he did know. They didn't have a chance now.

16

Watching the one rider returning with the other slung across his led horse, the man whose arm hung in a sling said loudly, "What did I tell you, John? That's what I mean! Tricky!"

As far as Dain could make out, Hollen betrayed no disappointment and no anger, although it was hard to tell in quick glimpses from the hut. One thing was certain, and that was that Hollen entertained no regret for the man he had sent to his death. The lean hawk face turned away from the rider and the led horse. It was some of the men who lifted the body off the horse and laid it out. The rider wiped sweat from his brow and swore.

Hollen lifted his reins. "The day ain't getting any shorter. If we're going down there, we better go!"

There was some half-hearted shifting about, but most of the reins stayed slack. Someone raised a querulous voice. "Show me where it's worth it. I ain't that hot on the prod just for the pleasure of it."

"Ain't no pleasure so far," put in another.

"An' no profit!"

If Hollen made any response, it was lost in the general rumble of agreement. Seemingly unperturbed, he said after they quieted, "If that's the way it is, I'll go down and finish it off m'self! See you later!" He touched his horse to a walk, holding an unwavering, heavy-lidded stare on the open door of the hut.

A man called after him, "John, hold it! Look east!"

Hollen drew rein, but didn't look east. He didn't take his eyes off the door. "What is it?"

"Well, look, dammit!"

Hollen bowed low, wheeled his horse swiftly, and rode back to the bunch. Then he looked east, as they were doing. They had taken Bledsoe's fieldglasses, and he borrowed them and looked again. He nodded, unexcited.

Some scurrying confusion ensued among the men, and the horses caught it and some of them reared, spooky-eyed, but Hollen raised a hand and talked quietly. In a minute the bunch paid heed to him, and soon his icy calmness penetrated and steadied them. He had them under control, where before he had exerted no effort to influence them, and nobody raised objection.

The disorderly haste dropped from them. They relaxed, nodding assent to what he was saying, and once a laugh went up, apparently at the expense of the man carrying the pink wrap on his saddle, who grinned and made a coy gesture that brought more laughter. A final word and they ranged into loose column and jogged up the slope, not filing in among the boulders for concealment, but traversing the highest and most open spaces.

Letting the column leave without him, Hollen walked his horse to the nearest boulder. He passed behind it and did not reappear on its far side.

"I guess it's Walker's crowd they saw coming in from the main road," Dain observed. "Can't think what else it could be." He shook his head, puzzled. "But what's Hollen up to, skulking behind that rock? Does he figure to stage a one-man rearguard action?"

Equally puzzled, Tresa gazed up the slope and her eyes widened. Following their direction, Dain saw what she saw — a woman in the departing column of riders, or what appeared to be a woman. But he realized that it couldn't be. The man carrying the pink wrap had donned it and removed his hat.

Usually the difference between a man and a woman rider could be detected, though they might be dressed much alike and at a distance. Something about the swing and carriage of the body, especially that of a young woman, told you for certain. But the loose pink wrap hid all that, and it so contrasted with the garb of the men that the illusion was startling. To all effect it was a woman, closely guarded by the men around her. A woman captive, being carried off.

Dain actually looked at Tresa to assure himself that she was with him. Then into their field of vision, restricted by the door, swept a second pack of riders, coming from the east. Dain picked out Walker's blocky figure in the lead. They had sighted the Bledsoe bunch and were hot after them. They quartered up the slope long before reaching the settlement.

"Oh, no!" Tresa cried out. "No!"

Dain cupped his mouth with his hands and shouted, and Tresa pealed scream after scream. That far off, Walker and his roughnecks couldn't hear them above the noise of their own hard riding. The white of bandages showed among them. They had evidently tried to bull through the main road to Travis City, and got mauled by Bledsoe's lookouts posted in the caprock.

The column ahead lifted to a run, and the pink wrap billowed out behind. Those riders vanished into the jumble of rocks, bound for the ambushers' paradise of the Breaks and the caprock, and presently the Walker faction vanished after them. Their sounds dwindled and died away, and Dain prepared to meet Hollen.

The killer walked out from behind the boulder with the air of a man who knew exactly what he was about. His steps fell unhurried and precise. He looked at the ground, only occasionally raised a glance to the hut, and picked his way with some care, avoiding rubble and weeds. He might have been taking an easy walk in town from hotel to saloon, in no doubt that his drink would be waiting for him, and that he would enjoy it.

At the halfway mark he stroked out his pair of guns, and this he did with languid grace, as though exposing them for some-one's admiring inspection. He carried them muzzles down, letting them swing with the natural swing of his arms, not at all menacingly. And that was the essence of menace.

Dain, not much on bluffing, made a stab at it. "Stop there, unless you want a bullet!" he sang out.

Hollen leveled his narrow, deeply set eyes under their heavy-lidded cowls. He said, "I guessed those two shots was your last. Now I know." His tone was patient. "If you had a shell left you'da fired it to get Walker's attention. He'da heard that." He came on at the same even pace.

Dain drew Tresa back. He picked up the heavy old flintlock and motioned Pancho to a corner. He set the lock, to try one more bluff.

There was the tingling wait. The soft crunch of Hollen's boots. The hawk-shadow on the sunlit ground, moving inexorably to the threshold. The legs, the body, the lean and cheek-sunken face of Hollen peering in the door.

Dain held the flintlock rifle waist high, lined at the body.

Hollen eyed it curiously. "So that's the whopper that put the jumbo-size hole in Bledsoe!" he commented. "It ain't loaded, or you'd shoot. Pull the trigger." He paused. "Pull the trigger, or I'll do it with a bullet!"

Dain pulled it. The lock clicked, a futile little snick of sound.

Hollen said, "Throw it down. I don't trust you with anything in your hand. Nor near it, 'memberin' that knife!" He looked at Tresa, huddled on her knees against the wall. He looked without pity at her lacerated feet, and tilted a gun at her and shuttled a look at Dain.

Dain tossed the old flintlock rifle onto the cot, and Hollen said, "You better! Who's that kid?"

"Just a boy who happened to be in here when we took over the place," Dain said.

Hollen stood over Sid. He rolled the prostrate body over with his foot. "Well, well! The Beau's about gone. Couldn't happen to a better prospect." He ran a glance around. "Nice little place you got here. No windows. Thick walls, an' blind on two sides. Right cozy."

It struck Dain that Hollen either was deliberately prolonging the agony or else was discovering within himself an unexpected streak of indecision now at the last minute. Hollen shot another look at Tresa. His left-hand gun crept up, tilted again at her, and Dain watched the trigger-finger and tensed set to jump him.

Tresa looked at the gun and up at Hollen. She would not beg for her life, would not grovel. She knelt only because her feet would not let her stand. Her lips moved in silent prayer while her eyes grew withdrawn and stoical. The gun lowered.

Pancho released a dry sob. Hollen turned on him savagely, his chill composure momentarily cracked, hawk face contorted and hooded eyes glaring. "Quiet, damn you!"

A paroxysm of rage such as that could do it. Dain watched him throttle the sudden convulsion and regain his self-command, and he thought: *Meet him on a cool level and don't spill his temper!*

Hollen scraped shut the sagging rear door and kicked the cot against it. He had sensed Dain's readiness to jump at him barehanded, and from then on he never removed his eyes or gun-point from him. He said, "Step out front ahead o' me! Walk till I tell you to stop!"

Dain figured it likely that he would stop when a bullet in the back stopped him. Still, if it entailed drawing Hollen out of the hut, that was a tiny something to the good. He walked a dozen paces into the plaza, his spine crawling, before Hollen behind him said, "Stop! Now turn round!"

Dain turned. He faced Hollen at five steps' distance. Hollen was not negligent, not the kind of man to put his guns within reach of the victim, especially Dain, whom he thoroughly distrusted. For a moment they stood like that, staring at each other, and for his life Dain couldn't figure what Hollen had in mind. A bullet in the back was so simple and easy for a man of Hollen's type, yet Hollen had not pulled trigger.

Hollen said to him reasonably, "Moore, you know as well as I do that if the girl went free she'd talk. She'd be questioned. She'd be bound to tell she was carried off by Bledsoe an' me an' the Beau. Bledsoe's dead. The Beau's dyin', I'm alive, so the kidnap charge would land on me — only me! It was a damfool thing to do, but it can't be helped now. Right?"

"Go on," Dain said. *Meet him on a cool level . . .*

"Half my life I been on the dodge," Hollen admitted. "My description's known on the books. I ain't goin' on the dodge again, not with that hangin' over me. I wouldn't stand a shake. They'd hunt me down like a mad dog! I came to the Panhandle for a fresh start. Now Bledsoe's dead I'm out of a job, but I can still make a go of it here — provided this kidnap charge ain't hung round my neck! That girl can't go free to talk, an' that's flat!"

"You took a shot at her," Dain mentioned. "Killed the burro she rode." He knew or thought he did, what was stalling Hollen, and he played it up. "Of course, I guess it's one thing to shoot at a woman at long range. Different when you're close up and have to see her die — bloody and broken, lying murdered at your feet."

But he was giving credit for squeamishness where little of it existed. The ghastly picture that he sketched failed in its intent. Hollen didn't change expression. He said coolly, "You'll know about that, because you're the one who's goin' to do it!"

A chill shook Dain. He forced his voice to a moderate pitch. "How d'you figure that?"

"If you don't," Hollen answered, "I'll shoot you where it hurts the most! You'll be a long time dyin'!"

"You intend shooting me anyway, don't you?"

Hollen shook his head. "Not if you do the job. No point in it. Let that kid see you do it. A witness. Then *you* go on the dodge, not me! I'll be in the clear. If you're

243

caught, what story you give won't matter."

"Suppose I tell you to go ahead and shoot?"

"Then I'll shoot you. Her an' the kid, too. I'd sooner I wouldn't have to. Not just for the reason you think, but because these things have a way of cropping out later sometimes."

"Murder will out, yes."

"Not always. All right, we've talked enough. Make up your mind!"

Fingering his jaw, Dain pretended to consider and to arrive at a decision. "Let me have a gun."

Hollen smiled for the first time, a grimace that down-creased his thin lips. "Am I crazy? Not on your life do you get a loaded gun! Nor a knife!"

"Then how —"

"Shouldn't be hard. That old rifle in there must be heavier'n a sledge hammer. One good clout in the head —"

"You butchering devil!" Dain ground out, discarding entirely his resolution to maintain coolness. "You unholy disgrace to mankind!"

Hollen speared his cold glare at him. "That's enough!"

"Enough? I could go on for a month and not begin naming you for what you are!

Scum! The lowest —"

His features savagely diabolic, Hollen dipped both his guns at Dain's middle. "Why, you — !"

A report crashed through his flow of profanity. He jerked as though kicked in the rear, and turned rigidly, his face draining white. At the front door of the hut, where he had dragged himself forward, Sid Beaugrand lay smiling dreamily up at him, smoke coiling from the gun that he held in both hands.

Hollen hammered a shot downward before Dain reached him. Dain rammed him with a shoulder and elbow, and Hollen fell and rolled over. He was striving to sit up when Dain kicked the guns from his hands. Dain went to Sid.

Sid whispered very slowly, "I — saved — out — one shell. Saved — it for — for . . ."

"Yes, Sid," Dain said. "Thank God you did." But he guessed Sid couldn't hear him.

In a little while he thought of Hollen and looked at him. Hollen lay on his back, staring up at the sky. He wasn't breathing.

Dain stepped across Sid into the hut. "You might get your folks here now," he told Pancho. Bone tired, he helped Tresa to the cot and sat on the edge.

"Maybe we can borrow clothes from them," he said to her. "Some kind of dress for you, and moc'sins. A shirt for me. Walker and the crowd will be coming back sooner or later. If they haven't run into a jackpot they can't get out of, and I don't want him calling us scarecrows. Reminds me I better take Hollen's guns and belts. And see the folks about giving Sid decent burial. They can have his guns."

"Will you never rest?"

"There'll be a time for it."

Walker didn't exactly call them scarecrows, but he used similar language about their appearance, although the people of San Rosario had done their best for them.

"My God!" he boomed, staring hard at Tresa. "You look like a squaw! Moore, is that how you take care of a young lady? What's wrong with her feet? Did you let her *walk* here?"

He laid his aggressive stare on Dain, surveying him up and down. "And did they kick you all the way here? You look it!"

Dain made no reply and didn't even glance his way. He knew what lay behind Walker's hectoring reaction, and felt more embarrassment for him than anger. Walker was seething inside at his own failure. He and his armed roughnecks had showed up in mid-afternoon, roaring into San Rosario like a liberating army out to save all Texas.

"Where's Bledsoe?" Walker demanded, and one of his own men, returning from scouting around, spoke up. "Layin' up there in the rocks. Dead. You musta had a helluva scrap here, Moore!"

"Where's that friend of yours, that Sid Beaugrand?" Walker wanted to know next.

"They've buried him," Dain said sparely.

"Where's —" Walker cast about — "where's Hollen?"

"You fell over him coming in. How far did you chase after that pink wrap?"

His square face brick-red, Walker snapped, "It got snarled in the wind and we saw it was a man wearing it. Thought for a while it must be Tresa. What happened to her clothes?"

But he didn't wait for an answer. His manner toward the people of San Rosario had been rough since he arrived, and now he thrust Don Porfirio aside, saying, "All right, let's get out of this rat hole!"

Dain sent Don Porfirio a look of apology, and the fat man nodded understandingly, although his soft eyes flashed. "Tresa isn't riding any saddle," Dain said. "Not today."

"Left my buggy and team along the main road," Walker responded. "One of you go get it. Hurry up!"

While waiting, Dain spoke to Don Porfirio and Pancho. "Good friends, you have done much for us."

"Little," murmured the boy.

"It is pleasant to do small favors for those from home," said the fat man, bowing first to Tresa, then to Dain. "We are too far from our own kind," he added pensively. "This place, it is not our *querencia . . .*"

Querencia, Dain mused. Where the heart lives. Quaint old term of affection for a favorite land, never forgotten. He met Tresa's eyes and saw the quick glisten of them. She too was remembering.

Walker's buggy, when it arrived, was the signal for departure. It was an expensive rig, with a leather top, drawn by a fast-stepping pair of matched bays. Nothing but the best for Milam Walker.

Dain lifted Tresa into the shaded back seat, where she curled up thankfully on the leather cushions. He took his seat beside Walker, who drove. Walker evidently realized that he had displayed unjustified resentment and over-stepped the bounds, and he tried another tack. He would not let down in front of his men, though, and he waved them on ahead, with some difficulty restraining the team of bays from racing them.

"We'll take over Bledsoe's stageline first thing," he stated. "We'll be taking over Travis City, and spread out from there. That means a big crew. New organization. New equipment."

He talked on, enlarging and expounding the picture of the future: Milam Walker, stagecoaching king of the whole Panhandle and the South Plains. Competition to be fought and ruthlessly eradicated until no other than the Walker stages plied the roads.

"Can't have that fat Mexican's brokendown coach cluttering up my roads, either. Small thing, I grant but it's the principle of it. Make an example of him!"

Growing aware of Dain's continued silence, he irritably inquired, "Well?"

"Don Porfirio is a friend of mine."

"There's no room for sentiment in my business, Moore!"

"Count me out of your business," Dain said evenly. "Tresa, too. We've quit."

"The devil you say!" Walker began loudly, then broke off to utter a short laugh. "Got an overdose of gunsmoke, eh?

I've seen it, other men. It'll wear off. You're my number-one man. You've built yourself a record and showed what you can do. You smashed Bledsoe. I admit it. No, no, you're too valuable for me to let go. With me, you've got a big career ahead of you. Top man!"

Dain said nothing, and in a few minutes Walker, sending him a side glance, remarked sharply, "Don't forget that little matter at Flint Corners. And that deputy from Travis City had a warrant for you. And while you haven't told me what-all you did last night, I can imagine it was plenty rough and illegal. The law could be mighty interested in you. Hold that thought in mind!"

Dain held it, and kept his silence all the rest of the way to Muleshoe Junction. Once he turned his head, and he and Tresa exchanged a long look in which there was silent communion. The buggy bowled smartly down the main street, Walker giving back nods to men who hailed him. His riders had got in ahead of him and put up their horses. The sun was low, and they stood about in the long shadows, telling their story to groups of listening citizens.

Walker slowed to turn in at the stageline

yard, and Dain finally spoke. "Drive on through town."

Walker's head came around. "What for?"

"Because if you don't," Dain told him quietly, "this gun is going off."

He held the gun on his lap, partly covered by his hands so that no casual passerby was likely to notice it. The gun was cocked, its muzzle directed point-blank at Walker's stomach.

Walker's breath whistled in his nostrils. "You don't dare."

"After what I had to do last night and this morning," Dan murmured, "it wouldn't bother me. Drive on south."

"So you solidly meant you quit, eh?"

"Sure did. Let the team out a bit."

The buggy cleared town and ran the straight road southward. Dain said, "Walker, stagecoaching isn't a young man's business. Do you think the railroad's going to stop forever at Travis City? No, when the traffic warrants it, the railroad will build on down, and then where's your stagecoaching business? You're getting old. It'll last you long enough. Not me. Beef will always find a market, and I'm a cattleman."

"I wish sometimes I'd never got into anything else." Walker spoke with none of his

251

previous bluster. "But old? Me?"

"Yes." It was Tresa who said that, and Walker sagged a little in his seat, shaking his head. "If you send the law after Dain, and they take him from me," she vowed, "I swear I'll come back and kill you!"

He drew the bays to a standstill. He turned and looked into her face for a full minute. "I believe it," he said heavily, "I thoroughly do believe it! But d'you really think I'd do it, girl? I'd have to send the law after you, too. I wouldn't do that. Show myself up for an old fool? Play Judas? Moore, damn you, what do you want from me? My life?"

"No, only to borrow the rig. We'll leave it somewhere south and one of your drivers can pick it up."

Walker handed him the lines and stepped down. "Where will you pair of hellions end up?"

"On our own *querencia*," Tresa said to him.

He understood the phrase. "I see." Some of his ordinary arrogance rose irrepressibly to the surface. "I tried wrong tactics on you, I don't know why. Yes, I do know. When a man's after a woman, it doesn't pay him to respect her too much or to fall too deep . . . I should've gone all out and —"

"You would've had to be younger," Dain interrupted him, purposely cruel, "by twenty years or more."

Walker gazed off. The boldness on his face faded with the fading light. He tapped his fingers on the front wheel of the buggy. "Moore, you've done a lot for me. I've gained by it. I don't agree the days of stagecoaching are numbered. Practically, I'm king of the Panhandle right now." Strain deepened his voice to a growl. "But I'd give all I own to be your age and go where you're going and have what you've got, damn you!"

He slapped the wheel and turned away. "Keep the rig and team. Wedding gift."

They looked after him, an aging man, thickset and baggy, all but ludicrous with his wild shock of gray hair and his stumpy legs, tramping stolidly back to Muleshoe Junction in the reddish glow of sundown. Walker would keep his word. He had that pride, and he had bigness, in spite of his faults and corrosive obsessions.

Dain shook the lines. The bays settled into a lively gait. "You know," he said, his thoughts dwelling on the future, "we've got that thousand dollars I won for bringing in the Travis City mail. We could get a pretty good price for this rig. It's way too fancy for us. That'll give us something to start on when we reach home. I doubt we'll ever grow rich, but we'll be where we want to be and live like we want to live. Is it enough, Tresa?"

"It's everything, Dain!" She was leaning forward, and she spoke close to his ear.

He felt strongly the tingle of her nearness, her vibrant femininity. On second thought he guessed he'd better not stop the team.

"We ought to send for Don Porfirio and his folks," he said a bit hurriedly. "I've an idea they're as homesick as us. We'll need help rounding up the cattle and horses my folks left running wild down there. Pancho would like that."

"Pancho is quite a boy."

"He'll do till a boy comes along," he agreed. And then, realizing how that might sound, he looked quickly around at her.

Although she had a high color that grew richer as he looked, her eyes met his steadily and directly. "Yes," she said, quite breathlessly, "that's right!"

She moved her face the inch or two closer, and while they kissed he pulled back on the lines. Couldn't risk spilling the rig in the dark. It was a long, long way down to Padre Island.

The employees of Thorndike Press hope you have enjoyed this Large Print book. All our Thorndike and Wheeler Large Print titles are designed for easy reading, and all our books are made to last. Other Thorndike Press Large Print books are available at your library, through selected bookstores, or directly from us.

For information about titles, please call:

(800) 223-1244

or visit our Web site at:

www.gale.com/thorndike
www.gale.com/wheeler

To share your comments, please write:

Publisher
Thorndike Press
295 Kennedy Memorial Drive
Waterville, ME   04901